Cooling Down

Cooling Down

Local Responses to Global Climate Change

Edited by
Susanna M. Hoffman, Thomas Hylland Eriksen,
and Paulo Mendes

First published in 2022 by
Berghahn Books
www.berghahnbooks.com

Library of Congress Cataloging-in-Publication Data

Names: Hoffman, Susannah M., editor. | Eriksen, Thomas Hylland, editor. | Mendes, Paulo, editor.
Title: Cooling Down: Local Responses to Global Climate Change / edited by Susanna M. Hoffman, Thomas Hylland Eriksen, and Paulo Mendes.
Description: New York: Berghahn Books, 2022. | Includes bibliographical references and index.
Identifiers: LCCN 2021039722 (print) | LCCN 2021039723 (ebook) | ISBN 9781800731899 (hardback) | ISBN 9781800734173 (paperback) | ISBN 9781800732988 (open access ebook)
Subjects: LCSH: Climatic changes—Social aspects—Case studies. | Climatic changes—Effect of human beings on—Case studies.
Classification: LCC QC903 .C6656 2022 (print) | LCC QC903 (ebook) | DDC 363.738/7452—dc23/eng/20211027
LC record available at https://lccn.loc.gov/2021039722
LC ebook record available at https://lccn.loc.gov/2021039723

British Library Cataloguing in Publication Data

A catalogue record for this book is available from the British Library

ISBN 978-1-80073-189-9 hardback
ISBN 978-1-80073-417-3 paperback
ISBN 978-1-80073-298-8 open access ebook

https://doi.org/10.3167/9781800731899

Contents

List of Illustrations vii

Acknowledgments x

Introduction
Scaling Down in Order to Cool Down 1
Thomas Hylland Eriksen and Paulo Mendes

Part I. Ways of Knowing 25

Chapter 1
Environmental Pluralism: Knowing the Namibian Weather in Times of Climate Change 29
Michael Schnegg

Chapter 2
How a Storm Feels: Storying Climate Change in the Eastern Himalayas 49
Alexander Aisher

Chapter 3
Who Is Perturbed by Ecological Perturbations? Marine Scientists' and Polynesian Fishers' Understandings of a Crown-of-Thorns Starfish Outbreak 65
Matthew Lauer, Terava Atger, Sally J. Holbrook, Andrew Rassweiler, Russell J. Schmitt, and Jean Wencélius

Chapter 4
Urban Transformations in the Hydric Landscapes of Belém, Brazil: Environmental Memories and Urban Floods 90
Pedro Paulo de Miranda Araújo Soares

Part II. Situations and Decisions 109

Chapter 5
Climate Change and Mitigation in Bangladesh: Vulnerability in Urban Locations 113
Tasneem Siddiqui, Mohammad Jalal Uddin Sikder, and Mohammad Rashed Alam Bhuiyan

Chapter 6
Localizing Climate Change: Confronting Oversimplification of Local Responses 131
Brian Orland, Meredith Welch-Devine, and Micah Taylor

Chapter 7
"The Times They Are A-Changin'" but "The Song Remains the Same": Climate Change Narratives from the Coromandel Peninsula, Aotearoa New Zealand 167
Paul Schneider and Bruce Glavovic

Chapter 8
Climate Change and East Africa's Past: Three Cautionary Tales 201
A. Peter Castro

Chapter 9
"Our Existence Is Literally Melting Away": Narrating and Fighting Climate Change in a Glacier Ski Resort in Austria 223
Herta Nöbauer

Part III. Politics, Policies, and Contestation 245

Chapter 10
Where Floods Are Allowed: Climate Adaptation as Defiant Acceptance in the Elbe River Valley 249
Kristoffer Albris

Chapter 11
Climate Resilience through Equity and Justice: Holistic Leadership by Tribal Nations and Indigenous Communities in the Southwestern United States 269
Julie Maldonado and Beth Rose Middleton

Chapter 12
The Return to What Has Never Been: A View on the Animal Presence in Future Natures 292
Guilherme José da Silva e Sá

Chapter 13
Emitting Inequity: The Sociopolitical Life of Anthropogenic Climate Change in Oaxaca, Mexico 313
Amanda Leppert and Roberto E. Barrios

Chapter 14
Disaster and Climate Change 339
Susanna M. Hoffman

Afterword
Toward Eco-Socialism as a Global and Local Strategy to Cool Down the World-System 363
Hans A. Baer

Index 377

Illustrations

Figures

3.1. Moorea island, main settlements, and its two research centers. © Matthew Lauer. 69

3.2. Strings of fish (*tui*) sold by roadside. © Terava Atger. 75

4.1. A drainage channel in the Una Basin, Belém, 2017. © Pedro Paulo de Miranda Araújo Soares. 102

6.1. Coastal Georgia showing the normal Mean Higher High Water line, the estimated surge extent of Hurricane Matthew, and survey and interview respondent locations. Map by Micah Taylor, USA 31.489946° N, -81.499712° W. Esri, HERE, Garmin, © OpenStreetMap contributors, and the GIS user community. Accessed December 2018. 133

6.2. Estimated surge extent of Hurricane Matthew in Savannah, Brunswick, and Richmond Hill. Map by Micah Taylor, USA 31.489946° N, -81.499712° W. Esri, HERE, Garmin, © OpenStreetMap contributors, and the GIS user community. Accessed December 2018. 139

6.3. Study conceptual design: main study components shaded and external factors in open boxes. © Brian Orland. 140

6.4. Intentions to move away versus (a) social norm, resistant to move, (b) perceived behavioral control, victim of circumstances. © Brian Orland. 157

6.5. Intentions to stay in place versus (a) social norm, expect to stay, (b) perceived behavioral control, seek advice. © Brian Orland. 158

7.1. Coromandel beachfront property development at Hahei. © Paul Schneider. 172

7.2. New Zealand Coastal Policy Statement 2010 decision context for coastal areas exposed to coastal hazards and climate change. Source: MFE 2017. 176

7.3. Newspaper articles covering coastal risk/climate change on the Coromandel since 2010. © Paul Schneider. 182

7.4. Beachfront property development at Te-Whanganui-o-Hei/Mercury Bay. © Paul Schneider. 184

7.5. Local newspaper articles reporting on emergency action taken to defer coastal erosion. © Paul Schneider. 185

7.6. Fifteen-year-old high school student Helena Mayer and Coromandel residents Nancy and Eric Zwaan protesting in front of the Thames-Coromandel District Council office in Thames. © Beaton, 2019. 187

7.7. Key events, publications, turning points, and policy actions affecting climate change action and adaptation on the Coromandel. © Paul Schneider. 189

9.1. The "occupationscape" of the Pitztal glacier ski resort, summer 2015. © Herta Nöbauer. 224

9.2. A worker uses a digger for removing some textiles in late summer 2015. © Herta Nöbauer. 236

13.1. *The History of Mexico*, mural by Diego Rivera in the Mexican National Palace. © Roberto E. Barrios. 335

Tables

6.1. County populations and populations at risk of coastal Georgia, sorted north to south. US Census, 2017. Source: American Community Survey, reported by Headwaters Economics, headwaterseconomics.org/par. Table by Brian Orland. 141

6.2. Interview and survey respondents versus census demographic characteristics. US Census, 2017. Source: American Community Survey, reported by Headwaters Economics, headwaterseconomics .org/par. Table by Brian Orland. 142

6.3. Reasons for living in coastal Georgia. ^{Q} = Quality of Life, ^{F} = Family Ties, ^{P} = Pragmatic. The superscript letter indicates the factor grouping. Data by Brian Orland. 144

6.4. Correlation table, major demographic characteristics. Data by Brian Orland. 145

6.5. Considerations for moving away. Factor groupings: ^{Q} = Loss of Quality of Life, ^{C} = Increased Costs, $^{J/F}$ = Job/Family Loss. Data by Brian Orland. 147

6.6. Regional and personal expectations regarding sea level rise and damaging storms. Data by Brian Orland. 148

6.7. Attitudes to adaptation responses. Factor groupings: [E] = Government Entitlement, [P] = Personal Responsibility, [G] = Government Appreciation. Data by Brian Orland. 150

6.8. Actions that might be taken. Factor groupings: [M] = Intention to Move, [S] = Stay in Place. Data by Brian Orland. 151

6.9. Interviewee intentions to stay or move away from the coast. Data by Meredith Welch-Devine. 153

6.10. Investigating social norms. Factor groupings: [R] = Reluctant to Move, [S] = Stay in Place, [C] = Climate Skeptic. Data by Brian Orland. 155

6.11. Investigating perceived behavioral control. Factor groupings: [C] = Personally in Control, [A] = Seek Advice, [V] = Victim of Circumstances. Data by Brian Orland. 156

6.12. Behavioral outcomes: intentions to move away. Bold indicates strong associations. Data by Brian Orland. 156

6.13. Behavioral outcomes: intentions to stay in place. Bold indicates strong associations. Data by Brian Orland. 157

Acknowledgments

A volume that proposes to present a set of chapters intended to illuminate the specifics of a complex problem takes the concerted effort of many people. We, the editors, in particular would like to offer our truly heartfelt thanks to all our contributors who worked first as unflagging ethnographers and then as ardent enthusiasts upon this endeavor. They come from a wide variety of backgrounds, work in a sweeping range of places, and have approached the book's purpose from a spectrum of standpoints, yet together they have produced a compendium of remarkable relevance and caliber. We would also like to thank the trio of peers who reviewed our manuscript. Their insightful comments both affirmed the importance of the work and helped improve it. We are sincerely indebted to our very concerned publisher, Marion Berghahn, and to our editors, Tom Bonnington and Anthony Mason, who guided the book to completion, along with Elizabeth Martinez and her team for their pinpoint editing and Andrew Esson for his striking cover art. We would also like to express our gratitude to our colleagues in anthropology and in every discipline who work in the crucial field of climate change for their essential involvement. Paulo Mendes would like to express his personal gratitude to his research center, Centro em Rede de Investigação em Antropologia (CRIA), and to the Portuguese research funding agency, Fundação para a Ciência e Tecnologia (FCT). Lastly, we would like to thank one another.

Introduction
Scaling Down in Order to Cool Down

Thomas Hylland Eriksen and Paulo Mendes

Perhaps the publication of Naomi Klein's influential *This Changes Everything* in 2014 marked a watershed in the sense that climate change was by then, as the author had come to realize, not just another human challenge to add to an already lengthy catalogue of ailments and injustices but *the* main problem facing humanity. Climate change has become the single most important lens through which phenomena such as inequality, displacement, indigenous issues, migration, corporate power, new political movements, environmental degradation, and racist exclusion must be viewed in order to obtain a full picture of any of them. This holds true whether the investigation is fueled by curiosity or activist concerns—or, as the case may often be, both. Extreme weather is now in the news every day, ranging from the massive 2019–20 forest fires in Australia to the European heatwaves in the same years and the simultaneous hailstorm in Guadalajara, Mexico, which deposited a meter-and-a-half layer of wet snow in the middle of summer in a city otherwise known for its dry and warm climate. In January 2020, a mild wind blew through Oslo, where temperatures reached eight degrees centigrade above zero, a far cry from the normal minus five degrees and at least half a meter of snow. Although the coronavirus pandemic led to a sudden slowing down of anthropogenic climate change, with air traffic plummeting by more than 90 percent in April 2020 compared to April 2019, it is in itself unlikely to have long-term effects on the underlying dynamics of climate change.

Global climate change seems abstract, difficult to understand, relate to, and deal with politically. It is well documented, yet it lends itself easily to conspiracy theories and alternative interpretations. It is a product of

Notes for this chapter begin on page 21.

modernity, which seriously questions a central tenet in the very modern project that has produced it, with growth and acceleration as key values. Notably, climate change leads to a profound questioning of the belief in a particular kind of progress based on the partnership between science and technology.[1] It also indicates the limitations of nationalism as a political project for the twenty-first century and reveals the starkness of global inequality *and* the need for humanity to act as one. The causes of climate change are also the causes of the unprecedented economic growth, comfortable middle-class living for a growing minority of humanity, and, in some places, the successful struggle against abject poverty. Accordingly, contemporary global civilization is caught in a double bind (Bateson et al. 1956) at two systemic levels: The individual benefitting from the modernity of fossil fuels and capitalist growth relies on a world economy that simultaneously provides them with comfortable lives *and* undermines the very conditions for those lives. The global economic system relies on accelerated growth (Eriksen 2016, 2018) of a kind that destroys its own foundations by using up nonrenewable resources and damaging the global ecology beyond repair.

It is difficult to imagine a more critical or prominent topic in the world today than climate change. Books on the topic range from popular science to the political, from the journalistic to the academic. Atlases and handbooks showing the scope of the issue have appeared. Research centers have been established, usually with an interdisciplinary element and often with a mixed basic and applied research mission. Major journals have been established, both specific to particular disciplines and those that are more wide-ranging. Important transnational institutions such as the United Nations have produced germane and overarching examinations, appraisals, and increasingly insistent policy recommendations. New terms, such as *the Anthropocene*—tailored to describe a new era for human life on Earth—have spread quickly (Steffen, Crutzen, and McNeill 2007), while the more recent concept *the Capitalocene* suggests that the overuse of resources, the relentless search for profitability, the translation of nature into quantifiable "resources," and the commitment to endless growth are not characteristics of humanity as such but of a particular phase in our history.

Attempts to describe and understand climate change generally fall into one or several of three categories: (1) descriptions of comprehensive worldwide happenings, such as sea level rise, temperature rise, desertification, and increasing storms; (2) warnings of dire consequences if measures are not taken; and (3) discussions of implications for development, industry, and socioeconomic policy. Virtually every scientific discipline at every major academic institution seems to have developed a section dedicated to the topic, and many institutions and professional organizations (such as the American Anthropological Association) have established commission task forces aiming to produce disciplinary statements with details, charts, and analytical breakdowns on the subject.

Even if massive human impact on climate is a recent phenomenon, the awareness that climate has an impact on human life is not new. As Dove (2013) reminds us in his historical reader on the anthropology of climate change, one of the founders of medical science, Hippocrates (b. 460 BCE), wrote a treatise called *Airs, Waters, Places* that argued for a connection between the climate, the environment, and the human condition. Much later, during the Enlightenment, the social theorist Montesquieu (1689–1755) saw a close relationship between climate and social institutions, temperament and social life. Dismissed by later social theorists as simplistic environmental determinism, similar ideas have nevertheless never quite disappeared. What is new in the current age is the recognition of humanity's impact on climate and the potentially catastrophic consequences for life on the planet in the future. In this area, anthropologists are making important contributions to knowledge.

Perspectives from Social Theory

Attempts to describe and understand climate change mainly fall into one or several of three categories: (1) descriptions of comprehensive worldwide processes, such as sea level rise, temperature rise, desertification, and extreme weather events; (2) warnings of severe consequences if measures are not taken; and (3) discussions of implications for development, industry, and policy.

In other words, the contemporary world of climate change and the Anthropocene, and that of global transformation in general, has not evaded the attention of academics, and this is also the case in the social sciences. In general social theory, Zygmunt Bauman (2000) and Ulrich Beck (2009) wrote important works about risk and unpredictability around the turn of the millennium, while Hartmut Rosa has devoted his research to social acceleration, with clear implications for climate (2016). The term Anthropocene was initially proposed by the atmospheric chemist Paul Crutzen (with Eugene Stoermer), who is also the coauthor of a much-cited article with his colleague Will Steffen and the historian John McNeill (Steffen, Crutzen, and McNeill 2007) on social aspects of climate change. Slightly earlier, the archaeologist Brian Fagan published several books about the significance of climate for human society (see Fagan 1999). Another archaeologist, Joseph Tainter, has produced important analyses of the causes of civilizational collapse in the past (1988), a perspective subsequently popularized by Jared Diamond (2005). Tainter's work shows ways in which contemporary societies can learn from archaeological research when faced with mounting or simmering crises. In his comments on the present, which make comparisons with the collapse of the Roman and Maya empires, climate change nevertheless comes across as just one factor in accounting for the decline of complex societies. The decisive cause, as Tainter sees it, will consist in decreased marginal returns on investments in energy (EROI), owing to population growth and subsequent intensification of food production with decreasing returns, coupled with growth in bureaucratic, logistic, and transport costs. Presently, resource shortages, a direct result of anthropoid dominance of the planet, may be a more acute problem than climate change in his view.

Since the late eighteenth century, we have been able to exploit unprecedented amounts of energy, at first in the shape of abundant and easily accessible coal deposits, and subsequently through the extraction of oil and gas for the sake of economic growth and the improvement of the human condition (Mitchell 2011). The fossil fuel revolution enabled humanity to support a fast-growing global population—it has increased sevenfold since the beginning of the fossil fuel revolution. Yet the cost of taking out fossil fuels grows as the low-hanging fruit is being used up. At the same time, production relying on fossil fuels has an inevitable element of destruction (Hornborg 2019), in a dual sense, since we are simultaneously eating up capital that it has taken the planet millions of years to produce and undermining the conditions for our own civilization by altering the climate and ruining the environment on which we rely.

Interdisciplinary collaboration is necessary in order to understand the full implications of climate change. While climate scientists adopt a birds-eye perspective on the planet and archeologists move their gaze back in time, anthropologists enter deeply into local realities in order to understand perceptions of and responses to climate change. The last couple of decades have produced a fast growing corpus of anthropological knowledge about climate change, much of which performs a double task in that it improves our understanding of society and may also be relevant for policy and action.

The Unique Contribution of Anthropology

Through its insistence on the primacy of local realities, anthropology builds its theoretical insights in dialogue with the social and cultural worlds studied and engaged with by researchers, who have spent years qualifying as specialist connoisseurs of the local knowledge and practices of the communities with which they work. In this, anthropology differs from other academic disciplines by developing theoretical insights, not exclusively through internal academic debates but by way of active engagement with local experiences and worldviews. The ethnographic method is not particularly expensive, but it is immensely time-consuming since the researcher has to get to know their collaborators personally rather than merely doing interviews (Shore and Trnka 2013). As a result, anthropologists tend to learn a lot about a few rather than a little about many, and herein lie both the strengths and the weaknesses of the ethnographic method. The strengths, when faced with systems of staggering scale such as the global climate system, have been demonstrated in a number of recent books, some taking on anthropogenic climate change explicitly (e.g., Crate and Nuttall [2009] 2016), others emphasizing the lessons that can be learned from indigenous people and their engagement with the environment (e.g., Hendry 2014). A collection of essays by Claude Lévi-Strauss (1983) is entitled *Le regard éloigné*, but what characterizes most anthropological work in the field is rather the view from below and from the inside. This gaze and methodology inevitably produces diversity rather than uniformity, displaying locally tailored solutions to the problems facing actual human beings rather than standardized options of the one-size-fits-all kind.

The plurality of perspectives presented through anthropological research effectively falsifies the TINA (There Is No Alternative) doctrine popularized by Margaret Thatcher in the 1980s by showing that, in fact, TAMA (There Are Many Alternatives). Yet it could be argued that the tendency toward myopia and provincialism haunts anthropological research for precisely the same reasons that it shines in its ability to produce a dazzling range of distinctive local knowledges. Faced with large-scale phenomena such as global capitalism and human ecological footprints traceable on a global canvas, anthropologists need help to fill in the blanks, lift their gaze from their local community, and challenge their own prejudices and assumptions. This is why interdisciplinarity must be part and parcel of an anthropology of climate change.

In a short position paper written for non-anthropologists, Jessica Barnes and coauthors (Barnes et al. 2013) list three kinds of knowledge that anthropology can contribute to the field: (1) ethnographic insight, (2) historical perspective, and (3) holistic view. These will be elaborated below.

Anthropologists are well positioned to make a difference and, perhaps, help mitigate effects, or even to propose deeper systemic change to combat climate change. A considerable, and growing, number of edited volumes on climate change by anthropologists have appeared since the turn of the millennium. Interest in the area has grown very rapidly, and to this development we now turn.

The Growth of Climate Anthropology

Although the study of climate change is recent in anthropology (as it is elsewhere), it has important precursors in the history of the discipline, especially in environmental anthropology and the anthropology of energy.

While mainstream British and French social anthropology in the mid-twentieth century was mainly preoccupied by research on social organization, politics, and ritual, American cultural anthropology tended to emphasize the study of symbolic meaning. However, in the United States, there was also a tradition, going back to the nineteenth century, of studying material culture, technology, and ecological adaptation. After World War II, Julian Steward (1955) championed human ecology, while Leslie White ([1949] 2005) studied technology and energy use from a social evolutionist perspective. These approaches ceased to wield influence in the discipline by the early 1980s, and especially White was criticized for not paying enough attention to power and symbolic meaning. Yet the emphasis on energy and ecology remains relevant for the current anthropology of climate change.

A different approach to ecology is represented in Gregory Bateson's work, which remains highly influential (Bateson 1972). As early as 1970, he identified three root causes to what he already then spoke of as the ecological crisis. The first was technological progress, the second was population increase, and for the third he pointed to a set of entrenched Western cultural values and ideas that place humanity in an unhealthy relation to the environment (what he speaks of as a flawed epistemology based on Cartesian dualism and individualism). What Bateson criticized was the idea that humans should strive to control the environment, along with the strong focus on the individual, the belief in economic growth, the assumption that we live within an infinitely expanding frontier, and the conviction that technology will solve any problem facing us. What Bateson calls a healthy ecology requires ecological flexibility and slow change, "a single system of environment combined with high human civilization in which the flexibility of the civilization shall match that of the environment to create an ongoing complex system, open-ended for slow change of even basic (hard-programmed) characteristics" (Bateson 1972: 502).

Whereas Bateson identified a central contradiction of contemporary civilization early on, he did not address climate change explicitly. His ex-wife Margaret Mead may in fact have been the first anthropologist to do so (Kellogg and Mead 1980), as she convened a conference about the atmosphere as early as 1975. Whereas climate change was not yet on the agenda—in fact, many scientists at the time believed that we were heading toward a new Ice Age rather than an overheated world—the conference took on smoke, smog, and other forms of atmospheric pollution as genuinely global challenges that needed to be dealt with politically.

By the 1990s, climate change (still spoken of as global warming) began to enter the political and research agenda more visibly. In anthropology, an early, important contribution was a four-volume work edited by Steve Rayner and Elizabeth Malone titled *Human Choice and Climate Change: An International Assessment* (1998), an interdisciplinary work with contributors from many countries. Another pioneering work was Ben Orlove's ethnoclimatological research in the Andes, which—among other things—showed how farmers used the influence of El Niño events on the visibility of the Pleiades to predict rainfall and temperature (Orlove et al. 2000). In the 1990s, the concern with climate change was nevertheless still marginal and peripheral in anthropology.

A decade later, this was about to change.

Coming from the anthropology of health, Hans Baer and Merrill Singer published *Global Warming and the Political Ecology of Health* (Baer and Singer 2009). The book investigates a particular aspect of climate change, namely its impact on water, nutrition, and the spread of disease. Unlike many other anthropological studies of climate change, this book strongly emphasizes that climate change affects different communities unequally owing to an economic system that produces inequality.

In the same year, Susan Crate and Mark Nuttall edited the widely cited and read *Anthropology and Climate Change* (Crate and Nuttall [2009] 2016), which was a pioneering, indeed groundbreaking, volume when it was published, with chapter authors working in different parts of the world. The main perspective in this book is interpretive, and the text explores local responses to, and perceptions of, climate change in a wide range of societies. It should nevertheless be mentioned that the societies that are the main contributors to climate change—the rich OECD countries, as well as China—are sparsely represented. This shortcoming is addressed in the second edition of the book (Crate and Nuttall [2009] 2016), as well as in the later edited volume *Cultures of Energy* (Strauss, Rupp, and Love 2013), but perhaps most consistently in Kari Norgaard's *Living in Denial* (Norgaard 2011). Based on fieldwork in a rural Norwegian community where erratic winters interfere with winter tourism, *Living in Denial* asks how it can be that people who are aware of, and experience the effects of, climate change continue to lead unsustainable lives.

A few years later, a very substantial anthropological literature dealing with different aspects of climate change had appeared, and professional interest in the field had skyrocketed. Whereas there was just a single panel at the Society for Applied Anthropology (SfAA) devoted to climate change in 2006, the number had increased to twenty a decade later. Crate and Nuttall sum up the growth and diversification of the field by stating that anthropologists today "are engaging [in] research that has a concern with resilience, vulnerability, adaptation, mitigation, anticipation, risk and uncertainty, consumption, gender, migration, and displacement. Anthropologists have developed significant work on the politics of climate change, inequality, health, carbon markets and carbon sequestration, and water and energy" (Crate and Nuttall [2009] 2016: 11).

Global Diversity

The body of knowledge that anthropologists have so far accumulated ranges from critical studies of the discourses and practices of carbon offsets (Dalsgaard 2013) to comparative studies of retreating glaciers (see Ben Orlove's website, https://glacierhub.org), in addition to a fast-growing number of ethnographies describing how communities deal with the local effects of climate change, in projects that look, in Kirsten Hastrup's evocative terms, at the drying lands, the rising seas, and the melting ice (Hastrup and Hastrup 2015). A political economy approach informed by anthropological reflexivity is provided, inter alia, in works by Harold Wilhite (2016) and Alf Hornborg (2019). Local responses to climate change are explored in Stensrud and Eriksen (2019), the relationship between health, capitalism, and climate has been analyzed by Hans Baer and Merrill Singer (2009), and the historical antecedents of current concerns with environmental change and climate are covered in Michael Dove's historical reader (Dove 2013). Anthropologists have also contributed some very significant ethnographic monographs on climate issues, ranging from Jessica Barnes's research on water in the Nile Delta (Barnes 2014) to Linda Connor's work on mining in Australia (Connor 2016).

Not all environmental anthropology has a focus on climate. Important research on topics such as deforestation, mining, waste, and toxins may be only tangentially related to climate. However, it is fair to say that the broader field of environmental anthropology is being renewed and reformulated because of the intensified attention to climate, as witnessed, for example, in the edited volume *The Angry Earth: Disasters in Anthropological Perspective* (Oliver-Smith and Hoffman 2000, 2020), where, in the second, revised, and updated edition of the book, nearly all contributors mention the atmospheric changes that have begun to affect the sites of their prior studies. It also deserves mentioning that the most famous living anthropologist without an anthropology degree, Bruno Latour, shifted his attention years ago to the causes and politics of climate change (Latour 2017). It is everywhere, and it is now; it is comprehensive, it brims with methodological implications, it buzzes with theoretical possibilities, and indeed, the fact of anthropogenic climate change may be about to to redefine the very foundations of anthropological (and other) research, and it also raises the question of what it entails to be a human being within a new existential framework. Climate change, the immediate cause of the coining of the neologism Anthropocene, may retrospectively be seen as a major game changer in intellectual and political life in general, and also in anthropological research. It is no coincidence that the increased interest in multispecies fieldwork, and the rise to prominence of the Deleuzian term *assemblage* (which transcends the human/nonhuman and material/symbolic barriers), have shaped the work of many anthropologists in the present century.

As opposed to attempts to create top-down solutions through international agreements, some of which have a perceptible element of magical thinking (Rayner 2016), the anthropological view from below and within provides a number of useful insights. First, an awareness of variation is essential to all anthropological research. The clunky distinction between developing and developed countries, and indeed the very category of the country, does not always fit the territory. The Seychelles is not "a place" in the same sense as China is "a place." There is no reason to assume that actions that have been proved successful in Namibia would work in Nepal. The challenges faced by Greenlanders confronting melting ice differ from those faced by Bangladeshis, who are challenged with intensified flooding, salination of the soil, and mudslides, or from those encountered by Sahelian nomads, who witness their pastures turn to dust.

Second, any successful social change has to begin with an appreciation of local life-worlds and has to be developed not for but with the people affected. Neither of these principles, commonsensically true to any working anthropologist, are reflected in the abstract, large-scale worlds of international negotiators. In other words, a reasonable conclusion is that climate change policy must be scaled down and perhaps built from the bottom and not from the top.

Comparison is a third asset. One of anthropology's main methods for generating knowledge and opening new theoretical horizons, as well as for stimulating the political imagination, comparison generates new ideas about human worlds. For example, a comparative approach shows that it is not self-evident that land can be subject to personal ownership and that "resource management" and "sustainability" are integrated in the taken-for-granted knowledge. It goes without saying, because it comes without saying, that in societies where "the economy" has not been disembedded from everyday life, making people accountable to their surroundings consists of ways that are unknown and perhaps unknowable to those who own and profit from property elsewhere.

The methodological and analytical holism on which anthropologists insist has often made their knowledge somewhat unwieldy and unmanageable for governments and development agencies, since it goes against the segmentation of worlds into separately manageable sectors that bureaucratic planning requires. Yet at this point in history, more holism may be precisely what is needed: in order to understand the refugee crisis in Syria, which began with the outbreak of civil war in 2011, the seven-year drought preceding the unrest needs to be taken into account; in order to explain the rise of ethnonationalism in Europe, the containerization of shipping and its role in catapulting Chinese exports to global omnipresence must be understood; and not least, the knowledge, usually contested, enabling people to navigate, interpret, and act upon the world must form an integral part of any project, whether academic or applied, concerning the human implications of climate change. Anthropology, its methods, and its knowledge are particularly well equipped to consider the local, to scale down, while pondering the weight of the global and its impacts on local worlds.

In spite of the thriving research and reporting activity in the field, this book is needed. By examining the real, practical assessments, solutions, and calls for concern that are happening on the minute, regional, parochial, and diverse levels of humans encountering a problem, it is an account from the half-forgotten backwaters of the contemporary, overheated world. It also presents chronicles from some of its centers. Like other anthropologists contributing to the field, we recognize the global dimension of climate change, but we also mean to show in what ways climate change is also always local and has to be understood as such, ecologically, socially, politically, culturally. While politicians until recently might write off concerns of urgency by calling for more research, it is by now abundantly clear that the natural science knowledge needed to act has been available for many years.

However, while we possess sufficient knowledge from the natural sciences, pointing to it, it is by no means evident that the human dimension of climate change is understood sufficiently well. A blunt question, interrogating the actual impact of the massive natural science knowledge now available, may be why so little is happening, since nearly all countries are signatories to a series of climate agreements beginning with the Kyoto

Protocol in 1997, which specifies the steps that need to be taken to mitigate the impact of changes that are already taking place. Later reports from the Intergovernmental Panel on Climate Change (IPCC) have further been increasingly insistent about the need to take action immediately. Yet, global emissions continue to rise and are nowhere near to reaching the targets agreed to in Kyoto and later affirmed. Indeed, emissions were, by 2013, more than 60 percent higher than in 1990 (Khokhar 2017).

To begin to explain this conundrum, we will now take a short excursus to Norway, which provides an interesting case not included in this book but one that the first author of this introduction knows well since he lives there.

The Cases of Norway and Portugal

The case of Norway is intriguing. On the one hand, the very concept of (ecological) sustainability was coined by an influential UN commission headed by the then Norwegian prime minister, Gro Harlem Brundtland, in 1987. Norway further markets itself as a clean and scenic tourist destination with vast areas of unspoiled nature. Indeed, nature as wilderness forms a central dimension of the collective Norwegian cultural self-understanding (Gullestad 1992).

On the other hand, through its massive exports of oil and gas, Norway may indirectly be responsible for as much as 3 percent of the global CO_2 emissions. At home, the country appears to have a better track record than many countries, in spite of the fact that the affluent Norwegians drive and fly often and are enthusiastic consumers of imported commodities. Most of the energy used in Norwegian households and industry comes from hydroelectric plants,[2] and the exported oil does not affect the domestic emission statistics. Yet, it is commonly known that Norway is a part of the problem, not of the solution, due to its considerable oil and gas exports. On this background, Norwegian governments—and in particular the center-left government that ran the country from 2004 to 2013—have in mainly two ways sought to balance out some of the detrimental effects of Norwegian oil and gas exports: (1) The directors of the Sovereign Fund, into which most of the state oil profits are invested, are concerned with ethical investments and have appointed an ethical council that oversees its activities, aiming to ensure that it does not invest in "unethical" products such as weapons and coal (!). More importantly, (2) the country commits itself to considerable investments in projects aiming to reduce carbon emissions elsewhere, notably in the Global South. The most familiar of these may be the UN-sponsored REDD Programme (Reducing Emissions from Deforestation and forest Degradation; the acronym REDD, incidentally, means *save* or *rescue* in Norwegian, but it can also mean *afraid*).

The irony is evident: Instead of implementing changes at home, such as reducing the rate of oil extraction or the level of consumption, Norway pays foreigners to change their behavior in order to reduce the impact of—inter alia—Norwegian oil exports. Prime Minister Jens Stoltenberg (resigned in October 2013), trained as an economist, argued that investing in climate-friendly activities in the Global South was far more cost-effective than spending similar sums in the expensive north.[3]

This duality in Norwegian policy, whereby social welfare and economic growth are closely associated with oil extraction whereas foreign investments and development assistance aim to reduce carbon footprint and environmental destruction, reveals a profound double bind (Bateson 1972). This is arguably the central contradiction in contemporary civilization, where growth in energy use and ecological sustainability are desired at the same time but rarely simultaneously achieved. Successive governments have pledged to fulfill their commitment to reduce emissions by 40 percent compared to the 1990 levels by 2030. So far (2021), emissions are slightly higher per capita than they were in 1990. So, one might ask, are they lying, or do they believe in miracles? Conveniently, political elites in many countries encourage their citizens (seen as consumers) to live more sustainably, perhaps to fly less and eat less meat. A consequence is that the citizens may eventually be blamed for the outcome of a global process on which the politicians did not themselves act.

Another small, geographically peripheral country in Europe is Portugal, where the second author lives. Located in the southwest corner of Europe, its consecutive governments have subsidized "green power" (mainly hydroelectric and windmills) heavily since 2005 but never canceled comparable financial support to fossil energy. In Lisbon, policies to "clean the air" are being implemented—interdicting older cars from circulating in the city center and increasing public transportation, mainly—while the same local and national authorities expand cruise terminals and airports, arguably to serve one of Portugal's main exports, tourism.

The Portuguese do not see themselves as being major global polluters. The circumstance of being a small country with a weak industry reinforces a narrative that places Portugal as a net recipient of climate change. Ursula von der Leyen, president of the European Commission, addressing the European Parliament on the "European Green Deal" in December 2019, said that "Portugal is one of the countries most affected by climate change. The loss of coast, hurricanes, floods, horrible forest fires have taken already a very high toll . . . [and it has] invested significantly [in clean energies] and it will close its last coal mine in 2023. . . . It already has a surplus in renewal energies. . . . From the Portuguese perspective the European Green Deal is about energy infrastructure, is about interconnectivity [ways to sell renewal energy] and is about adaptation to climate change." These words kindly pair with the discourse and policies of different Portuguese governments: renewal energies are the Portuguese contribution to lower CO_2 emissions and are at the same time a commodity that may be exported. Economic growth is paramount and almost unconditional. The Portuguese national debt and ways to succeed, as a citizen and collectively, postpone strong climate change policies unless they conform to enrichment and economic growth too. The same could be said for many other countries—changes are urgent, but we may say the exchange of goods and mainly the transfer of money is more urgent yet.

The Puzzling Lack of Climate Action

The kind of change of which we are talking is more difficult to achieve than it may superficially seem. Across the world, lives are entangled with things, policies, and everyday activities that contribute to climate change, and while changing ideas may seem feasible, changing infrastructures requires time and investment of a different order. This is the world as we know it, an overheated world that has shifted into a higher gear in its movement toward greater profits, greater prosperity, and more of everything. One could look at anything from groceries in Western supermarkets to the factory that produced the concrete for the house in which the typical member of the global middle or upper class lives. Or we could lift our gaze to a higher scale and consider the phenomenal growth of the Port of Shanghai since the beginning of this century and the container ship technology that has reduced the price of transport by more than 90 percent since the 1960s (Eriksen 2016). Neither the films you watch on Netflix nor the smartphone you depend on for payments and communication are climate neutral.

One explanation for why so little is happening is *path dependence*, a technical term for systemic habit. Most of us affluent northerners act as we are accustomed to, perhaps with a few symbolic tweaks, such as composting kitchen waste before getting into a plane to speak about climate change in another country. On a larger scale, the electrification of the Norwegian oil platforms is touted as a great victory for the climate cause, conveniently failing to mention what the climate-neutral platforms actually extract and produce.

Changing habits is difficult, especially if you feel that things are getting better, which is the case regarding consumption and well-being in much of the world (Rosling 2018). This is why the contemporary youth protest movement, led by the Swedish teenager Greta Thunberg, is interesting. Adolescents have invested very little in the existing system and, thereby, possess far more flexibility when it comes to endorsing and practicing radical change.

Secondly, *temporality* is significant. Everybody lives in many temporalities, long and short. The meeting tomorrow morning that you need to prepare for has a short—indeed urgent—temporal horizon, as do your children's immediate needs. In the medium term, we plan for our own and the next generation's future, investing perhaps in a house, saving money for our children's studies, borrowing money for a pilgrimage (to Mecca, Lourdes, or Varanasi), a holiday, or a vehicle. Yet, in the long term, we shall all be dead. The question is, thus, to what extent are human beings capable of adjusting their behavior on the basis of events that will (or may) take place when our grandchildren's generation is on the brink of retirement? Evolution has not equipped us with a capability for this kind of global maximum (i.e., accepting a reduction in well-being in the short term in order to improve it in the long term), and it is uncertain whether we are actually able to change our behavior.

Thirdly, *spatiality* is similarly important. As with the case of temporality, human beings typically live most of their lives at a small scale, even if they are fully integrated into large-scale or indeed transnational or global systems. More than half of the text messages we send are addressed to between three and five persons. What matters most to most people is that which is near at hand and the people into whom we have invested our love and commitment or to whom we owe an intangible debt. This is a fundamental insight from anthropological research. On the other hand, abstract ideologies like nationalism and abstract religions like Islam and Christianity show that human solidarity can be extended to higher scales. Yet, it is uncertain to what extent most people will modify their actions, particularly to the detriment of people close to them, for the sake of lofty ideals or abstract communities populated by people they will never meet, such as their great-grandchildren's children.

Fourthly and finally, the problem of *affluence and excessive success* was never addressed in our evolutionary history. Evolution adapted us for a life in scarcity, competition, dangers, and threats, requiring instant gratification and local maxima. Shifting the focus, with the help of cultural practices, values, and knowledge, to a situation where there is too much and not too little will not be easy.

These four problems have not been properly addressed by climate scientists or politicians, even the most well-intentioned of them. Anthropology cannot give an unequivocal answer suitable for every budget, climate zone, or way of life, but we can offer some ideas, not drawn on discussions with other intellectuals but developed in close dialogue with other people's experiential worlds. This is where the anthropological form of knowledge production differs from nearly every other scientific endeavor. Ethnographies are shaped and created through the interaction of researcher and participants, not by asking particular questions to the latter. For an ethnography to be credible, it has to give a realistic and truthful rendering of local views, knowledges, lives, and experiences. In other words, if the political, economic, and technological elites agree that local perspectives need to be integrated into climate policy, the kind of knowledge represented in anthropology is indispensable.

The solutions offered in mainstream political discourse are typically of two kinds. One family of solutions holds out "green technology" and "green growth" as the only feasible way to deal with the issues. Pointing out that we are currently a global population of seven and a half billion (and counting), who all need food, shelter, and the right to a good life, advocates for this view, who include most politicians and corporate leaders, look to electric cars, solar power, large-scale tree planting, bans on plastic bags, and similar sustainable economic practices for solutions. They argue that a sustainable world will continue to require large amounts of energy in order to avoid famine and human suffering on an unimaginable scale and that the green transition requires huge investments. One of the heroes in this narrative is Norman Borlaug, the main architect behind the Green Revolution, which enabled food production in many countries to increase manifold thanks to extensive mechanization, new and more productive strains of cereal, and chemical fertilizer. The other narrative, supported by many intellectuals and researchers, argues that this kind of solution is short term, produces severe side effects and a loss of flexibility, and is incompatible with fundamental ecosystem needs. An early proponent of this holistic, ecological way of thinking was William Vogt, whose *Road to Survival* ([1948] 2010), a precursor to Rachel Carson's *Silent Spring* and, many would argue, the starting point for the radical environmental movement, claimed that the finiteness of Earth's resources should serve as a guide for political strategy and action (see Mann 2018 for an assessment). Often associated, and rightly so, with neo-Malthusian pessimism (see, e.g., the influential Club of Rome report *The Limits to Growth,* Meadows et al. 1972), Vogt and many of his followers advocate a reduction in the global population, while technological optimists have so far proved that the world is capable of feeding a population that has trebled in size since the publication of Vogt's book (Rosling 2018).

Ever since Marx and Engels argued against (indeed ridiculed) Malthus's warning, published on the cusp of the Industrial Revolution, Malthusianism has been proved wrong in the industrialized parts of the world. Technological and logistical advances have enabled increasing proportions of humanity to grow and prosper for two hundred years since the onset of the fossil fuel revolution just after the French and American Revolutions. However, the fact of anthropogenic climate change, resulting from the kind of accelerated change and economic growth that could be described as global overheating (Eriksen 2016, 2018), may yet prove Malthus right. The fact that natural resources have now been acknowledged to be finite, and that contemporary civilization is undermining the conditions for its own existence by being too successful for its own good in the short term, prompts a rethinking of the conditions of human life, its parameters, and its limitations.

The Primacy of the Local

As late as the 1990s, environmental concerns were a slightly countercultural specialty, inside and outside the academy. The philosopher Arne Johan Vetlesen, who has recently engaged with current anthropological approaches to the culture/nature divide (Vetlesen 2019), points out that during his studies in Oslo and Frankfurt in the 1980s environmental questions were never ever broached (Vetlesen 2015). Eriksen could echo his view from his vantage point across the university square in Oslo. In the anthropology they were taught at the time, environmental questions were associated with classic studies of human adaptation (often deterministic, often with a ring of cultural evolution and its assumed stages) or with distinctively unfashionable anthropologists like Leslie White and Marvin Harris, the latter often dismissed as a vulgar materialist, the former merely as dated. Neither have any visible influence on the field today.

The situation has changed radically in just a few decades. Research money, prestige publications, and professional profiles now often include an environmental interest, sometimes using the term Anthropocene and often mentioning climate change as a professional concern. In this book we are, in other words, adding our voices to a chorus that has very quickly become lively and multivocal. While contributing to shifting the gaze and acknowledging the need for an interdisciplinary, multiscalar, and multitemporal approach that highlights some of the shortcomings of the ethnographic method, we in the volume to follow insist on cultivating, and indeed advertising, the virtues of classic anthropological method in the present endeavor. Through the method of participant observation, we offer a perspective based on experience, from within and from below. We draw on a century of holistic research on integrated life-worlds that make sense on their own terms (if not necessarily on those of modern scientists) and that are continuously evolving and changing. Culture is not a thing, it is a process. Yet we do not deny that there is a need for historical, statistical, and macrosociological data to produce a full picture.

Climate change is not a catastrophe as this term is commonly understood, that is, in the rapid-onset sense. Unlike the coronavirus pandemic, it is incremental, a slowly creeping process in the slow-onset disaster sense, gradually altering ecologies and livelihoods in ways that differ significantly both physically and culturally, as has been acknowledged by anthropologists before (Oliver-Smith and Hoffman 2020). Although its effects are only now being felt in tangible and often dramatic ways, climate change has been advancing for years. In addition, the changes are often subtle, not always even steady as they effect many different locales. Furthermore, they differ from place to place, sometimes involving erosions, sometimes flooding, sometimes aridity, sometimes crop loss and flora and fauna changes, insect infestations, suddenly intolerable heat, or massive storms. The consequence is that while governments may increasingly issue national rules to counter the effects, edicts to reduce carbon emissions and the use of plastic, implement sustainable energy, and so on, most actual climate change maneuvers, dealings, assessments, adaptations, and countermeasures are taking place at local scales, as they must. Central actors in these efforts are the groups and peoples inhabiting the multifarious locales of the world, and although their experiences and responses reveal that although the ultimate cause of changes in their environments is global climate change, this may not be how the changes are understood. They may be perceived as enigmatic, divine, or routine events that have simply increased in size, intensity, and frequency or have mysteriously morphed. Yet human memory is frail, and most of us are mainly concerned with getting by day to day or, at best, year to year. The temporal and spatial scales of living are out of sync with the large scale and long term of planetary processes.

Changes in the climate may take place without many noticing them until the livestock begin to die, the fields are inundated by seawater, or the soil dries out because of the disappearance of glacier meltwater or rain. Science and erudite forbearance may play no part at the level of the concrete. Nonetheless, it is people on the local levels who are the ones actually coping or adapting to the changes and raising their voices to protest that the changes are not being heeded or dealt with.

The Contribution of This Book

These are the three areas this book deals with: (1) "Ways of Knowing"; (2) "Situations and Decisions"; and (3) "Politics, Policies, and Contestation."

Climate change is already perceptible, although it is not always understood as such. From Thailand to Queensland, many tend to blame weather (not climate), the vagaries of nature, or higher powers rather than the long-term effects of corporate extractive capitalism and complacent government policies when extreme events become more frequent and more intense. However, people in various locations respond to the changes in weather and other patterns by adjusting their practices, migrating, trying out alternative livelihoods, or working discursively or politically to change their circumstances or the underlying causes of their problem. For this reason, the term *resilience* has become a key concept in social research on climate change.

As shown above, a number of volumes on climate change already exist in anthropology. What is unique to this book is its dual emphasis on *regions* and *themes*. The book hopefully shows the importance of ethnographic detail in coming to terms with climate change, showing simultaneously that this is a planetary problem that affects people everywhere, that it is responded to very differently in different parts of the world, and that it requires a broad range of solutions anchored in local circumstances. Just as the main mission of anthropology in the last century has been to document and make sense of human diversity, this book shows variation. Dealing with the loss of snow in the Austrian mountains and its consequences for skiers is quite different from dealing with flooding in the Elbe Valley (to take a neighboring country) and calls for different kinds of stratagems. As we have pointed out, there is frequently no general agreement about the appropriateness of particular solutions, especially at the point where politics takes over from cultural analysis.

In order to provide an appropriate frame for the present book, we could also approach it like this. In a programmatic article written for non-anthropologists, Barnes et al. (2013) identify three areas where anthropology may be in a privileged position to contribute to research on climate change: (1) cultural values and political relations; (2) a historical awareness connecting the present to the past; and (3) the holistic perspective on human life connecting culture and society to its broader ecological context. To this, we add a fourth: if it is anything at all, anthropology is the study of cultural *diversity*, and the niche distinguishing this book from many others consists in the breadth of its ethnography, which indicates that the problems, effects, and solutions of climate change vary considerably. If there is a takeaway lesson for policymakers here, it must be that one size does not fit all, which is to say that climate change is a global phenomenon that stems from a relatively short number of causal factors (commonly referred to under the umbrella term *emissions*), but the ways to fight it have to be localized.

Our empirical cases range from the US Southwest and Southeast, Germany, and Austria to Bangladesh, Mexico, Namibia, New Zealand, and Portugal. In the first part, "Ways of Knowing," the main focus is on differing perceptions of climate change. Starting by distinguishing between climate and the weather, Michael Schnegg points out that Namibian pastoralists in fact possess considerable knowledge about the latter but lack concepts about the former. Like most of us, perhaps. He also proposes the concept *environmental pluralism* to designate the diverse sources of knowledge about weather and the environment. In the next chapter, Alexander Aisher takes us across the globe to Arunachal Pradesh in the Eastern Himalayas, where concerns with weather are no less prominent than in semiarid Namibia but are played out differently in a very wet climate; here the major North Indian rivers originate and cosmological explanations are invoked to make sense of "strange weather" such as sudden storms. Matthew Lauer and coauthors provide a third, locally grounded lens through which to view climate change in their account of the diverging perceptions of the spread of the crown-of-thorns starfish in French Polynesia. It had been known for a long time by local fishermen, who did not accord it much importance as it did not interfere directly with fishing; however, scientists, who represented different knowledge interests, understood the prevalence of starfish in a different light: as destructive to coral and indicative of climate change. In the final chapter of this section, Pedro Paulo de Miranda Araújo Soares traces the transformation of the Amazonian city of Belém from a "tropical paradise" to a flood-prone, profit-generating, ecologically precarious city seen as a success through the eyes of planners but not from the point of view of residents or ecologists.

The second part, titled "Situations and Decisions," focuses on changes in the physical world resulting from conscious, if sometimes misguided, decisions at a political level. Tasneem Siddiqui, Mohammad Jalal Uddin Sikder, and Mohammad Rashed Alam Bhuyian's chapter, focused geographically near Aisher's field site but socially and culturally a world apart in low-lying, Bangla-speaking, Muslim Bangladesh, presents findings from research on migration into Bangladeshi cities. The migrants come from ecologically vulnerable places (with flooding, land scarcity, and land grabbing) and go to ecologically vulnerable places (with poor hygiene and housing, and so on). Although population growth and vagaries of nature may be invoked, the analysis makes it clear that the situation is a result of policy decisions and anthropogenic climate change. Brian Orland, Meredith Welch-Devine, and Micah Taylor, in the following chapter, investigate the reluctance of people in the US state of Georgia to migrate following a devastating hurricane, quite the contrary of the Bangladeshi situation, where many are prone to leave owing to erratic weather. A similarity is nevertheless that the poor are more likely to leave than the affluent. Moving yet again to a different continent and a different local context, Paul Schneider and Bruce Glavovic describe responses to erosion and environmental degradation in the Coromandel peninsula, a popular holiday destination, in Aotearoa New Zealand's North Island. Compounding the complexity of the locality is the fact that it is inhabited partly by people of European descent, partly by Maori. In the following chapter, A. Peter Castro, who has worked in different East African countries, presents three "cautionary tales" from three countries—Kenya, Somalia, and Ethiopia—which all indicate conflicts of interest and power struggles where, alas, advocates for the environment tend to lose out, beginning with logging activities in the Kenyan highlands as early as the turn of the last century. Shifting the attention to problems of the affluent, Herta Nöbauer then shows how ski resorts in Austria are developing technological solutions to the increasingly erratic snowfall and retreating glaciers by building artificial, climate-independent slopes and tracks.

The third and final part, "Politics, Policies, and Contestations," begins with Kristoffer Albris's analysis of resilience and reconstruction following a devastating flood in the Elbe Valley, Germany. Here, in a setting comparable to that of neighboring Austria, adaptation rather than calls for systemic change is the main response to perceived climate change. Julie Maldonado and Beth Rose Middleton, in the next chapter, detail how Native American tribes in the Southwest struggle, as they have since the beginning of settler colonialism, to retain autonomy and their livelihoods in the face of encroaching industrial capitalism, and how the unpredictability of the weather has exacerbated their problems. Loss is also the topic of the next chapter, by Guilherme José da Silva e Sá, who provides an account of a "rewilding" project in Portugal that is an attempt to restore a natural habitat to an imagined pristine condition by introducing species that may have thrived there before the Anthropocene. The thin membrane separating humanity from nature in the modern constitution becomes visible in this way, and rewilding is also a reminder that the boundary between nature and culture is now wholly managed by humans. Returning to the theme of knowledge but supplementing it with an analysis of the political forces creating a particular, volatile relationship between humans and nature, Amanda Leppert and Roberto E. Barrios explain Meso-American historical perspectives on the environment and those in the contemporary situation. Susanna M. Hoffman's chapter, finally, identifies some of the human drivers of seemingly natural disasters, connecting them to the general processes of climate change.

The local issues differ; local understandings vary; the opportunities presenting themselves to the affected people are hugely different between the ski slopes of the Austrian alps and the shanties of Bangladeshi cities. Nonetheless, at the same time, there is a pattern that connects them, that of globalization in the Anthropocene, which is not merely about labor migration, or consumer goods, or social media, or mining jobs, or outsourcing and a growing scalar gap between decision-makers and the people decisions are made about. It is invariably about the entanglement of everybody with everybody. Ironically, this turns out to be the crisis that requires a truly global conversation about our common destiny, and in this area, anthropologists can make a significant difference. Time is running out, and at the time of this writing, it is still easier to imagine the end of the world than the end of capitalism.

Thomas Hylland Eriksen is professor of social anthropology at the University of Oslo and external scientific member of the Max Planck Society. His research has focused mainly on identity politics and globalization, but he has published widely on other topics as well, in recent years with a focus on accelerated change ("overheating") in the realm of the economy, identity, and environment. Among his recent books in English are *Fredrik Barth: An Intellectual Biography* (2015); *Overheating: An Anthropology of Accelerated Change* (2016); *Boomtown* (2018); the coedited, with Elisabeth Schober, *Identities Destabilised* (2016); and the coedited, with Astrid B. Stensrud, *Climate, Capitalism and Communities* (2019).

Paulo Mendes is a professor of social anthropology at Universidade de Trás-os-Montes e Alto Douro (UTAD) and researcher at Centro em Rede de Investigação em Antropologia (CRIA). His research has focused on fishing villages and on the (inter)relation of environment and place making. The books *Se o Mar Deixar* (1996) and *O Manda é que Manda* (2013) address that matter. The latter, titled in English *The Sea Commands,* is forthcoming from Berghahn Books. More recently he edited with Humberto Martins a book on the personal experience of fieldwork, *Envolvimento e Experiência de Trabalho de Campo* (2016).

Notes

1. A brief history of notions such as "progress," "economic development," or "sustainable development" would translate systems of thought and ways of doing that would highlight simultaneously the appropriation and exploration of nature resources, modes of production, and shared concerns that are key for the under-

standing of anthropogenic forces behind climate change. In one word, *growth* still is the main goal, and although it may be measured in economical charts, its value is also moral, if not mainly so. Concomitantly "the faith in technology," in the unstoppable technological progress, contributes to deferring in time, if not suspending, changes and policies that could mitigate climate change more rapidly.
2. The production of "clean energy" is in itself a never-ending puzzle. Hydroelectric dams are big sources of methane and CO_2; wind turbines use sulfur hexafluoride (SF_6), a potent greenhouse gas; solar energy relies heavily on mining and metallurgical industries and produces large amounts of toxic waste (mainly tetrachloride). Though these energy sources are reportedly less harmful to the environment than fossil fuels, controversies remain; see, e.g., the views of James Lovelock (2007) on nuclear energy.
3. At the same time, it is not uncommon to find affluent persons traveling around the world in private jets while supporting financially "green projects" such as electric cars, forestation, or "transition communities."

References

Baer, Hans, and Merrill Singer. 2009. *Global Warming and the Political Ecology of Health: Emerging Crises and Systemic Solutions*. Walnut Creek, CA: AltaMira Press

———. 2018. *The Anthropology of Climate Change: An Integrated Critical Perspective*. 2nd ed. London: Routledge.

Barnes, Jessica. 2014. *Cultivating the Nile: The Everyday Politics of Water in Egypt*. Durham, NC: Duke University Press.

Barnes, Jessica, et al. 2013. "Contribution of Anthropology to the Study of Climate Change." *Nature Climate Change* 3: 541–44.

Bateson, Gregory. 1972. "The Roots of Ecological Crisis." In Bateson, *Steps to an Ecology of Mind*, 494–98. New York: Ballantine.

Bateson, Gregory, D. D. Jackson, J. Haley, and H. Weakland. 1956. "Toward a Theory of Schizophrenia." *Behavioral Science* 1: 251–64.

Bauman, Zygmunt. 2000. *Liquid Modernity*. Cambridge: Polity.

Beck, Ulrich. 2009. *World at Risk*. Cambridge: Polity.

Connor, Linda. 2016. *Climate Change and Anthropos: Planet, People and Places*. London: Routledge.

Crate, Susan A., ed. 2019. "Storying Climate Change." Special issue of *Practicing Anthropology* 41(3).

Crate, Susan A., and Mark Nuttall, eds. (2009) 2016. *Anthropology and Climate Change: From Encounters to Actions*. Walnut Creek, CA: Left Coast Press.

Dalsgaard, Steffen. 2013. "The Commensurability of Carbon: Making Value and Money of Climate Change." *HAU: Journal of Ethnographic Theory* 3(1): 80–98.

Diamond, Jared. 2005. *Collapse: How Societies Choose to Fail or Succeed*. New York: Viking.

Dove, Michael R., ed. 2013. *The Anthropology of Climate Change: An Historical Reader*. Chichester: John Wiley & Sons.

Eriksen, Thomas Hylland. 2016. *Overheating: An Anthropology of Accelerated Change*. London: Pluto.

———. 2018 *Boomtown: Runaway Globalisation on the Queensland Coast*. London: Pluto.

Fagan, Brian. 1999. *Floods, Famines, and Emperors: El Niño and the Fate of Civilization*. Cambridge: Cambridge University Press.
Gausset, Quentin, Jens Hoff, and Simon Lex, eds. 2019. *Building a Sustainable Future: The Role of Non-state Actors in the Green Transition*. London: Routledge.
Gullestad, Marianne. 1992. *The Art of Social Relations*. Oslo: Universitetsforlaget.
Hastrup, Kirsten, and Frida Hastrup, eds. 2015. *Waterworlds: Anthropology in Fluid Environments*. Oxford: Berghahn Books.
Hastrup, Kirsten, and Karen Fog Olwig, eds. 2012. *Climate Change and Human Mobility: Global Challenges to the Social Sciences*. Cambridge: Cambridge University Press.
Hendry, Joy. 2014. *Science and Sustainability: Learning from Indigenous Wisdom*. London: Palgrave Macmillan.
Hornborg, Alf. 2019. *Nature, Society and Justice in the Anthropocene: Unravelling the Money-Energy-Technology Complex*. Cambridge: Cambridge University Press.
Kellogg, William W., and Margaret Mead, eds. 1980. *The Atmosphere: Endangered and Endangering*. Turnbridge Wells, UK: Castle House Publications.
Khokhar, Tariq. 2017. "Global Emissions Rose 60 Percent between 1990 and 2013." World Bank: Atlas of Sustainable Development Goals. Retrieved 12 October 2020 from http://blogs.worldbank.org/opendata/chart-global-co2-emissions-rose-60-between-1990-and-2013.
Klein, Naomi. 2014. *This Changes Everything: Capitalism versus the Climate*. New York: Simon & Schuster.
Latour, Bruno. 2017. *Down to Earth: Politics in the New Climatic Regime*. Cambridge: Polity.
Lévi-Strauss, C. 1983. *Le Regard éloigné*. Paris: Plon. (English edition: *The View from Afar*. New York: Basic Books 1985.)
Lovelock, James. 2007. *The Revenge of Gaia: Why the Earth Is Fighting Back—And How We Can Still Save Humanity*. London: Penguin.
Mann, Charles C. 2018. *The Prophet and the Wizard*. New York: Simon & Schuster.
Meadows, Donella H., Dennis L. Meadows, Jørgen Randers, and William W. Behrens III. 1972. *The Limits to Growth*. New York: Universe Books.
Mitchell, Timothy. 2011. *Carbon Democracy: Political Power in the Age of Oil*. London: Verso.
Norgaard, Kari M. 2011. *Living in Denial: Climate Change, Emotions, and Everyday Life*. Cambridge, MA: MIT Press.
Oliver-Smith, Anthony, and Susanna M. Hoffman, eds. 2020. *The Angry Earth: Disaster in Anthropological Perspective*. 2nd ed. London: Routledge.
Orlove, Ben, J. C. H. Chiang, and M. A. Cane. "Forecasting Andean Rainfall and Crop Yield from the Influence of El Niño on Pleiades Visibility." *Nature* 403: 68–71.
Rayner, Steve. 2016. "What Might Evans-Pritchard Have Made of Two Degrees?" *Anthropology Today* 32(4): 1–2.
Rayner, Steve, and Elizabeth Malone, eds. 1998. *Human Choice and Climate Change*. Columbus: Battelle Press.
Rosa, Hartmut. 2016. *Resonanz: Eine Soziologie der Weltbeziehung*. Berlin: Suhrkamp.
Rosling, Hans, with Ola Rosling and Anna Rosling Rönnlund. 2018. *Factfulness: Ten Reasons We're Wrong about the World—And Why Things Are Better Than You Think*. London: Sceptre.
Shore, Cris, and Susana Trnka, eds. 2013. *Up Close and Personal: On Peripheral Perspectives and the Production of Anthropological Knowledge*. Oxford: Berghahn Books.
Steffen, Will, Paul J. Crutzen, and John R. McNeill. 2007. "The Anthropocene: Are Humans Now Overwhelming the Great Forces of Nature?" *Ambio* 36(8): 614–21.

Stensrud, Astrid B., and Thomas Hylland Eriksen, eds. 2019. *Climate, Capitalism and Communities*. London: Pluto.
Steward, Julian. 1955. *Theory of Culture Change*. Urbana: University of Illinois Press.
Strauss, Sarah, Stephanie Rupp, and Thomas Love, eds. 2013. *Cultures of Energy: Power, Practices, Technologies*. Walnut Creek, CA: Left Coast Press.
Tainter, Joseph A. 1988. *The Collapse of Complex Societies*. Cambridge: Cambridge University Press.
Vetlesen, Arne Johan. 2015. *The Denial of Nature: Environmental Philosophy in the Era of Global Capitalism*. London: Routledge.
———. 2019. *Cosmologies of the Anthropocene: Panpsychism, Animism, and the Limits of Posthumanism*. London: Routledge.
Vogt, William. (1948) 2010. *Road to Survival*. Whitefish, MT: Kessinger.
White, Leslie. (1949) 2005. *The Science of Culture: A Study of Man and Civilization*. Clinton Corner, NY: Percheron Press.
Wilhite, Harold. 2016. *The Political Economy of Low Carbon Transformation: Breaking the Habits of Capitalism*. London: Routledge.

Part I

Ways of Knowing

The ubiquity of climate change is undeniable. However, as the ways in which it is lived, interpreted, and understood vary locally, so do the social narratives that explain it. While by and large (hard) sciences, technology, mathematical models, and algorithms dominate public discourse, there are places across the world where this is not the case. In the geographies where scientific accounts do not prevail, the definition often speaks of weather, not climate; reports to long-term cultural lore or embedded belief systems; elides historical depth and accumulating knowledge to cite obscure long-term cycles; or speaks of personal experiences and not overall global scales. It is commonly said that "climate change does not respect borders." In turn, also, national and international organizations, and the sciences that inform them, often do not respect other forms of knowledge—although some authors consider that this may be changing. Ways of knowing are also ways of forgetting: forgetting past events, forgetting forms of interpretation, forgetting shared worldviews and cosmogonies.

The ethnographies in the first section of *Cooling Down: Local Responses to Global Climate Change* provide a good example of how anthropology is able to contribute to the study of climate change, not only by invoking other forms of knowledge but also by reflecting on how other societies and cultures are dealing with the phenomenon. In the forthcoming pages we travel through different ontologies and geographies: Namibia, the Himalayas, Polynesia, and Brazil. All chapters in this section further address concomitantly these issues: culture-situated cosmogonies and *Westernized* (as M. N. Srinivas named it long ago) ways of knowing, their juxtaposition and relations of power and authority.

Michael Schnegg takes us to Fransfontein, a small community in northwestern Namibia. The *ǂnūkhoen*, hunter-gatherers, nowadays predominantly shepherds, refer to the most drastic periods of drought they face as *ǂû-i ǀkhai*, times of "no food." More than a long period without precipitation, drought is understood as a phenomenon centered on its sociobiological effects. However, after a long fieldwork, Schnegg, himself a cattle farmer who has lost half of his herd to drought, is challenged by different local ways of understanding and narrating the climatic phenomenon. Gendered winds, deities, and climatology coexist locally and are often expressed collectively by the same person as sources of explanation for both recurrent and extraordinary phenomena. Consequently, traditional and modern ecological knowledge coexist. This concurrence of narratives based on tradition and religion crossing with scientific knowledge suggests to Schnegg a notion of *environmental pluralism*, "where a person uses different, ontologically and epistemologically contradicting knowledge systems to explain environmental phenomena."

Alexander Aisher takes us to the Eastern Hymalayas, to Talum, a village mainly inhabited by Nyishi people who depend on subsistence cultivation for their survival. He writes, "place-based communities in the region have already noticed more erratic rainfall," and while many local perceptions of climate change in the Himalayas are now being validated by scientific evidence, as all anthropologists know, storytelling, traditional ecological knowledge (TEK), and the ethnographies about it are as important as the data collected upon the thousands of meter-deep polar ice drills.

Matthew Lauer and his coauthors follow this line of enquiry unequivocally and question how different ways of knowledge, namely TEK and science, approach marine life (and its eradication) in French Polynesia. On the reefs surrounding Moorea, an island twenty kilometers west of Tahiti, an outbreak of crown-of-thorns starfish was being understood by marine scientists as the cause of a major coral depletion. At the same time, local fishing communities were aware of the outbreak and the coral loss, but did not consider them significant and did little to respond to them, as the authors mention. The gigantic and poisonous starfish has been a matter of vivid discussions among the numerous scientists associated with "one of the most studied tropical coral reef systems in the world." The multidisciplinary team of scientists of this chapter pose questions such as, "Who notices changes to Moorea's coral reefs, and how can it be judged if they are noteworthy?" and, "What should be done if a perturbation and its effects are identified?" It is noteworthy that French Polynesia, although being a relatively highly subsidized colony, remains a colony. Questions of "knowledge colonization" gain particular relevance here, if not else for symbolic related issues. Nonetheless, as the starfish overshadows the coral reef, scientific knowledge overlays local ecological knowledge and local everyday practices deeply rooted in time.

Concerning the Una Basin, Belém, Brazil, Pedro Paulo de Miranda Araújo Soares depicts a "modernization" process: "In the name of progress, modernity, and ultimately development, both nationally and internationally funded economic projects attracted a massive influx of investments and migrants to Belém between the 1960s and the late 1980s." During those decades, the population of Belém more than tripled its size, and the banks of the Amazon Delta region became densely populated, crowded with highly polluting factories and subject to never-ending works carried out on drainage, basic sanitation (sewage and solid waste management), and water supply. The main project turned what was once a clean river flowing through the city into a concrete-embanked polluted drainage ditch. Soares centers his research on the impacts of the policies mainly on what concerns the management of floods and introduces the concept of environmental memory. Soares's notes on the development of the sanitary conditions of Belém are particularly striking. A large number of deaths in Brazil, specifically in the cities of the Amazon Delta, are certainly related to the schemes of the government, but also to centuries of devastation of the Amazon River forest, and ideas of progress, economic growth, and personal success. Yet, despite the horrific pollution caused by the capitalistic development programs that is now constantly exacerbated by climate change flooding, the inhabitants of the even more marginalized neighborhoods along what was once the fresh river continue to treat the putrid stream as if it were still fresh. They interpret the befouled floodwaters as per the memory of "what once was," not "what is now." Perhaps surprisingly, or not so much, Pedro Soares tells us as well that the Amazon cities are not being considered in the Brazilian climate change policies and reports.

Ways of knowing are ways of living and, therefore, different ways of participating. The obliteration of other forms of knowing, namely those based on TEK, is a major loss to the study of climate change and to forms of mitigation and disaster prevention. Back in 2004, several people of the Andaman Islands became a world news sensation: although isolated and with no modern earthquake alert technologies, they managed to survive a tsunami by "climbing up the mountains" hours before it made landfall. Reportedly, their TEK establishes links between a number of natural events, such as earth tremors, sea retrocede, change in behavior of other animals, tsunamis, and the effects of climate change, and even affords ways of dealing with it.

Relying on ethnographies of different epistemologies and ontologies, anthropology may not only report "case studies" but also contribute to the proposition of different ways of doing, therefore to different ways of fixing that finally may configure other ways of knowing and inhabiting shared environments.

Chapter 1

Environmental Pluralism

Knowing the Namibian Weather in Times of Climate Change

Michael Schnegg

Introduction

At present, human knowledge about the weather and climate is undergoing a period of transformation. From one perspective, global warming modifies local weather patterns: winds are changing, soils are drying, permafrost is thawing, storms are increasing, and melting glaciers are contributing to rising sea levels. These phenomena challenge local understandings if, for example, they invalidate weather predictions that held true in the past (Brüggemann and Rödder 2020; Ehlert 2012; Krupnik and Jolly 2002; Green, Billy, and Tapim 2010; Roncoli, Ingram, and Kirshen 2002; Schnegg, O'Brian, and Sievert 2021). From another perspective, newly developing scientific knowledge of climate change that is spreading around the globe promotes new ways of seeing nature. If, as Hulme explains, "the idea of climate works to stabilize cultural relationships between humans and their weather" (Hulme 2015: 10), then the globalization of climate change knowledge can alter the way people understand the weather as well (Jasanoff 2010; Paerregaard 2013; Pettenger 2016).

Following both transformations, the aim of this chapter is to explore whether and how global climate change and scientific knowledge alter local understandings of the weather in Namibia. In doing so, this chapter focuses on perceptions. In my related work, I have developed a phe-

Notes for this chapter begin on page 43.

nomenological perspective and argued that perceptions are not the only way of accessing the world.[1] On the other side of the coin are practices through which we enact the environment. As I have shown, people switch between perceiving and practicing, which helps to explain how different ways of accessing the world create different ways of knowing the environment and, possibly, also different worlds (Schnegg 2019, 2021a, 2021b). While this duality of practicing and perceiving provides a very effective approach to explore different layers of knowing, it first requires an in-depth understanding of the epistemic structures (i.e., perceptions) that this chapter provides.

The anthropological literature has shown convincingly that scientific and indigenous ways of knowing the environment are in part based on distinct ontological assumptions about how the world works (Antweiler 1998; Berkes 2008). In environmental science, climate is defined as the "average weather." Thus, while weather describes the conditions of the atmosphere over a relatively short period of time, climate refers to long-term averages of daily weather, described in terms of the mean values and variability of specific indicators. To accumulate knowledge about both the weather and climate, environmental scientists rely on a number of epistemological assumptions. They often *assume* knowledge to be independent of contexts, allowing explanations gleaned from one set of specific circumstances to be applied to other, similar contexts (DeWalt 1994; Schnegg 2014).[2] Moreover, environmental scientists *assume* that many patterns are beyond direct human observation and require the aid of scientific instruments for study. The aggregation from weather to climate follows formal rules that transcend local meaning systems (Jasanoff 2010). While, in these scientific terms, human activities can affect climate, these relationships are understood as global aggregates both in terms of cause and in terms of humanity's ability to mitigate the harmful consequences of human-induced climate change (Schnegg 2019, 2021a).

Therefore, on a more abstract level, scientific climate knowledge is grounded in a set of principles: science allows for the separation of cause and effect on a large temporal scale, e.g., burning coal on an industrial scale one hundred years ago may have effects on the climate-influenced weather today. Science also allows for the separation of cause and effect spatially. Emissions from industrialized countries can contribute to extreme weather events in less industrialized countries. Additionally, scientific climate knowledge does not recognize natural forces, such as the wind or water, as agents incorporating a design. Moreover, scientific knowledge is not based on immediate sensory experiences but on longitudinal measurement and observation. Finally, while scientists can predict local weather patterns a few days in advance, predictions of changes in climate and average weather conditions can be made on an extended timescale.

Indigenous understandings of the weather differ in many regards. They are often embedded in holistic worldviews that connect the land to the air and water, the earth to the sky, plants to the animals, and people to spirits (Cochran et al. 2013; Paerregaard 2013; Antweiler 1998).[3] Thus, reasoning includes diverse aspects of nature, and weather results from the interaction of these components, to which humanlike agency is often attributed (Roncoli, Ingram, and Kirshen 2002). Moreover, given its importance as the source of weather and its uncontrollability, the sky and the dynamic events that occur there are unsurprisingly animated frequently in human thought (Donner 2007; Ingold 2006). Storms punish; lightning frightens. Across many cultures, weather-related phenomena are associated with specific supernatural powers (Bierlein 1994). Given this, weather predictions are often based on an interpretation of the intentions of the supernatural world and how these are present in the behavior of the elements of nature, including birds, plants, animals, winds, cloud patterns, and the movements of the moon and stars (Elia, Mutula, and Stilwell 2014; Orlove et al. 2010; Roncoli, Ingram, and Kirshen 2002; Nyong, Adesina, and Osman-Elasha 2007; Ifejika Speranza et al. 2010; Lefale 2010; King, Skipper, and Tawhai 2008).

On an abstract level, many indigenous explanations of the weather share a number of ontological and epistemological principles as well. First, they are usually integrated into human moral concerns, thereby establishing a concrete and relatively short-term temporal link between human-induced causes and weather-related effects. For example, when humans do something immoral or careless, they are more or less immediately punished by harmful weather events. To calm and placate disturbed natural agencies, people may take various actions, including offering ritual sacrifices. Secondly, these causal linkages are local: "*our*" behavior shapes "*our*" weather (Friedrich 2018; Rudiak-Gould 2014; Schnegg, O'Brian, and Sievert 2021). Third, nonhuman agents, including the elements of nature, also influence the weather. Sometimes their interactions are mediated by supernatural powers (Rayner 2003). Fourth, local weather-related reasoning does not always rely on causal explanations but may reflect a more fatalistic view. Fifth, epistemologically, laypeople's tacit knowledge is typically laden with emotion and sensitive to the context in which it is applied (Gorman-Murray 2010; Vannini et al. 2011).

While these differences between indigenous and scientific knowledge are relatively well established in anthropology, what happens when those epistemologies collide is much more controversial. There are at least three different answers that can be proposed to this question.

Established Approaches to Linking Different Ways of Knowing

In one view, indigenous knowledge will (and should) be overcome, giving way to scientific truth. Studies conducted in the *public understanding of science* (PUS) paradigm begin from the premises of an information-deficit model. They attribute skepticism regarding scientific knowledge to a lack of understanding resulting from a lack of information. For example, a large number of studies show that common explanations of climate change implicate the hole in the ozone layer, while experts eschew this connection (Kempton 1991; Löfstedt 1991; Rayner 2003; Bostrom et al. 1994; Thompson and Rayner 1998). A conclusion in this research has been that this "misinterpretation" can be corrected through better communication. In general, research that is conducted in the public understanding of science paradigm is based on a sender-receiver communication model that has been criticized as being one-dimensional and in part naïve (Hulme 2009; Jasanoff 2010; Kearney 1994; Weingart, Engels, and Pansegrau 2000).

In another view, indigenous knowledge is increasingly repressed by dominant discourses. This *sociocultural approach* is not restricted to knowledge but encompasses norms, values, actors, and their social networks. As Jasanoff states, "Without human actors . . . even scientific claims have no power to move others" (Jasanoff 2004: 36). Instead of assuming a linear transfer of climate change knowledge from sender to receiver, the sociocultural model explores how social fields, including science, politics, the media, and the general public, negotiate climate change socially and culturally (Krauss 2012; Jasanoff 2010; Rudiak-Gould 2012). To study this, a discourse-centered analysis is typically used to deconstruct how dominant actors shape forms of knowing and, eventually, reality. Theoretically, the sociocultural approach is often grounded in science and technology studies (STS) (Jasanoff 2010; Weisser et al. 2014; Pettenger 2016).

In a third view, indigenous and scientific knowledge are less mutually contradictory than is often assumed (Gagnon and Berteaux 2009; Nyong, Adesina, and Osman-Elasha 2007; Huntington et al. 2004; Herman-Mercer, Schuster, and Maracle 2011; Weatherhead, Gearheard, and Barry 2010; Roncoli, Ingram, and Kirshen 2002; Risiro et al. 2012, Orlove et al. 2010; Lefale 2010; Kalanda-Joshua et al. 2011). Indigenous knowledge offers new insights for science, and both perspectives can be integrated (*integrated approach*) to effectively tackle the environmental problems we face.[4] For example, Gearheard and colleagues (2010) have argued in their comparison of indigenous and scientific interpretations of changing wind patterns among Inuit hunters in the Canadian Arctic that similarities in observations and interpretations of long-term patterns can strengthen confidence in the conclusions, while differences can lead to new questions for further investigation. While many authors have pointed to this approach as an opportunity for collaboration (Nyong, Adesina, and Osman-Elasha 2007; Huntington, Suydam, and Rosenberg 2004; Huntington 2000; Green, Billy, and Tapim 2010; Ifejika Speranza et al. 2010), others take a more cautious view (Nadasdy 2003; Chanza and De Wit 2013).

Although the three approaches differ significantly, they are similar in their focus on the interactions between distinct epistemologies and forms of knowing when asking: (1) how one "wins over the other" (public understanding of science, sociocultural approach), or (2) how they complement and stimulate one another (integrated approach). In doing so, they assume that an actor typically has (only) one way of knowing, and also that this way of knowing typically differs between the scientist and the nonscientist. The framework I offer overcomes this view and argues that people can (but also may not) combine plural ways of knowing about the environment. I propose to address this as an *environmental pluralism.*

Environmental Pluralism

Anthropological research in the fields of medicine and law has shown convincingly that the introduction of new principles of knowing does not automatically lead to the replacement of existing ones. In medical anthropology, it is widely agreed that people recognize different, often contradictory, interpretations of the body, its functioning, and ways of diagnosing and treating its illnesses. People differentially draw upon these understandings at different times (Pelto and Pelto 1997). In a similar vein, legal anthropologists have shown that multiple normative frameworks often coexist and that people actively choose between them depending on context (Merry 1988; Benda-Beckmann, Benda-Beckmann, and Wiber 2006). While the ideal of plural normative orders and knowledge is applied successfully in other social fields, a careful study of how pluralism might also be relevant for knowledge of the environment has not been adequately considered.

Environmental pluralism describes a situation, where a person uses different, ontologically and epistemologically distinct knowledge systems to explain environmental phenomena. It brings into focus the role of climate change in introducing new knowledge about the natural world that can result in multiple—even contradictory—ways of knowing the environment based on different epistemological and ontological assumptions (Schnegg 2019, 2021a; Schnegg, O'Brian, and Sievert 2021).

From the literature, there are some indications that environmental pluralism exists. For example, Ehlert (2012) reports that wet rice farmers in

Vietnam combine both traditional and modern means of weather forecasting to make farming decisions. Among the farmers she studied, short-term weather predictions are typically based on "reading the water," while the longer-term future is judged with meteorological knowledge transmitted by radio and loudspeakers set up by the state (Ehlert 2012). Similarly, Paerregaard (2013) has shown that climate change discourses introduced new ways of understanding human-environment interactions in the Peruvian community she studied. At the same time, people deny that the changes they observe could be caused by factors outside the community itself (Paerregaard 2013). Equally, for the Iñupiat in western Alaska, their local discourse on weather change and the scientific discourse are separate discourses. Unlike the examples of hybrid ways of explaining, where different discourses are mainly compatible, Iñupiat knowledge and daily observations of the environment are far from matching the generalized scientific knowledge. At the same time, they can readily apply both ways of knowing depending on their situation (Marino and Schweitzer 2009: 212).

Against this background of the existing literature, I discuss my case study, which reveals in more detail how environmental pluralism emerges and exists.

Being in Northwestern Namibia

Fransfontein is a community of roughly 250 households in the arid environment of northwestern Namibia, a region also referred to as Kunene. The communal pastures surrounding it are dotted with small settlements of five to twenty homesteads each. They cluster around drilled boreholes that provide water to humans and their livestock. The majority of people consider themselves as ǂnūkhoen (or Damara people). ǂnūkhoen is a Khoekhoegowab word and literally translates as "black people."[5] Most likely, the name was given to the ǂnūkhoen by strangers. Before contact with German colonizers in the late nineteenth century, the ǂnūkhoen were presumably hunter-gatherers, with significant contributions to their economy coming from small-scale trading.

With forced integration into a colonial system that began around the turn of the twentieth century, and which was mainly achieved through land and grazing taxes, the need to produce for the market grew. Taxation was a major force that led to the spread of pastoralism among the Damara people, and today all Damara households own livestock. However, during the middle of the twentieth century the reduction of land through the Odendaal Plan meant that it became impossible to make a living from the land alone.[6] Livelihoods began to diversify, leading to new subsistence patterns that involved labor markets and the state. While some combine pastoralism with wage labor in the local economy, others link pastoralism with state welfare and wage labor in the national economy through strategies such as migration (Greiner 2011).

Since German colonial rule, Fransfontein has had a school, and since the 1970s, all children attend at least primary school. The classes are taught in Khoekhoegowab, and a few students proceed from the secondary school to university. The radio is the most important means of public communication. A radio station (NBC Nama-Damara) that broadcasts in Khoekhoegowab can be received throughout the entire area, where only approximately two hundred thousand Khoekhoegowab speakers in total reside. Much of the information aired is personal and relates to family matters, e.g., upcoming funerals, weddings, and things "lost and found." The radio is also the most important source for people to get to know about climate change and its scientific causes and explanations. Moreover, in Fransfontein, some households have access to electricity and own a television. However, many people, especially the older residents, have never seen a meteorological map or the perspective on the weather that such maps entail.

In the arid environment of northwestern Namibia, precipitation varies between one and three hundred millimeters per year with marked variation in both time and space. As in the rest of Namibia, the precipitation increases from west (the Atlantic coast) to east and is coupled with a high evaporation rate. Every seven to ten years the amount of precipitation is so low that scientists refer to it as a drought. These interannual fluctuations recur regularly (at least they have done so for as long as reliable data has been recorded), and meteorologists associate the cyclic ups and downs with El Niño and La Niña phenomena. Droughts are the most dramatic weather-related events, and if asked how the weather shapes their lives, the people in Fransfontein are likely to respond that "droughts" pose the most severe challenge for them all. In Fransfontein, one word often used to refer to a drought is *ǂû-i ǀkhai*, which literally translates as "no food." Thus, for the people, drought is more than a lack of precipitation; it is a sociobiological phenomenon largely focused on its effects. The period from 2013 until the time of writing is considered a drought in Kunene, both by scientists and by the local people (Schnegg and Bollig 2016). This situation made it comparatively easy for me to talk to people about potential explanations.

I first came to Kunene in 2003 when my wife and colleague Julia Pauli and I were looking for a place to conduct a community ethnography. We lived in Fransfontein for more than a year, and we have returned many times since. In 2010, I started a comparative research project to explore how the notion of community-based natural resource management (CBNRM) transforms social forms of water sharing in northwestern Namibia (Schnegg 2016), and I have returned more regularly to Fransfontein since then. During this engagement and my associated preoccupation with water and the water cycle, it became clear to me just how detailed the knowledge is that people in Fransfontein have about the weather. The data presented and analyzed here was mostly collected after 2015 as part of a larger project on different ways of knowing the environment (Schnegg 2019). In total, I taped interviews with diverse people that covered a broad range of topics, including the weather, the forces that influence it, its changes, and its causes.

Different Ways of Knowing Weather Change

The concept of environmental pluralism directs a researcher to pay attention to whether and how actors combine different ways of knowing. It puts into focus different forms of knowledge that may coexist in various degrees of integration, from complete independence to significant overlap. Different ways of knowing tend to coexist separately because their foundational epistemologies offer largely mutually contradictory explanations that cannot be subsumed under one framework. In the Namibian case, indigenous, religious, and scientific bodies of knowledge exemplify different epistemologies that are sometimes hard or even impossible to integrate. To explore the hypothesis that knowledge coexists in plural forms, I follow three cases in my analysis. Through these case studies I show how and to what degree people in Fransfontein combine different forms of knowing to make sense of weather events (i.e., drought) in times of climate change.

Indigenous Discourses

The following episode explains the meaning of this translation quite well.[7]

One day in late February 2015, I was sitting in front of my hut in ǁgamo!nâb (literally translated "place with no water inside"), preparing for the interviews I was planning to conduct in the late evening. ǁgamo!nâb is a small farming community about seven kilometers outside of Fransfontein, situated uphill. The surrounding land is relatively flat, and the views are endless. The temperature had risen to more than thirty-five degrees Celsius by ten o'clock in the morning and I knew it would not get cooler again until seven in the evening. Like most other people in ǁgamo!nâb, I spent my day in the shade.

Sitting under a tree, I could see the clouds forming toward the east, an estimated two hundred kilometers away from us. While they were approaching us, the temperature rose to more than forty degrees Celsius. Just around noon, my neighbor Robert passed by to ask for a cup of sugar and some tea. While chatting, we soon turned our conversation to the impressive cloud formation approaching from inland and began to discuss whether it would rain that day. Around that time of the day and year, almost any conversation turns to the subject of rain. Robert, about sixty years old, has spent all his life in that area. Like most people there, he is a pastoralist and keeps a small number of goats, sheep, and cattle. When I asked Robert if he thought it would rain that day, he replied, "Yes, Michael, don't you feel that she is not blowing so strongly? She will let him in." I did not understand. So he explained further: "You know, around this time of the year, the female wind, *huriǂoab*, comes every morning from the coast and searches for her male lover *tūǂoab*, far inside the land."[8] He pointed toward the east, where the clouds had begun to form. "There, the two meet, and only if they agree will they jointly return and bring the rain." He was right. We received some soft rain later that day (Schnegg 2019).

The particular episode is singular, and yet it already points to some general principles: two winds, *huriǂoad* and *tūǂoab*, bring the rain. Both winds are animated in human thought and have a gender and a personality. While the westerly wind is female, the easterly wind is male. To bring the rain the two must interact and, more importantly, agree. During the morning hours, she (*huriǂoab*) comes and searches for the easterly wind (*tūǂoab*) farther east of Fransfontein, where the clouds eventually form as they interact. If she continues to blow too strongly, the clouds will not reach Fransfontein. Only if she stops and lets him in will they arrive.

Similarly to Robert, most people I consulted framed this interplay of wind and rain using the metaphor of a love affair. According to Helga, a woman in her late sixties, the two propose to each other. To use Helga's words: "She goes down there and takes the male wind. Then they both come along this way, and that will bring the rain. They are a couple."[9] Later in the course of the interview, she makes additional reference to the engagement ceremony. Engagement ceremonies were most likely the precursor to Christian marriages among the Damara and are still an integral part of the splendid marriages people celebrate (Pauli 2011, 2019). The events span over three days, and each day the family of the groom has to ask for the bride in a nightlong ceremony. On the first two days the groom's family is sent home, and only on the third day is marriage agreed upon. As with human engagements, only if both the male and the female wind reach an agreement do they decide to come together and move toward the west, pushing the clouds, and eventually bringing the rain to Fransfontein.

If the winds fail to agree for longer periods of time, the result is a drought.

Religious Discourses

Robert and I were still sitting in front of my hut looking at the sky and the heavy rainclouds that were forming in the north, where the descendants of European settlers own big commercial farms. While we talked, Robert explained.

Michael: And is there drought on the white farms as well?
Robert: No, there is no drought on these farms.
Michael: Why not?
Robert: Maybe the rain, because God is also white, because the rain always prevails there, this side. They have long hair, and God is also white; I saw in the Bible.
Michael: So He favors some people?
Robert: Yes, God favors the white people, because He is having, he is white, and He is having long hair like them.

According to Robert, and many others I talked to, a Christian God is the ultimate cause of things. German missionaries have Christianized people in this area since the late nineteenth century, and most consider themselves as Lutheran Protestants or as belonging to one of the quickly spreading Pentecostal churches. For many people today, God has many things under His control. Or, as Robert put it, "The rain is not in our hands; it is in God's hands, and this is why in some years it rains and some it does not. It can even happen that it will not rain for three years." When asked whether God would make the winds agree, he denied it. For him, the two domains are not intertwined.

This example shows how a different context, the political and economic inequality between European settlers and indigenous people, triggered a very different explanation of the lack of rainfall than the first: one being religious and the other being predominantly indigenous. This explanation relates to colonialism, Christianization, and God. Christian beliefs and indigenous knowledge come from different epistemological frameworks and are not fully integrated in one worldview. They become meaningful and are applied in different contexts.

A second case further exemplifies Christianity as a relevant knowledge domain. Charles is a well-respected elderly man in the community who worked for the so-called second-tier administration under colonial rule. Today he is an admired local leader, and many people seek out his advice.

When we talked, Charles explained: "The ongoing drought is a punishment from God; all we can do is pray." Charles related the drought to supernatural forces and eventually to humans who have done something wrong. They are punished. This is a widespread concern, and many people gave similar responses when I asked them about the current situation in which rains fail to come.

However, Charles, like others, did not see God as the single cause. In the course of our conversation, we touched on the political situation and the changes since independence in 1990 as well. In this context, Charles gives a second reason for the ongoing drought: overstocking. Remember, drought means "no food," and overstocking is a plausible cause. In the area around Fransfontein, there are two factors that drive stocking rates up: in-migration from large herd owners (mostly from the north of Kunene) and so-called part-time pastoralists. Part-time pastoralists are farmers who combine well-paid jobs in the urban centers with pastoralism in the rural hinterland (Schnegg, Pauli, and Greiner 2013). Part-time pastoralists have financial capital to invest, and their herds often become very large. To put it in Charles's words: "As we can observe, the leader and the government are doing nothing. In the past, they reduced the number of cattle so that it matches the carrying capacity of the land. If the number of livestock gets over the carrying capacity, you are destroying the land."

Charles paints a positive picture of the apartheid regime, in which he was a member of the local government. Admittedly, at that time, the state was much more active in controlling and regulating people's herding decisions. The state and the Ministry of Agriculture actively relied on scientific knowledge about the management of rangelands and the model of a "carrying capacity," the maximum number of livestock that can be held in a given area, as an appropriate management tool. With the postapartheid government and the weakening of the state, drought also occurred as a man-made phenomenon. This tendency to explain drought as a result of the social and economic ruptures after independence has also been observed by Sullivan among the ǂnūkhoen people with whom she worked (Sullivan 2000, 2002). This case shows once more that in two different contexts, alternative explanations are appropriate. In the first instance, this is a religious model, in which God is the ultimate power and—only a short while later—it is a political one, in which the loss of state control is the ultimate cause.

Scientific Discourses

Charles explained that both people and God cause drought, and he is equally aware that the ongoing destruction of nature plays a critical role in changing weather patterns. He continued: "As the countries are developing, some gases go up in the air, and this can have negative effects on the clouds and the weather. There is also a cyclone, which has a negative effect on the SADC [Southern African Development Community] area. For those reasons, there is no rain this year." Charles made a reference to development and emissions that cause climate change. Like many people in the area, he has accumulated scientific knowledge through listening to the radio, and he relates this knowledge to his observations about the weather. Thus, in another context, he refers to a third model to make sense of the ongoing drought.

In a similar vein, Hanna reasoned about the rains and, again, eventually, the effects of climate change.[10] Hanna is about thirty years old and teaches science in the primary school of Fransfontein. She did not grow up in the area, originating instead from a community some three hundred kilometers south, which is mostly inhabited by a different ethnic group, the Ovaherero. Damara and Ovaherero people speak different languages, and many hold prejudices about each other. These prejudices derive from the influence of German missionaries who have promoted the (historically wrong) notion that the Damara people were slaves of the Ovaherero in the past. Today, many Damara are likely to consider Ovaherero to be arrogant and overly proud, while many Ovaherero would respond that Damara are lazier than one should be. It is in this interethnic context that Hanna's explanation must be placed, when she explains why the effects of climate change are different here and there.

Michael: Do you know why it's not raining anymore?

Hanna: People are cutting too many trees in this area. This contributes to climate change, which causes this drought. The trees store CO_2. It's true.

Michael: You said the other day that in the community where your parents live it is raining nicely; how can this be?

Hanna: It rains nicely, very nice.

Michael: So why? Do you have an explanation?

Hanna: There, people are not the same; people are not building their houses with trees. Us, we only build with bricks. So we don't cut down the trees.

Hanna considered deforestation responsible for the drought in Fransfontein, yet at the same time the drought was not really an issue in Omatjete, where she was born. While the people in her home community respect trees and do not cut them down, the people she knows in Fransfontein do not possess the same knowledge. Her argument refers to a global discourse about deforestation, which is seen as a key driver of climate change in scientific debates. At the same time, she localizes this discourse to explain the differences she observes.

However, for Hanna, like all my informants, God too plays His role. When she reflects on the differences between people in both places, she finds that the inhabitants of Fransfontein are too lavish and lazy as well. They would not even send their children to school if not coerced, and the parents would not care. Educated in a very religious family, she is confident that this is against God's will and that He punishes those who misbehave. A drought, as the Bible says, is a common means of punishment, and people should take this seriously by changing their behavior.

Again, just as in the two cases discussed before, we find a reference to different epistemological frameworks and explanations for one phenomenon, the drought. In different contexts, varying epistemological and ontological frameworks are applied to explain why the rains fail to come. As in the cases analyzed before, Hanna confirms the notion of an emerging environmental pluralism rather well.

In brief, Robert, Charles, and Hanna all explain the occurrence and existence of drought as the most dramatic weather-related phenomenon they experience in particular ways. However, there are parallels too. In their views, and following the meaning "no food," *ǂû-i ǀkhai* can have two distinct causes: first, low precipitation, and second, animals that overpopulate an area and reduce the grazing to such a degree that the animals will die. People attribute the latter to the grazing management of the current government and the inequality that was introduced through colonialism and that still continues today. When it comes to explaining why there is no rain, the ǂnūkhoen people see the interaction of two winds that are animated in human thought to be one cause. Moreover, the lack of rain is also explained as God's will. And, with "climate change," a new explanatory model for the lack of precipitation has been introduced. It links changing weather patterns to changes outside the local realm. While those different explanations coexist in society, they also exist simultaneously for many individuals.

Discussion: Patterns of Pluralism

To explain the drought, one of the most salient weather phenomena in northwestern Namibia, individuals refer to various knowledge domains, i.e., indigenous, religious, and scientific ways of knowing. None of the people I interviewed referred to one cause alone, and many, like the three informants I introduced, use plural epistemological frameworks.

While Robert's understanding of the drought is deeply rooted in the dominant ǂnūkhoen cultural model, which explains weather as an interaction between different winds that have their own intentions, he equally acknowledges that God is the ultimate cause to explain why settlers of European descent receive much more rain than the people of Fransfontein do. Charles's knowledge is rooted in a ǂnūkhoen worldview as well. However, as someone who occupied a prominent position in the government, responsible for rangeland management for many years, he extends his reasoning into the scientific domain. Drought, as the inability to provide food, is therefore also explained as a failure of the "new" political system to enforce scientific truth. Thirdly, God plays a crucial role in Charles's reasoning. In the last case study, Hanna connects the two domains of scientific and religious knowledge.[11] While on the one hand she draws upon deforestation as an explanatory model derived from the climate change discourse, she refers to God on the other. Hanna's case also reveals an interesting gradient between environmental pluralism and different, merging epistemologies. Although she clearly refers to climate science knowledge to explain climate change by relating it to deforestation and burning trees, she localizes that global scientific model. In so doing, she integrates a circulating global model into the local worldview, in which local agents make the weather. Her reasoning underlines the importance of moral discourses that frame knowing about the weather to a large degree.

Conclusion

The analysis reveals that people combine explanations from different epistemic sources to make sense of weather change. At the same time, the analysis does not explore the nature of this coexistence in detail. It remains a challenge for future work to show how bounded or coherent those discourses remain and under what circumstances a discourse is more likely to become dominant. Equally, while the cases I present indicate that knowledge systems are applied in specific situations and contexts to understand aspects of weather-related phenomena, the analysis does not systematically establish when and under what conditions one system is more likely to be applied than others (Schnegg 2021b).

Both types of analysis are methodologically challenging and require following the same individual across contexts of space and time. A strict confirmation of an environmental pluralism hypothesis would be that the actors perceive or enact the world differently depending on the situation they find themselves in. While I have provided selected evidence to support this, much more research is required here (Schnegg 2019, 2021a, 2021b).

What are the theoretical and methodological lessons learned? Since climate change knowledge about the weather has been challenged and is transforming, the concept of environmental pluralism opens up space for exploring the coexistence of epistemologically and ontologically different knowledge systems. The empirical evidence from Namibia confirms the existence of environmental pluralism and proves useful for understanding current transformations and for assessing the consequences they have.

Acknowledgments

Without the continuous support of numerous people and communities in Kunene, this research would not have been possible. I thank all of them for being-in-their-world with me, especially Jorries Seibeb. Julia Pauli and Edward Lowe have shaped my thinking about knowing since the beginning of my career, and I am extremely grateful for the challenges and the advice they have given to me. In addition to them, Lena Borlinghaus-ter Veer, Inga Sievert, Coral O'Brian, João Baptista, Thomas Friedrich, and the editors of this volume have offered critical and extremely constructive comments to earlier drafts of this chapter. Moreover, I am deeply indebted to Sylvanus Job, who opened the door to his language, Khoekhoegowab, and the world behind it to me. The results presented here are a product of the research projects LINGS (Local Institutions in Globalized Societies, http://lings-net.de/) and CliSAP (Integrated Climate System Analysis and Prediction), and I am grateful to the DFG (Deutsche Forschungsgemeinschaft) for the continuous and generous financial support.

Michael Schnegg is a full professor of social and cultural anthropology at the University of Hamburg, and he has conducted extensive ethnographic fieldwork in Mexico and Namibia since 2000. His current research interest focuses on understanding how people perceive, value, and govern nature in an age of unprecedented globalization.

Notes

1. As a matter of fact, this chapter was drafted in early 2016, prior to my other writing on the same theme.
2. This assumption has been challenged by a great many authors since Husserl's original work in the *Krisis* (Husserl 1976). Today especially the fields of science and technology studies (STS) systematically demonstrate that in fact the knowledge acquired in scientific investigations of climate change is context dependent (Grundmann and Rödder 2019; Hulme 2016; Jasanoff and Martello 2004; Wynne 1995).

3. Throughout the chapter, I use the terms *local knowledge, lay knowledge,* and *indigenous knowledge* interchangeably to denote what Berkes defined as "a cumulative body of knowledge, belief, and practice, evolving by accumulation of TEK and handed down through generations through traditional songs, stories and beliefs. [It concerns] the relationship of living beings (including human) with their traditional groups and with their environment" (Berkes 1993: 3).
4. Within the political arena, the *Intergovernmental Panel on Climate Change Synthesis Report* stresses the value of indigenous, local, and traditional knowledge as a major, and still largely unexplored, resource for adapting to climate change (IPCC 2015: 80; Nakashima et al. 2012; Martello 2001).
5. Khoekhoegowab is a "Khoisan" language of the Khoe-Kwadi family with four (primary) click sounds (ǂ, palatal; ǁ, lateral; ǀ, dental; !, alveolar) that function like other consonants. The region around Fransfontein is multiethnic and multilingual. In this chapter, I restrict myself to the largest ethnic group but hope to broaden my focus in future work.
6. The term Odendaal Plan refers to a commission and the report it published in 1963. The report recommended, among other things, the establishment of so-called "homelands" to foster South Africa's racist apartheid politics. Homelands were to become the settlement areas of people with specific ethnic classifications.
7. I reported this encounter in Schnegg (2019) as well.
8. In Khoekhoegowab, the suffix for male nouns is "b" and for female nouns "s." Words can be taken out of their normal gender context to indicate that this particular instance is atypical, e.g., atypically shaped or atypically strong. The word for wind is *ǂoab*, which is male. If the wind were to be referred to with a female ending, this would imply that it was unusual, which is not done. It is likely that this can explain the use of the male "b" in connection with the female wind *huriǂoab*.
9. Interview, Fransfontein area, 21 January 2015.
10. This episode has partly also been reported in Schnegg (2019).
11. On another day, she, like Charles, mentions indigenous explanations as well.

References

Antweiler, Christoph. 1998. "Local Knowledge and Local Knowing: An Anthropological Analysis of Contested 'Cultural Products' in the Context of Development." *Anthropos* 93 (4/6): 469–94.

Benda-Beckmann, Franz von, Keebet von Benda-Beckmann, and Melanie Wiber. 2006. *Changing Properties of Property*. New York: Berghahn Books.

Berkes, Fikret. 1993. "Traditional Ecological Knowledge in Perspective." In *Traditional Ecological Knowledge: Concepts and Cases,* edited by Julian T. Inglis, 1–9. Ottawa: International Program on Traditional Ecological Knowledge and International Development Research Centre.

———. 2008. *Sacred Ecology*. 2nd ed. New York: Routledge.

Bierlein, John F. 1994. *Parallel Myths*. New York: Ballantine Books.

Bostrom, Ann, M. Granger Morgan, Baruch Fischhoff, and Daniel Read. 1994. "What Do People Know about Global Climate Change?" *Risk Analysis* 14(6): 959–70. doi: 10.1111/j.1539-6924.1994.tb00065.x.

Brüggemann, Michael, and Simone Rödder, eds. 2020. *Global Warming in Local Discourses: How Communities around the World Make Sense of Climate Change.* Cambridge: Open Book Publishers.

Chanza, Nelson, and Anton De Wit. 2013. "Epistemological and Methodological Framework for Indigenous Knowledge in Climate Science." *Indilinga African Journal of Indigenous Knowledge Systems* 12(2): 203–16.

Cochran, Patricia, Orville H. Huntington, Caleb Pungowiyi, Stanley Tom, F. Stuart Chapin III, Henry P. Huntington, Nancy G. Maynard, and Sarah F. Trainor. 2013. "Indigenous Frameworks for Observing and Responding to Climate Change in Alaska." *Climatic Change* 120(3): 557–67. doi: 10.1007/s10584-013-0735-2.

DeWalt, Billie. 1994. "Using Indigenous Knowledge to Improve Agriculture and Natural Resource Management." *Human Organization* 53(2): 123–31. doi: 10.17730/humo.53.2.ku60563817m03n73.

Donner, Simon D. 2007. "Domain of the Gods: An Editorial Essay." *Climatic Change* 85(3–4): 231–36. doi: 10.1007/s10584-007-9307-7.

Ehlert, Judith. 2012. "We Observe the Weather Because We Are Farmers: Weather Knowledge and Meteorology in the Mekong Delta, Vietnam." In *Environmental Uncertainty and Local Knowledge: Southeast Asia as a Laboratory of Global Ecological Change,* edited by Anna-Katharina Hornidge and Christoph Antweiler, 119–43. Bielefeld: Transcript.

Elia, Emmanuel F., Stephen Mutula, and Christine Stilwell. 2014. "Indigenous Knowledge Use in Seasonal Weather Forecasting in Tanzania: The Case of Semi-arid Central Tanzania." *South African Journal of Libraries and Information Science* 80(1): 18–27. doi: 10.7553/80-1-1395.

Friedrich, Thomas. 2018. "The Local Epistemology of Climate Change: How the Scientific Discourse on Global Climate Change is Received on the Island of Palawan, the Philippines." *Sociologus* 68 (1): 63–84. doi: 10.3790/soc.68.1.63.

Gagnon, Catherine A., and Dominique Berteaux. 2009. "Integrating Traditional Ecological Knowledge and Ecological Science: A Question of Scale." *Ecology and Society* 14(2): 19.

Gorman-Murray, Andrew W. 2010. "An Australian Feeling for Snow: Towards Understanding Cultural and Emotional Dimensions of Climate Change." *Cultural Studies Review* 16(1): 22(1): 60–81. doi: 10.5130/csr.v16i1.1449.

Green, Donna, Jack Billy, and Alo Tapim. 2010. "Indigenous Australians' Knowledge of Weather and Climate." *Climatic Change* 100(2): 337–54. doi: 10.1007/s10584-010-9803-z.

Greiner, Clemens. 2011. "Migration, Translocal Networks and Socio-economic Stratification in Namibia." *Africa: Journal of the International African Institute* 81(4): 606–27. doi: 10.1017/S0001972011000477.

Grundmann, Reiner, and Simone Rödder. 2019. "Sociological Perspectives on Earth System Modeling." *Jounal of Advances in Modeling Earth Systems* 11(12): 3878–92. doi: 10.1029/2019MS001687.

Herman-Mercer, Nicole, Paul Schuster, and Karonhiakt'tie B. Maracle. 2011. "Indigenous Observations of Climate Change in the Lower Yukon River Basin, Alaska." *Human Organization* 70(3): 244–52. doi: 10.17730/humo.70.3.v88841235897071m.

Hulme, Mike. 2009. *Why We Disagree about Climate Change: Understanding Controversy, Inaction and Opportunity*: Cambridge: Cambridge University Press.

———. 2015. "Climate and Its Changes: A Cultural Appraisal." *Geo: Geography and Environment* 2(1): 1–11. doi: 10.1002/geo2.5.

———. 2016. *Weathered: Cultures of Climate.* London: Sage Publications.

Huntington, Henry P. 2000. "Using Traditional Ecological Knowledge in Science: Methods and Applications." *Ecological Applications* 10(5): 1270–74. doi: 10.1890/1051-0761(2000)010[1270:UTEKIS]2.0.CO;2.
Huntington, Henry P., Terry Callaghan, Shari Fox, and Igor Krupnik. 2004. "Matching Traditional and Scientific Observations to Detect Environmental Change: A Discussion on Arctic Terrestrial Ecosystems." *Ambio*: 18–23.
Huntington, Henry P., Robert S. Suydam, and Daniel H. Rosenberg. 2004. "Traditional Knowledge and Satellite Tracking as Complementary Approaches to Ecological Understanding." *Environmental Conservation* 31(3): 177–80. doi: 10.1017/S0376892904001559.
Husserl, Edmund, ed. 1976. *Die Krisis der europäischen Wissenschaften und die transzendentale Phänomenologie. Eine Einleitung in die phänomenologische Philosophie, Husserliana: Edmund Husserl. Gesammelte Werke*. Den Haag: Martinus Nijhoff.
Ifejika Speranza, Chinwe, Boniface Kiteme, Peter Ambenje, Urs Wiesmann, and Samuel Makali. 2010. "Indigenous Knowledge Related to Climate Variability and Change: Insights from Droughts in Semi-arid Areas of Former Makueni District, Kenya." *Climatic Change* 100(2): 295–315. doi: 10.1007/s10584-009-9713-0.
Ingold, Tim. 2006. "Rethinking the Animate, Re-animating Thought." *Ethnos* 71(1): 9–20. doi: 10.1080/00141840600603111.
IPCC (Intergovernmental Panel on Climate Change), ed. 2015. *Climate Change 2014: Synthesis Report: Contribution of Working Groups I, II and III to the Fifth Assessment Report of the*. Geneva: Intergovernmental Panel on Climate Change.
Jasanoff, Sheila. 2004. "Heaven and Earth: The Politics of Environmental Images." In *Earthly Politics: Local and Global in Environmental Governance*, edited by Sheila Jasanoff and Marybeth Long Martello, 31–52. Cambridge, MA: MIT Press.
———. 2010. "A New Climate for Society." *Theory, Culture & Society* 27(2–3): 233–53. doi: 10.1177/0263276409361497.
Jasanoff, Sheila, and Marybeth L. Martello, Peter M. Haas, and Gene I. Rochlin, eds. 2004. *Earthly Politics: Local and Global in Environmental Governance*. Cambridge, MA: MIT Press.
Kalanda-Joshua, Miriam, Cosmo Ngongondo, Lucy Chipeta, and F. Mpembeka. 2011. "Integrating Indigenous Knowledge with Conventional Science: Enhancing Localised Climate and Weather Forecasts in Nessa, Mulanje, Malawi." *Physics and Chemistry of the Earth, Parts A/B/C* 36(14–15): 996–1003. doi: 10.1016/j.pce.2011.08.001.
Kearney, Anne R. 1994. "Understanding Global Change: A Cognitive Perspective on Communicating through Stories." *Climatic Change* 27: 419–41. doi: 10.1007/BF01096270.
Kempton, Willett. 1991. "Lay Perspectives on Global Climate Change." *Global Environmental Change* 1(3): 183–208. doi: 10.1016/0959-3780(91)90042-R.
King, D. N. T., A. Skipper, and W. B. Tawhai. 2008. "Māori Environmental Knowledge of Local Weather and Climate Change in Aotearoa—New Zealand." *Climatic Change* 90(4): 385–409. doi: 10.1007/s10584-007-9372-y.
Krauss, Werner. 2012. "Localizing Climate Change: A Multi-sited Approach." In *Multi-sited Ethnography: Theory, Praxis and Locality in Contemporary Research*, edited by Mark-Anthony Falzon, 149–64. Burlington, VT: Ashgate Publishing Limited.
Krupnik, Igor, and Jolly, Dyanna, eds. 2002. *The Earth Is Faster Now: Indigenous Observations of Arctic Environmental Change*. Fairbanks, AK: Arctic Research Consortium of the United States.
Lefale, Penehuro F. 2010. "Ua 'afa le Aso Stormy Weather Today: Traditional Ecological Knowledge of Weather and Climate; The Samoa Experience." *Climatic Change* 100: 317–35. doi: 10.1007/s10584-009-9722-z.

Löfstedt, Ragnar E. 1991. "Climate Change Perceptions and Energy-Use Decisions in Northern Sweden." *Global Environmental Change* 1(4): 321–24. doi: 10.1016/0959-3780(91)90058-2.

Marino, Elizabeth, and Peter Schweitzer. 2009. "Talking and Not Talking about Climate Change in Northwestern Alaska." In *Anthropology and Climate Change: From Encounters to Actions*, edited by Susan A. Crate and Mark Nuttall, 209–217. New York: Routledge.

Martello, Marybeth L. 2001. "A Paradox of Virtue? 'Other' Knowledges and Environment-Development Politics." *Global Environmental Politics* 1(3): 114–41. doi: 10.1162/152638001316881430.

Merry, Sally E. 1988. "Legal Pluralism." *Law & Society Review* 22(5): 869–96. doi: 10.2307/3053638.

Nadasdy, Paul. 2003. "Reevaluating the Co-management Success Story." *Arctic* 56(4): 367–80. doi: 10.14430/arctic634.

Nakashima, Douglas J., Kirsty G. McLean, Hans D. Thulstrup, Ameyali R.Castillo, and Jennifer T. Rubis. 2012. *Weathering Uncertainty: Traditional Knowledge for Climate Change Assessment and Adaptation.* Paris: UNESCO.

Nyong, Anthony, Francis Adesina, and Balgis Osman-Elasha. 2007. "The Value of Indigenous Knowledge in Climate Change Mitigation and Adaptation Strategies in the African Sahel." *Mitigation and Adaption Strategies for Global Change* 12(5): 787–97. doi: 10.1007/s11027-007-9099-0.

Orlove, Ben, Carla Roncoli, Merit Kabugo, and Abushen Majugu. 2010. "Indigenous Climate Knowledge in Southern Uganda: The Multiple Components of a Dynamic Regional System." *Climatic Change* 100(2): 243–65. doi: 10.1007/s10584-009-9586-2.

Paerregaard, Karsten. 2013. "Bare Rocks and Fallen Angels: Environmental Change, Climate Perceptions and Ritual Practice in the Peruvian Andes." *Religions* 4(2): 290–305. doi: 10.3390/rel4020290.

Pauli, Julia. 2011. "Celebrating Distinctions: Common and Conspicuous Weddings in Rural Namibia." *Ethnology: An International Journal of Cultural and Social Anthropology* 50(2): 153–67.

———. 2019. *The Decline of Marriage in Namibia: Kinship and Social Class in a Rural Community.* Bielefeld: Transcript.

Pelto, Pertti J., and Gretel H. Pelto. 1997. "Studying Knowledge, Culture, and Behavior in Applied Medical Anthropology." *Medical Anthropology Quarterly* 11(2): 147–63. doi: 10.1525/maq.1997.11.2.147.

Pettenger, Mary E. 2016. *The Social Construction of Climate Change: Power, Knowledge, Norms, Discourses.* London: Routledge.

Rayner, Stephen. 2003. "Domesticating Nature: Commentary on the Anthropological Study of Weather and Climate Discourse." In *Weather, Climate, Culture*, edited by Benjamin S. Orlove and Sarah Strauss, 277–90. New York: Berg.

Risiro, Joshua, Dominic Mashoko, Doreen T. Tshuma, and Elias Rurinda. 2012. "Weather Forecasting and Indigenous Knowledge Systems in Chimanimani District of Manicaland, Zimbabwe." *Journal of Emerging Trends in Educational Research and Policy Studies* 3(4): 561–66.

Roncoli, Carla, Keith Ingram, and Paul Kirshen. 2002. "Reading the Rains: Local Knowledge and Rainfall Forecasting in Burkina Faso." *Society & Natural Resources* 15(5): 409–27. doi: 10.1080/08941920252866774.

Rudiak-Gould, Peter. 2012. "Promiscuous Corroboration and Climate Change Translation: A Case Study from the Marshall Islands." *Global Environmental Change* 22(1): 46–54. doi: 10.1016/j.gloenvcha.2011.09.011.

———. 2014. "Climate Change and Accusation: Global Warming and Local Blame in a Small Island State." *Current Anthropology* 55 (4): 365–386. doi: 10.1086/676969.
Schnegg, Michael. 2014. "Epistemology: The Nature and Validation of Anthropological Knowledge." In *Handbook of Methods in Cultural Anthropology*, edited by H. Russel Bernard, 21–53. Lanham, MD: Rowman & Littlefield.
———. 2016. "Lost in Translation: State Policies and Micro-politics of Water Governance in Namibia." *Human Ecology* 44(2): 245–55. doi: 10.1007/s10745-016-9820-2.
———. 2019. "The Life of Winds: Knowing the Namibian Weather from Someplace and from Noplace." *American Anthropologist* 121(4): 830–44. doi: 10.1111/aman.13274.
———. 2021a. "Ontologies of Climate Change: Reconciling Indigenous and Scientific Explanations for the Lack of Rain in Namibia." *American Ethnologist* 48(3): 260–73. doi: 10.1111/amet.13028.
———. 2021b. "What Does the Situation Say? Theorizing Multiple Understandings of Climate Change." *Ethos* 49(2): 194–215. doi: 10.1111/etho.12307.
Schnegg, Michael, and Michael Bollig. 2016. "Institutions Put to the Test: Community-Based Water Management in Namibia during a Drought." *Journal of Arid Environments* 124: 62–71. doi: 10.1016/j.jaridenv.2015.07.009.
Schnegg, Michael, Julia Pauli, and Clemens Greiner. 2013. "Pastoral Belonging: Causes and Consequences of Part-Time Pastoralism in North-Western Namibia." In *Pastoralism in Africa: Past, Present and Future*, edited by Michael Bollig, Michael Schnegg, and Hans-Peter Wotzka, 341–62. Oxford: Berghahn Books.
Schnegg, Michael, Coral I. O'Brian, and Inga J. Sievert. 2021. "It's Our Fault: A Global Comparison of Different Ways of Explaining Climate Change." *Human Ecology* 49: 327–339. doi: 10.1007/s10745-021-00229-w.
Sullivan, Sian. 2000. "Getting the Science Right, or Introducing Science in the First Place? Local 'Facts,' Global Discourse—'Desertification' in North-West Namibia" In *Political Ecology: Science, Myth and Power*, edited by Sian Sullivan and Philip A. Stott, 15–44. London: Edward Arnold.
———. 2002. "How Can the Rain Fall in This Chaos? Myth and Metaphor in Representations of the North-West Namibian Landscape." In *Challenges for Anthropology in the "African Renaissance": A Southern African Contribution*, edited by Debie LeBeau and Robert J. Gordon, 255–65. Windhoek: University of Namibia Press.
Thompson, Michael, and Steve Rayner. 1998. "Risk and Governance Part I: The Discourse of Climate Change." *Government and Opposition* 33(2): 139–66. doi: 10.1111/j.1477-7053.1998.tb00787.x.
Vannini, Phillip, Dennis Waskul, Simon Gottschalk, and Toby Ellis-Newstead. 2011. "Making Sense of the Weather: Dwelling and Weathering on Canada's Rain Coast." *Space and Culture* 15(4): 361–80. doi: 10.1177/1206331211412269.
Weatherhead, Elizabeth, Shari Gearheard, and Roger G. Barry. 2010. "Changes in Weather Persistence: Insight from Inuit Knowledge." *Global Environmental Change* 20(3): 523–28. doi: 10.1016/j.gloenvcha.2010.02.002.
Weingart, Peter, Anita Engels, and Petra Pansegrau. 2000. "Risks of Communication: Discourses on Climate Change in Science, Politics, and the Mass Media." *Public Understanding of Science (Bristol, England)* 9(3): 261–83. doi: 10.1088/0963-6625/9/3/304.
Weisser, Florian, Michael Bollig, Martin Doevenspeck, and Detlef Müller-Mahn. 2014. "Translating the 'Adaptation to Climate Change' Paradigm: The Politics of a Travelling Idea in Africa." *Geographical Journal* 180(2): 111–19. doi: 10.1111/geoj.12037.
Wynne, Brian. 1995. "Public Understanding of Science." In *Handbook of Science and Technology Studies*, edited by Sheila Jasanoff, Gerald E. Markle, James C. Peterson and Trevor Pinch, 361–88. Thousand Oaks, CA: Sage Publications, Inc.

Chapter 2

How a Storm Feels

Storying Climate Change in the Eastern Himalayas

Alexander Aisher

> But the sea is rising. . . . Don't you think we should keep silent just to enjoy this rather sinister moment?
>
> —Albert Camus, *The Fall* (1956: 30)

Introduction: Refuge and Vulnerability

Storytellers across the ages have called upon the power and chaos of storms to highlight human vulnerability. Storms abound in Western literature, as metaphors for emotional or spiritual upheaval, chaos, destruction, and sometimes transformation. From the magically invoked storm of Shakespeare's *The Tempest* to the dry, sterile thunder without rain of T. S. Eliot's *The Wasteland*, storms have been, and continue to be, "great revealers" (Garcia-Acosta 2002). Who can watch *King Lear* and not feel in the storm that bears down upon Lear his cognitive disintegration and the parallel disintegration of society? Storms can also reveal human helplessness in the face of the powerful ecological sovereignties that stand behind our human presence in the world. Within the history of religion, gods of storm and thunder abound: Zeus. Thor. Lei Gong. Indra.

> The wind came back with triple fury, and put out the light for the last time. They sat in company with the others in other shanties. . . . They seemed to be staring at the dark, but their eyes were watching God. (Hurston [1937] 1986)

Notes for this chapter begin on page 61.

As concepts, vulnerability and refuge weave through each other. It takes vulnerability to transform the destructive agent of a storm, which itself is but a hazard, into a *disaster*—and this vulnerability is socially and culturally produced (Oliver-Smith and Hoffman 2020). Vulnerability finds a counterpoint in the concept of refuge: a place of safety, somewhere that life is secure.

Scholarship on refugia is in its infancy and only now emerging as a major focus of research in conservation biology (Birks 2015). Most refugia have borders that separate what exists inside from what lies outside. Defined by borders—walls, barricades, membranes, the unseen border of values—refugia condition the entry into themselves of forms, forces, and actors that lie outside. Within conservation biology scholarship, *ecological refugia* are places that remain intact, even when areas around them are disturbed. These are places where life-forms can survive and even flourish during periods of intense existential disturbance; places from which, under the right conditions, they may again emerge (Turner 2005). *Climate refugia* harbor life-forms and genetic material needed to repopulate disturbed sites (Keppel and Wardell-Johnson 2012).[1] Sadly, as the multispecies scholar Donna Haraway observes, "Right now, the earth is full of refugees, human and not, without refuge" (2015: 160).

In his masterpiece *The Poetics of Space* (1964), the philosopher of the imagination Gaston Bachelard explores the house as a symbol of human security and refuge and well-being in a turbulent world. For Bachelard, storms *make sense of* the house as refuge, revealing its power to protect those who dwell within it against forces that besiege it. This quality of refuge extends across the shimmering border between humans and other species.

> Our consciousness of wellbeing . . . should call for comparison with animals in their shelters. . . . Physically, the creature endowed with a sense of refuge, huddles up to itself, takes to cover, hides away, lies snug, concealed. (Bachelard [1958] 1964: 37)

Like other local impacts of climate change, storms can threaten the fragile border between humans and ecological sovereignties. Even as storms provoke anxiety, they may call upon humans to "upframe" (Kohn 2013) their perceptions—to look beyond the individual storm to the powerful ecological sovereignties that exist beyond it. Storms also force those who experience them *back to place*.

Disasters arise at the nexus of potentially destructive agents and vulnerable human populations, neither of which are static. Some disasters come as "lightning bolts," but most are slow onset and arise as outcomes of long-running processes: they have *genealogies*, how they came to be what they are. The storm that came at dusk to Talum village, the ethnographic focus of this chapter, might appear to be a lightning bolt, but in fact it was centuries in the making.

Methods and Background

The following study is based on data gathered in the state of Arunachal Pradesh between 2001 and 2003. This state, located in the extreme northeast of India, is a core part of the Indo-Burma "biodiversity hotspot" (Myers et al. 2000), and represents one of the most biologically diverse terrestrial ecosystems on Earth (Thompson 2009). With a wide altitudinal range (100–7,090 meters), the state includes five major climatic zones—alpine, temperate, subtemperate, subtropical, and tropical. Inhabited by twenty-four major indigenous ethnic groups, self-identified "tribes," most communities depend directly on subsistence shifting cultivation for their livelihoods (Singh, Pretty, and Pilgrim 2010), informed by a rich heritage of biocultural knowledge systems (Singh et al. 2015).

While the Nyishi, whose biocultural knowledge is at the center of this chapter, reside in several districts of Arunachal Pradesh, this chapter is based on data collected in the upland district of Kurung Kumey—named after the two principal rivers that flow through it—one of the most remote districts in the state, and indeed anywhere in India. At the time of my doctoral fieldwork, no villagers had heard of global warming. While seven hundred villages dotted the valleys of this district, consisting of nearly fifteen thousand households, the focus here is on just one village. To preserve its anonymity, I call this village Talum.

The way communities model disasters matters. In his definition of *genealogy*, which evolved into a method, Michel Foucault included investigation into those elements of phenomena that "we tend to feel [are] without history" (1980: 139). It is impossible to attribute definitively any individual storm event to climate change; this form of causal analysis is fraught with statistical difficulties. However, it is possible to aim for something resembling the life history of a storm and to articulate its complex emotional form. Most storytellers do not aim for a continuous trail so much as a sequence of discrete footprints (Berger 1982: 284–85). So too for this genealogy of a storm. The movement shall be back from the present into a collectively remembered past, a movement "upriver" to the ontological sources of a storm-that-feels.

Climate Change in the Water Tower of Asia

In this moment, our species stands unequally, collectively, in all our diversity, before the ecological sovereignties of climate. Like some hunters in stories told by Nyishi storytellers, we inhabit the liminal silence between human action (our own, others'), and drastic outcomes in the more-than-human realm. There was a time when the novelist Albert Camus could write, "Nature is still there, however She contrasts her calm skies and her reasons with the madness of men" (1955: 137). That time has passed. We are together now—we always were.

The concept of adaptation is at the heart of contemporary scholarship on global climate change, as is the recognition that it occurs, so often, "inside" communities (Adger, Lorenzoni, and O'Brien 2009: 338). For good reason, climate ethnographers have begun to press into the foreground how climate change is experienced (Roncoli, Crane, and Orlove 2008; Strauss and Orlove 2003), often through those diverse conceptual and expressive instruments, like stories and songs, through which communities have always engaged places (Basso 1996: 53). Climate change is a multispecies event with its own distinctive sociality, and as multispecies scholar Anna Tsing notes, "We have a lot to learn about how humans and other species come into ways of life through webs of social relations" (2013: 28).

Across the emerging subfield of multispecies ethnography, a new wave of scholarship has started to foreground the sensory, embodied, and affective quality of interspecies encounters—the more-than-human *becomings* that occur when species meet. They are focusing the ethnographic lens upon "new kinds of relations emerging from nonhierarchical alliances, symbiotic attachments, and the mingling of creative agents" (Kirksey and Helmreich 2010), new intersubjectivities (Candea 2010), new forms of life that come into being in the intimate "contact zones" (Haraway 2008) where nonhuman vitalities blend with (apparently) other-than-human realities (Kirksey and Helmreich 2010), and points of contact where the border dividing "Nature" from "culture" shimmers and species in contact with each other co-create more-than-human forms of sociality (Tsing 2013). Through this lens, landscapes come into view that are "enactment(s) of multiple conjoined histories" (Tsing 2013: 34). It is as a *social* event produced by encounters between myriad species that climate change can threaten and violate social, moral, or religious norms (Crate 2008; Roncoli et al. 2008).

The unfolding story of climate change in the Himalayas carries global significance. Forming a 2,400-kilometer-long and 150- to 400-kilometer-wide chain of high mountains, deep valleys, and elevated plateaus, the Himalayas exert a profound influence on the climate of the Indian subcontinent and upon the Tibetan plateau (Nandargi and Dhar 2011). Forming a barrier to the southwest monsoon winds carrying humidity from the Indian Ocean northward toward the Tibetan plateau, they drive warm air upward, forcing moisture to condense and fall as heavy rain across the foothills and adjoining plains of India. As such, they directly affect the Indian monsoon system, upon which 20 percent of the human species depends. Here too is gathered 116,000 square kilometers of glacial ice, the source of ten of Asia's largest rivers, responsible for providing water for around 1.3 billion people. In this, "The Water Tower of Asia" (known also as the "Third Pole"), climate change is a story of things known precisely, lesser-known things, known unknowns, and the lurking presence of unknown unknowns.

The climate of the Himalayas varies with elevation, and climate scientists know that many terrestrial animal and plant species have already shifted their ranges and seasonal activities (IPCC 2014). They know a warmer climate means that a greater proportion of total rainfall will come from heavy precipitation events like blizzards and rainstorms (Cullen 2011), which is significant in a region given to sudden changes in the weather, cloud bursts, high winds, snowstorms, and flash floods (Nandargi and Dhar 2011).

Climate change in the Eastern Himalayas will have profound consequences for the well-being of hill communities as well as those downstream in Assam and Bangladesh (Sharma et al. 2009). The state of Arunachal Pradesh is already one of the wettest places on Earth (Roy 2005). Models predict that climate change here will be a story of increasing extremes, with more severe weather, droughts, heatwaves, and floods (Sharma et al. 2009). Already, according to current models, monsoon wet spells are getting wetter, and dry spells are becoming more frequent (Nature 2014). Glacial melting may also have ecological knock-on effects, including more frequent glacial lake outburst floods, as meltwater from snow and ice stored in high-elevation wetlands and lakes breach with devastating results. Of significance to the ethnographic subject matter of this chapter, climate change in the uplands may also bring more frequent and more severe cyclonic storms and monsoon depressions, and resulting landslides, debris flows, and flash floods.

In the Eastern Himalayas, climate change is also a story of changing interspecies dynamics. This includes changing breeding and migration patterns of birds and fish (Cruz et al. 2007), seasonal insect emergence, and disruption of pollinator relationships and predator-prey relationships (Xu et al. 2009). Models predict heightened extinction rates among species with narrow geographic and climatic ranges (Sharma et al. 2009).

Some of these changes are already underway. Ahead of detailed scientific data collection, indigenous and place-based communities in the region have already noticed more erratic rainfall (Singh et al. 2010), changes in snowfall pattern and intensity (Yadav and Kaneria 2012), earlier budburst and flowering of plants, and emergence of new agricultural pests and weeds (Chaudhary and Bawa 2011). Some report changing monsoon regimes, degrading permafrost, melting Himalayan glaciers, and shifting tree lines (Xu et al. 2009). Others have observed more frequent and more intense extreme weather events, including tropical cyclones and thunderstorms (Cruz et al. 2007).

At the intersection of indigenous environmental knowledge and rural livelihoods, villagers in Solukhumbu District in Nepal, the westernmost extent of the Eastern Himalayas, have already reported reduced snowfall and increasing difficulty in predicting the timing of rains and snows (Sherpa 2012). In West and East Siang Districts of Arunachal Pradesh, several hundred kilometers to the east, Adi tribal communities are also reporting more frequent weather anomalies, increasingly erratic rainfall, and more frequent flood events, with increasing soil erosion and increasing presence of crop pests, threats that will be aggravated by erosion of biocultural knowledge among younger generations in some indigenous groups (Singh et al. 2010).

No one knows how the story of climate change in the Eastern Himalayas will unfold, but human and nonhuman communities in the tribal state of Arunachal Pradesh appear to be particularly vulnerable. This is because climate change intersects with already existing heavy pressures on biodiversity, including species overexploitation through hunting (Aiyadurai and Velho 2018; Yadav and Kaneira 2012), high population growth rates across the state (Census of India 2011), weak infrastructure (Sharma et al. 2009), inadequate access to services (Committee on Himalayan Glaciers 2012), and the fragile mountain ecologies that define this region. These are likely to amplify the impacts of climate change.

A picture emerges of a region of outstanding biological and cultural diversity, where communities—human and other—are extremely vulnerable to the impacts of climate change. For this reason, the story of climate change must be told at a *human* scale, at the scale of communities who are *inside* these changes, and at the scale of the multispecies assemblages of which they are a part. As a villager comments in the event described below, as they chant away an approaching storm, "I do not want to see Dojung move."

A Storm at Dusk

The following event occurred at 4:15 P.M. on 17 May 2003 in Talum village, near Koloriang, high in the uplands of Arunachal Pradesh. It was dusk, and a dozen villagers had congregated at the second hearth of the largest longhouse in the village. Several hours earlier, the wind had begun to blow from an unfamiliar direction. Now it was whipping up through the longhouse floor, shoving flames to the side, throwing shadows across the split bamboo walls. With no electricity in this village (like most in the uplands at this time), the body of the longhouse was dark except for the flames of two fires. Several villagers had asked the hunter Tarido, an accomplished hunter (*nyigum*) who was able to commune with hunting spirits, to discover the cause of the gathering storm.

The hunter nudges a battered aluminum tin closer to the flames. Sitting on his haunches, he scrutinizes its contents. Inside, an egg turns in the boiling water, gathering signs. He waits patiently for several minutes, then removes the egg with his fingers and slowly unwraps the shell. Turning it in his hand, he inspects the white of the egg, then pulls it apart, searching for telltale marks and indents: traces of spirit influence. In front of the gathered villagers, he pops the yolk in his mouth and shifts it around with his tongue. Finding what he is searching for, he spits the embryo into his upturned palm, and moves it around with a finger, scrutinizing it. Shaking his head, without looking up, he states flatly, "Someone must have done something wrong."

Earlier that afternoon, the wind began to blow from the dry sunlit *nyobia* side of the valley: an unfamiliar direction. Those who noticed it said it was *karfoonum*: strange, unexpected, out of place. They didn't know if it was also *siru*, a message from spirits. That's why they asked him to perform the oracle. The strangeness of the wind raised questions.

A sudden blue flash illuminates the outline of the rough wooden doorway of the longhouse. A deep growl of thunder sends a pig scurrying under the longhouse, as several chickens scamper up from the muddy ground into the refuge of the roost. The hunter thrusts his chin toward the other side of the valley. "It is coming from the *nyobia* side. People there must have done something." The young second wife of my host swings her infant from her back to her breast, frowns, and snaps up at the roof, "Go to Yapup village! We don't know anything!"

"Someone must have killed a child of the spirit Dojung," the adult son of the old hunter Takar shouts across the fire. "Yes!" another villager calls. "Someone over there!" The man looks up at the roof and shouts, "Swallow whoever stole your child!" Across the fire, one of the hunters of the village, a young man who oscillates back and forth between the village and Koloriang township half a day's walk away, fixes his hunting mentor Tarido with a steady gaze. "Do it! Chant the clouds away! Tell them to go!"

Voices rise in agreement. For a moment, the hunter crouches still beside the fire, then he inhales deeply, leans into the fire, and begins to chant: "We humans who live here, we have done nothing wrong! We have done nothing wrong!" Jabbing a finger at the other side of the valley, he chants, "We are not your target! Go there to the people who swallowed your child! Whoever killed your child, go there! There, to that side! Sniff out the culprit there! We know nothing here! Go there! Go there!"

Another flash of lightning, and another deep rumble of thunder tumbles through the air of the valley, through the taught body of the longhouse and the villagers gathered there. From the shadows, a voice rises. Tirey, the youngest son of the *nyubu* shaman-priest of this village who also oscillates between the village and Koloriang, leans into the firelight. His eyes dart between the faces as he announces he has something they should know. Earlier that day, a group of his friends came to him in Koloriang and told him that they were planning to trek up above Koloriang to the mountain lake at the source of the Payu River. Like him, they knew it was a spirit-lake, *sinyuk*, set aside for the powerful mountain spirit, Dulu-Kungu Dojung. But they said they wanted to dam-fish there, where the fish were abundant. They asked him to come with them, but he declined. Instead, he walked back to the village.

For a moment, the villagers sit in silence around the fire, taking in what the young man has said. The hunter Tarido is the first to react. He jumps up and strides out through the low doorway. I grab my bag with my audio recorder and follow him outside, finding him crouched beside the skull-rack on the resting platform. Perched on the edge of the platform, facing into the valley, a silhouette against the darkening line of hills, he shouts over to me, "Dojung sent the storm! I do not want to see Dojung move!"

Crouching in the gathered dusk, he takes up the chant, to the storm, to the spirit Dojung: "Do not come to this place! Do not swallow our village! Do not come to this place! Do not swallow our village!" Another flash of lightning illuminates the clouds pushing like an army over the tropical forests of the valley, and again the thunder shakes the air of the valley. "No one here killed your children! Go there, to the sunlit side! Go there! Attack whoever swallowed your children there! No one here killed your children here! Do not come here!"

I look down at my digital recorder lying upon the split bamboo floor of the platform, buzzing in its little open universe, and in that moment the batteries fail, and the storm and village dissolve back into the antimatter of unrecorded time.

Genealogy of a Vengeful Storm

Re-: back. *Fugere*: flee. Refuge. In oral histories across the uplands, storms like this usually end in disaster. They rip open the fragile leaf thatch of the longhouse and tear it from its hardwood and bamboo frame, or uproot the entire structure and wash it and its terrified inhabitants down to the river below. In such stories, domestic animals living beneath the longhouse, tokens of human wealth, flee back to the forest and take refuge there (Aisher 2016). But the storm that came that dusk to Talum village did not conform to this pattern: no animals fled "back" to the forest, no human wealth was retrieved. Instead, the storm turned away. But questions remain.

The storm was alive. Villagers gathered in the longhouse felt it sniffing out a human culprit. From their perspective, this storm wasn't just metaphorically angry or vengeful: it *was* angry, furious, filled with desire for revenge. It was a site of more-than-human sociality: a storm-that-feels.

But there was more to it than that. To articulate how the storm felt for Nyishi villagers gathered in the longhouse that night we must track back through key moments in its genealogy, moments through which it became what it was. A genealogy of this storm-that-feels, and its emotion, can proceed by piecing together fragments of its past and placing them in service to the present; a methodological pathway that could lead to a deeper understanding of extreme events in general.

Understanding this storm demands an appreciation of the "innate" fury of storms, how they assault human senses. However, beneath its phenomenal surface, this storm was rich with stories and histories. Tirey's confessionary account to villagers traced a line back from the storm, to events that (supposedly) occurred earlier that day: a group of young men strode up through the forests above Koloriang, to a mountain lake at the source of the Payu River, to dam-fish there.

From the standpoint of villagers gathered in the longhouse, this was, in the language of the 1992 Earth Summit in Rio de Janeiro, "dangerous human interference." Why? Because for Talum villagers, mountain lakes like this were "mother-places" (*aaney-nyoku*): safe havens for spirit-owners of the wild. As such, they had to be approached with great care. For villagers gathered in the longhouse that night, such lakes should not have been approached at all: humans who visit them invite danger and misfortune. Such lakes, and their indwelling spirits, neither need nor desire human presence. From the standpoint of Dojung, such lakes were a refuge. They were safe havens from human predation. Fishing and hunting in their vicinity was forbidden. Even uttering the name "Dojung" in forests around them risked angering powerful master spirits.

So it was that the young men's plan to fish there amounted to an *invasion* of a refuge. In the Nyishi uplands, in the moment this storm came, stories abound of villagers who inadvertently wandered too near to such lakes and suffered bleeding from the mouth, nose, and ears, symptoms of spirit-attack. The power of such lakes is inseparable from the rumors that surround them. The landscape of the uplands is a shared landscape, suffused with diverse claims, human and other. Inevitably, perhaps, tensions existed between these diverse stakeholders. As villagers said, humans and spirit-owners of the wild were like "two hands holding the same object"; the land was akin to "a bead on a thread." Stories set them apart.

To understand how the storm felt that night, we must understand that the atmosphere in the village matters. Based on overhunting and declining wildlife—conceived as declining wealth of Dojung spirits—there was a sense in the village of mutual *distrust* between humans and Dojung spirits. Villagers well knew that hunters who neglected the feelings of spirit-owners of wildlife played a dangerous game. For all they took, or stole, many paid "the cost of a child." In this at times hostile landscape, this forest of mirrors suffused with multiple perspectives, simply witnessing something strange, unexpected, or out of place could be a precursor to spirit-attack (Aisher 2016).

With a strange hint of nostalgia, elderly respondents in Koloriang recalled the long era of clan warfare, cattle theft and longhouse burnings that continued on for centuries in the uplands. That world of surprise attacks, wife capture, abduction of enemies, and retrieval of blood-price—part of a clan warfare imaginary—lived on in uplanders' perceptions of spirits (Aisher 2020). Perhaps it was an appropriate imaginary for a landscape as fragile as that. Tales abounded of human actions and their consequences: a "reciprocal environment" (Bird-David 1990) that could switch to a *taking environment* and bite back, through hunting failure or crop failure, or death of domestic animals, or death of family members—or a storm. On the night of the storm, the wind blew from an unfamiliar direction, and this triggered villagers' collective discussion that led to Tirey's confessionary account of the young men fishing in the spirit-lake.

Across many stories, storms were media of revenge. They primed villagers to perceive the danger in the strange wind that blew that afternoon from an unfamiliar direction. That's why they gathered in the longhouse and why the hunter performed the oracle. Through their stories, they knew human folly could trigger revenge. The hurt that Dojung felt the evening of the storm had its origin also in an ancient story recounting the separation of humans and spirits. The storm's anger was the anger a person feels when agreed borders have been encroached. The villagers gathered there knew Dojung inhabited the hills long before humans came, and they knew Dojung bore the weight, and the cost, of the human presence.

Like climate change itself, storms can serve as triggers for those who encounter them, to "upframe" (Kohn 2013: 78) their perceptions to a realm of powerful ecological sovereignties that seem to stand behind these phenomena. The storm also triggered an exchange of perspectives between villagers and the storm. Like a mirror, it urged them to reflect on how Dojung felt about *them*. For a moment, it forced them to return to the value of refuge—for them, but also for Dojung.

Neat naturalistic classifications can erase these features of indigenous experience. As Venkatesan reminds us, "Ontologies, theories of being and reality, have histories (and genealogies)" (2010: 154). So too for this storm. Oral narratives, the central method of communicating knowledge in the uplands, offer a pathway into the atmosphere of mutual mistrust between humans and spirits. A hunter from the village sees a monkey with a necklace of white beads climbing into the crown of a *sangrik* fig tree. Unable to stop himself, he shoots at it with his shotgun. On his deathbed two years later, he admits to villagers that whenever he passed that tree after that event, he would hear a strange, high-pitched sound emanating from its crown. Everyone in Talum knew aggressive spirits dwelled in that spirit-tree. Everyone agreed that these spirits were responsible for the hunter's death.

Knowledge uncertainty suffused villagers' response to the storm. It underpinned their request to the hunter to perform an oracle and chant away the storm. They knew, through their stories, individual actions often have collective consequences. Like all storms, this storm had a genealogy: how it came to be what it was. It also possessed a genealogy of emotion: how it came to feel as it felt. And in feeling as it did, it revealed a deeper value: the value of refuge. And it also revealed what it means to invade a refuge and how this feels both for humans and for spirits of the wild.

Over the coming era of climate change, storms will continue to be sites of more-than-human sociality. As great revealers, they will continue to urge humans to upframe their perceptions, from individual storms to the ecological sovereignties that surround the human village.

Conclusion: Feeling into Climate Change—Why Stories Matter

In this era of anthropogenic climate change, extreme events will continue to remind humans of the need for *conviviality* with powerful ecological sovereignties and the more-than-human sources of human flourishing. Like an angry storm passing over a village, some will offer a counterpoint to *solastalgia*, that sense of loss that can come with dramatic change or deterioration of a once familiar landscape. Some will force those communities who encounter them to reacquaint themselves with the landscape and recognize once again the necessity for exchange with the other-than-human sources of human flourishing. In part, this is what it means to return to place.

As the Intergovernmental Panel on Climate Change recognizes, to deal with climate change we need new "convincing physically-based storyline[s]" (Stott 2016: 1518). Global climate change resists being condensed to a single story of molecules, particles, and other elements—in part because it is the story of countless communities facing what the weather brings. In the language of the poet Louis MacNeice, climate change is "incorrigibly plural." As Cullen puts it, "If climate is impersonal statistics, weather is personal experience" (2010). The personal and impersonal need to be reconnected. Stories offer a crucial bridge. Recognizing landscapes as *multispecies assemblages* (see Tsing 2013) also helps to draw us closer to their potent more-than-human agency. By drawing an inert background into a living foreground, such a recognition can facilitate the telling of new stories. As Le Carré puts it, "The cat sat on the mat is not the beginning of a story, but the cat sat on the dog's mat is" (Barber 1977).

Exploring how global warming intersects with local realities resonates with current work on the anthropology of emotion. As a leading scholar in the field, Andrew Beatty argues that the particularity of emotion runs counter to the dominant focus in anthropology and much climate science upon systems, groups, collectives, and cultures. The very multidimensionality of emotion renders it a casualty of any description that is too general, in part because "the occasion, expression and meaning of emotion are personal and particular, there being no such thing as a general emotion" (Beatty 2014: 555). This holds true for the storm that came to Talum village. We won't get anywhere by asking "how do storms feel?" because storms in general, like humans in general, don't feel anything at all.

To get back to personal and felt experience, we need to ask: "How did *this* storm feel to *these* villagers on *this* night?" To answer this, we need to ground our accounts in the stories of those who encounter them. As Beatty notes, "Emotions are not the creation of a moment. They participate in manifold relationships formed over periods of time" (2010). Only detailed narrative accounts that honor the particularity of individual lives—accounts that include the "plots and players, the people who inhabit the roles" (Beatty 2013), not all of whom are human—can get back to emotions in all their rich complexity. Sometimes it is only through story that humans can express how the human world appears a little way out from the human shore.

From my perspective now, several years after leaving Talum village, the storm is above all a story of two refuges, not one: the refuge of a longhouse and the refuge of a spirit-lake. It speaks to me now of humanity's need to identify and conserve refugia. As Bachelard reminds us, well-being, human and other, "takes us back to the primitiveness of the refuge" ([1958] 1964: 91). In the challenging time ahead, story may yet play a crucial role in conserving those places that humans co-create best through acts of setting them aside.

Acknowledgments

I wish to thank the British Academy and Economic and Social Research Council for funding the ethnographic research in the Eastern Himalayas that forms the basis of this chapter. I extend my deep and abiding gratitude to the Nyishi residents of Kurung Kumey District for their generosity, hospitality, and wisdom. Special gratitude goes to the hunters of Talum village for sharing their stories, and to Bengia Chongpi, Bengia Amit, Pige Ligu, and Bengia Takio for their friendship and extraordinary support in the field.

Alexander Aisher is based at the Department of Anthropology of the University of Sussex, UK. As an environmental anthropologist and ecological designer, he has a deep and abiding interest in multispecies ethnography, wildlife conservation, climate change, indigenous environmental knowledge, storytelling, and adaptation. In 2003, he became the first anthropologist in over forty years to conduct long-term ethnographic fieldwork in the internationally contested Protected Area and "biodiversity hotspot" of Arunachal Pradesh in the Eastern Himalayas. In 2007, he was awarded a British Academy Postdoctoral Fellowship to explore community-led environmental decision-making. In 2018, he returned to Arunachal Pradesh, site of his doctoral fieldwork, to offer policy recommendations on conservation of wildlife, oral history, and indigenous identity.

Note

1. For some scholars, the term *biocultural refugia* captures the quality of those places that "not only shelter species, but also carry knowledge and experiences about practical management of biodiversity" (Barthel et al. 2013: 2–3).

References

Adger, Neil, Irene Lorenzoni, and Karen O'Brien, eds. 2009. *Adapting to Climate Change: Thresholds, Values, Governance*. Cambridge: Cambridge University Press.

Aisher, Alexander. 2016. "Scarcity, Alterity and Value: Decline of the Pangolin, the World's Most Trafficked Mammal." *Conservation and Society* 14(4): 317–29.

———. 2020. "Fieldwork's Return: Troubled Steps Towards a Multispecies Imaginary." In *Uncanny Landscapes*, edited by Jon Michell and Karis J. Petty, special issue of *Material Religion* 16(4): 491–509.

Aiyadurai, Ambika, and Nandini Velho. 2018. "The Last Hunters of Arunachal Pradesh: The Past and Present of Wildlife Hunting in North-East India." In *Conservation from the Margins*, edited by Umesh Srinivasan and Nandini Velho, 69–93. New Delhi: Orient Blackswan.

Bachelard, Gaston. (1958) 1964. *The Poetics of Space*. Translated by M. Jolas. Boston: Beacon Press.
Barber, Michael. 1977. "John le Carré: An Interrogation." *New York Times*, 25 September. Retrieved 1 May 2020 from http://www.nytimes.com/books/99/03/21/specials/lecarre-interrogation.html.
Barthel, Stephan, Carole Crumley and Uno Svedin. 2013. "Bio-cultural Refugia: Safeguarding Diversity of Practices for Food Security and Biodiversity." *Global Environmental Change* 23(5): 1142–52.
Basso, Keith. 1996. *Wisdom Sits in Places: Landscape and Language among the Western Apache*. Albuquerque: University of New Mexico Press.
Beatty, Andrew. 2010. "How Did It Feel for You? Emotion, Narrative, and the Limits of Ethnography." *American Anthropologist* 112(3): 430–43.
———. 2013. "Current Emotion Research in Anthropology: Reporting the Field." *Emotion Review* 5(4): 414–22.
———. 2014. "Anthropology and Emotion." *Journal of the Royal Anthropological Institute* 20: 545–63.
Berger, John. 1982. *Another Way of Telling*. New York: Vintage Books.
Bird-David, Nurit. 1990. "The Giving Environment: Another Perspective on the Economic System of Gatherer-Hunters." *Current Anthropology* 31: 189–96.
Birks, H. John B. 2015. "Some Reflections on the Refugium Concept and Its Terminology in Historical Biogeography, Contemporary Ecology and Global-Change Biology." *Biodiversity* 16(4): 196–212.
Camus, Albert. 1955. *The Myth Of Sisyphus and Other Essays*. Translated from the French by Justin O'Brien. New York: Random House.
Candea, Matei. 2010. "'I Fell in Love with Carlos the Meerkat': Engagement and Detachment in Human-Animal Relations." *American Ethnologist* 37(2): 241–58.
Census of India. 2011. *District Census Handbook: Kurung Kumey*. Itanagar: Directorate of Census Operations.
Chaudhary, Pashupati, and Kamaljit S. Bawa. 2011. "Local Perceptions of Climate Change Validated by Scientific Evidence in the Himalayas." *Biology Letters* 7: 767–70.
Committee on Himalayan Glaciers, Hydrology, Climate Change, and Implications for Water Security. 2012. *Himalayan Glaciers: Climate Change, Water Resources, and Water Security*. Washington, DC: The National Academies Press.
Crate, Susan A. 2008. "Gone the Bull of Winter? Grappling with the Cultural Implications of Anthropology's Role(s) in Global Climate Change." *Current Anthropology* 49(4): 569–95.
Cruz, Rex. V., H. Harasawa, M. Lal, S. Wu, Y. Anokhin, B. Punsalmaa, Y. Honda, M. Jafari, C. Li, and N. Huu Ninh. 2007. "Asia: Climate Change 2007; Impacts, Adaptation and Vulnerability." In *Contribution of Working Group II to the Fourth Assessment Report of the Intergovernmental Panel on Climate Change*, edited by M. L. Parry, O. F. Canziani, J. P. Palutikof, P. J. van der Linden, and C. E. Hanson, 469–506. Cambridge: Cambridge University Press.
Cullen, Heidi. 2011. *The Weather of the Future: Heat Waves, Extreme Storms, and Other Scenes from a Climate-Changed Planet*. Repr. ed. New York: Harper Perennial.
Foucault, Michel. (1975) 1991. *Discipline and Punish: The Birth of the Prison*. London, Penguin.
———. 1980. *Language, Counter-memory, Practice: Selected Essays and Interviews*. Ithaca, NY: Cornell University Press.

Garcia-Acosta, Virginia. 2002. "Historical Disaster Research." In *Catastrophe and Culture: The Anthropology of Disaster*, edited by Susanna M. Hoffman and Anthony Oliver-Smith, 49–66. Santa Fe: School of American Research.

Haraway, Donna. 2008. *When Species Meet*. Minneapolis: University of Minnesota Press.

———. 2015. "Anthropocene, Capitalocene, Plantationocene, Chthulucene: Making Kin." *Environmental Humanities*: 159–65.

Hurston, Zora N. (1937) 1986. *Their Eyes Were Watching God*. London: Virago Press.

IPCC (Intergovernmental Panel on Climate Change). 2014. *Climate Change 2014: Impacts, Adaptation and Vulnerability. Working Group II Contribution to the Fourth Assessment Report of the IPCC*. Cambridge: Cambridge University Press.

Keppel, Gunnar, and Grant W. Wardell-Johnson. 2012. "Refugia: Keys to Climate Change Management." *Global Change Biology* 18: 2389–91.

Kirksey, Eben S., and Stefan Helmreich. 2010. "The Emergence of Multispecies Ethnography." *Cultural Anthropology* 25(4): 545–76.

Kohn, Eduardo. 2013. *How Forests Think: Toward an Anthropology beyond the Human*. Berkeley: University of California Press.

Mishra, Charudutt, M. D. Madhusudan, and Aparajita Datta. 2006. "Mammals of the High Altitudes of Western Arunachal Pradesh, Eastern Himalaya: An Assessment of Threats and Conservation Needs." *Oryx* 40(1): 29–35.

Myers, Norman, Russell A. Mittermeier, Cristina G. Mittermeier, Gustavo A. B. da Fonseca, and Jennifer Kent. 2000. "Biodiversity Hotspots for Conservation Priorities." *Nature* 403: 853–58.

Nandargi, S., and O. N. Dhar. 2011. "Extreme Rainfall Events over the Himalayas between 1871 and 2007." *Hydrological Sciences Journal* 56(6): 930–45.

Nature. 2014. "Climate Change: Monsoon Wet Spells Get Wetter." *Nature* 509: 11.

Oliver-Smith, Anthony, and Susanna M. Hoffman, eds. 2020. *The Angry Earth: Disaster in Anthropological Perspective*. 2nd ed. London: Routledge.

Roncoli, Carla, Todd Crane, and Benjamin S. Orlove. 2008. "Fielding Climate Change in Cultural Anthropology." In *Anthropology of Climate Change: From Encounters to Actions*, edited by Susan A. Crate and Mark Nuttall, 87–115. Walnut Creek, CA: Left Coast Press.

Roy, N. C. 2005. *Arunachal Pradesh Human Development Report*. Itanagar: Department of Planning, Government of Arunachal Pradesh. Retrieved 15 December 2007 from http://hdr.undp.org/en/reports/national/asiathepacific/india/name,3398,en.html.

Sharma, Eklabya, Nakul Chettri, Karma Tse-ring, Arun B. Shrestha, Fang Jing, Pradeep Mool, and Mats Eriksson. 2009. *Climate Change Impacts and Vulnerability in the Eastern Himalayas*. The International Centre for Integrated Mountain Development (ICIMOD). Retrieved 1 May 2015 from http://lib.icimod.org/record/8051.

Sherpa, Pasang Y. 2012. "Sherpa Perceptions of Climate Change and Institutional Responses in the Everest Region of Nepal." PhD diss., University of Washington, Seattle.

Singh, Ranjay K., Jules Pretty, and Sarah Pilgrim. 2010. "Traditional Knowledge and Biocultural Diversity: Learning from Tribal Communities for Sustainable Development in Northeast India." *Journal of Environmental Planning and Management* 53(4): 511–33.

Singh, Ranjay K., Ramesh C. Srivastava, Chandra B. Pandey, and Anshuman Singh. 2015. "Tribal Institutions and Conservation of the Bioculturally Valuable 'Tasat' (Arenga obtusifolia) Tree in the Eastern Himalaya." *Journal of Environmental Planning and Management* 58(1): 69–90.

Stott, Peter. 2016. "How Climate Change Affects Extreme Weather Events." *Science* 352(6293): 1517–18.

Strauss, Sarah, and Benjamin S. Orlove. 2003. "Up in the Air: The Anthropology of Weather and Climate." In *Weather, Climate and Culture*, edited by Sarah Strauss and Benjamin S. Orlove, 3–14. Oxford: Berg.

Thompson, Christian. 2009. "The Eastern Himalayas—Where Worlds Collide." Retrieved 1 September 2016 from https://wwfin.awsassets.panda.org/downloads/eh_new_species.pdf.

Tsing, Anna L. 2013. "More-than-Human Sociality: A Call for Critical Description." In *Anthropology and Nature*, edited by Kirsten Hastrup, 27–42. London: Routledge.

Turner, Nancy J. 2005. *The Earth's Blanket: Traditional Teachings for Sustainable Living*. Seattle: University of Washington Press.

Venkatesan, Soumhya. 2010. "Introduction." In "Ontology Is Just Another Word for Culture: Motion Tabled at the 2008 Meeting of the Group for Debates in Anthropological Theory, University of Manchester." *Critique of Anthropology* 30(2): 152–200.

Xu, Jianchu, R. Edward Grumbine, Arun Shrestha, Mats Eriksson, Xuefei Yang, Yun Wang, and Andreas Wilkes. 2009. "The Melting Himalayas: Cascading Effects of Climate Change on Water, Biodiversity, and Livelihoods." *Conservation Biology* 23(3): 520–30.

Yadav, Pramod K., and Manish Kaneria. 2012. "Shifting Cultivation in North-East India." Retrieved 15 December 2007 from http://www.academia.edu/3067626/ShiftingCultivationinNorth-EastIndia.

Chapter 3

Who Is Perturbed by Ecological Perturbations?

Marine Scientists' and Polynesian Fishers' Understandings of a Crown-of-Thorns Starfish Outbreak

Matthew Lauer, Terava Atger, Sally J. Holbrook, Andrew Rassweiler, Russell J. Schmitt, and Jean Wencélius

Introduction

Anthropological work focusing on local accounts of climate change has blossomed in recent years (Carey 2010; Crate 2011, Rudiak-Gould 2013a). In regions ranging from the arctic to small tropical islands, research has revealed how local people detect, understand, and interpret the local effects of global climate shifts (Krupnik and Jolly 2002; Mimura et al. 2007). This body of research builds on a long history in anthropology examining indigenous or local ecological knowledge (LEK) (Berkes, Colding, and Folke 2000). Beginning at least in the 1950s with Conklin's path-breaking work in the Philippines (1954), researchers began describing the rich and detailed compendium of knowledge held by indigenous people pertaining to local flora, fauna, and ecology.

LEK studies exploring marine and coastal ecosystems have tended to lag behind terrestrially focused research (Lauer 2017). In fact, the first detailed accounts of marine LEK began in the 1980s, almost thirty years after Conklin's work in the Philippines. It was a natural scientist, R. E. Johannes (1981), not an anthropologist, whose seminal work on Palauan fishers

Notes for this chapter begin on page 84.

brought the first comprehensive documentation of fisher knowledge and inspired a generation of researchers. His study in Micronesia revealed that islanders had greater depth of knowledge about some ecological processes, such as fish spawning aggregations, than did marine scientists. Moreover, on some Pacific islands, local kin groups continue to manage marine resources through long-standing practices, such as temporary closures and cohesive ridgetop-to-reef ecosystem management, and in certain cases have sustained limited island resources for generations (South et al. 1994). These knowledges and practices now serve as a foundation for contemporary marine resource management systems in many parts of Oceania (McMillen et al. 2014; Jupiter et al. 2014).

More recently, local islander knowledge about climate change, especially sea level rise, has captured attention both in the academic community and throughout the wider public (Rudiak-Gould 2013a). Low-lying atolls and the people who inhabit them are suffering the first effects of rising oceans, and studies have documented how island peoples are adapting, migrating, and interpreting these changes (Lazrus 2012). In addition to sea level rise, the degradation of coral reefs, especially in the Pacific region, has garnered much attention. Climate scientists have shown that coral reefs were one of the first ecosystems to begin to respond to climate-induced stresses, such as rising ocean temperatures, and in the coming decades will undergo major shifts (Hughes et al. 2003). While LEK research has detailed how Pacific Islanders can accurately detect the ecological effects of rapid perturbations such as tsunamis, LEK also develops around slower shifts that arise over many decades, such as expanding seagrass meadows (Aswani and Lauer 2014; Lauer and Aswani 2010; Lauer and Matera 2016). Of course, the decline of coral reef ecosystems threatens not only biodiversity but also the life-worlds of the Pacific peoples who depend on them as cultural and economic resources and as a source of cosmological inspiration. As Tongan anthropologist Epeli Hau'ofa eloquently expresses, Pacific peoples have a deep connection with the ocean: "The sea is . . . a major source of our sustenance, and is something we all share in common . . . the ocean is in us" (Hau'ofa 2000).

Although respect for LEK as a viable and accurate knowledge base has increased among the wider scientific community, there continues to be much debate about how to characterize knowledge production in nonscientific contexts and how scientific and nonexpert knowledge should relate (Goldman 2007; Klenk et al. 2017; Jasanoff 2004; Wynne 1996; Agrawal 1995; Bohensky and Maru 2011; Watson-Verran and Turnbull 1995; Goldman, Nadasdy, and Turner 2011). The predominant model has been to assume that science knowledge can serve as a neutral arbiter by which to judge the validity of all other accounts (Davis and Ruddle 2010). Most of the local ecological studies about climate change as well as earlier LEK research explicitly or implicitly accepted scientific knowledge as a means of legitimizing nonexpert ecological observations. It was through the validation of LEK by science that many non-anthropologists have become convinced that LEK is not inferior or deficient compared to expert knowledge. This is itself evidence that the relationship between science and indigenous or non-Western knowledge continues to be asymmetric. As postcolonial scholars have made clear, modern technoscientific knowledge has a dubious history not only of validating racist, sexist, and exploitative treatment of marginalized groups within Western society itself and the Global South more generally (Haraway 1991; Said 1979) but also of neglecting and denigrating local specialist knowledge (Hobart 1993). In response, some indigenous peoples have positioned their knowledge politically against official and scientific claims as a means to bolster their authority and autodetermination (Brosius 2006).

These issues become particularly salient when there is *disagreement* between the scientific community and local people about ecological dynamics. In a well-documented case on the island of Tuvalu, islanders attributed increased erosion, saltwater intrusion, and flooding to climate change–induced sea level rise, even though the scientific research community had not determined that sea level rise was responsible (Connell 2003). Scientific knowledge appeared to be ignored by the Tuvaluan government, which attempted to link many of the island's environmental problems with sea level as a means to blame the international community and seek compensation. This strategy of ignoring scientific knowledge for political gain underpins disagreements between experts and nonexperts that are now rampant, such as debates about climate change, vaccinations, GMOs, and pesticide use (Oreskes and Conway 2011). Yet, even when political motives are less salient and science practitioners avoid overt marginalization of nonexpert knowledge, the sociology of science literature highlights the problems of assuming science can serve as the benchmark to judge validity because it, like all knowledge systems, imposes subtle yet critical epistemic commitments and normative concepts (Latour 1999; Jasanoff 2004).

In this chapter, we enter into these debates by focusing on local fishers and marine scientists' characterizations of climate change–related coral loss on the island of Moorea, French Polynesia. Moorea is an interesting and illuminating case because there are rich bodies of both scientific and fisher knowledge about the same ecosystem. Fishing is central to Moorea households, and fresh reef fish caught locally are consumed nearly every day. In addition, activities in the ocean and the marine environment are central to Polynesians' cultural identity, everyday life, and way of being.

At the same time, Moorea is one of the world's centers of tropical coral reef research. The island is home to two prominent research centers that have accumulated a wealth of marine science observations in the past half century. The existence of both local and scientific knowledge enables a side-by-side comparison of how different knowledges are produced, received, intermingled, challenged, packaged, as well as entangled with political, economic, spiritual, and social processes.

We focus on an outbreak of crown-of-thorns starfish (*Acanthaster planci*) that, as measured by marine scientists, led to the destruction of 95 percent of the coral on the outer reefs of the island from 2008 to 2010. Crown-of-thorns starfish (COTS) are one of the most studied organisms on tropical coral reefs (Pratchett et al. 2017). These coral-eating organisms are well known across the Indo-Pacific for sudden, massive population booms where huge aggregations rapidly damage large areas of coral. Importantly, COTS outbreaks appear to be exacerbated by climate change–induced ocean acidification and warming (Kamya et al. 2017; Uthicke et al. 2015). During the Moorea outbreak, scientists characterized it as one of the most intense and devastating starfish population booms ever recorded by the coral reef science community (Trapon, Pratchett, and Penin 2011; Adam et al. 2011; Adjeroud et al. 2009). Reports from the marine science community and local NGOs advocated for the removal of starfish (Lagouy 2007; Lison de Loma, Chancerelle, and Lerouvreur 2006: 13). Following these recommendations, the local government supported and financed an eradication campaign on Moorea where starfish were removed and burned on the beaches. Marine science research monitoring the recovery since the outbreak has revealed that coral cover has returned to pre-disturbance levels in many areas (Holbrook et al. 2018), though the species composition has recovered in some reef regions but not others (Adjeroud et al. 2018). Local fishers, for their part, although well aware of the outbreak and of the coral-eating behavior of the crown-of-thorns starfish, did not view the outbreak as a major event warranting action. In what follows, we explore these contrasting standpoints and their broader implications for LEK research. We ask a seemingly basic set of related questions: Who notices changes to Moorea's coral reefs, and how can it be judged if they are noteworthy? Who notices the effects these changes have on coral reef fish? And what should be done if a perturbation and its effects are identified?

Moorea

Moorea (figure 3.1) is a triangular-shaped volcanic island with sharp mountain peaks that jut up abruptly from the coastline. A barrier reef

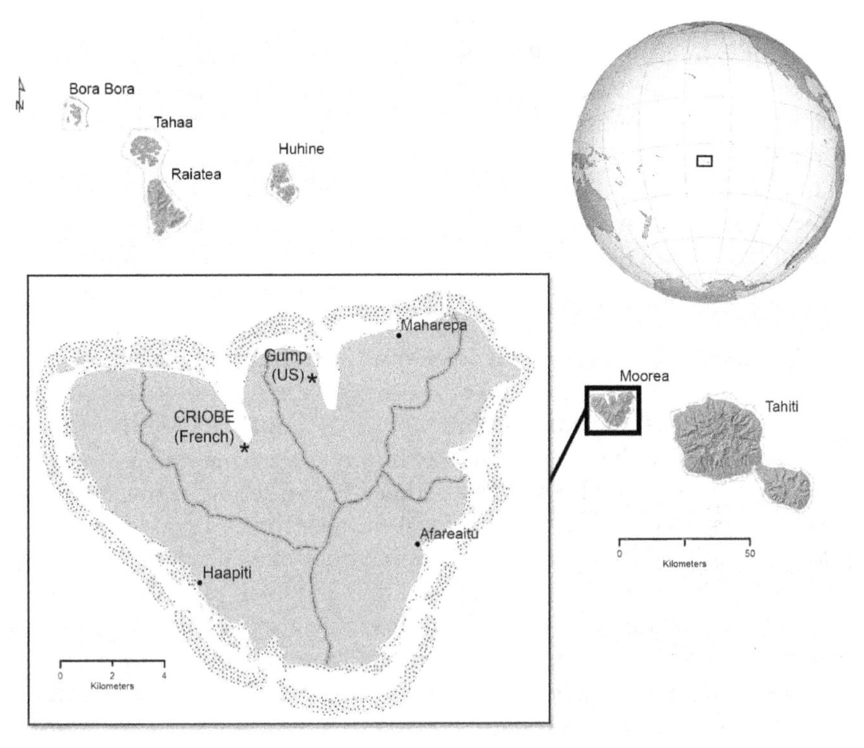

Figure 3.1. Moorea island, main settlements, and its two research centers. © Matthew Lauer.

rings the island, and ten main reef passes allow open ocean water to circulate through the shallow (less than ten meters deep) lagoons. The island is part of the Society Islands group in French Polynesia and is twenty kilometers west of French Polynesia's most populous and largest island, Tahiti. Moorea's close proximity to an international airport in Papeete on Tahiti across the channel has fueled tourism, and the island is now one of the most visited destinations in French Polynesia. With several large international hotels, numerous family-run guesthouses, and rental houses, tourism dominates the local economy. Attracted by employment in the tourist industry, immigrants from other islands in French Polynesia have swelled the island's population to over seventeen thousand inhabitants that reside in five administrative districts.

Despite centuries of major socioeconomic and cultural change associated with European colonization and the more recent effects of globalization, fishing continues to be a central part of Moorean life (Leenhardt et al. 2016). Over three-quarters of households have a member active in the fishery, and the consumption of fresh reef fish is high, with 67 percent of households reporting that they eat fresh reef fish at least three times per week. Unlike other Pacific island nations, Moorea's local communities are not as dependent on marine resources for food security or income. This is related to French Polynesia's status as a semiautonomous territory of France, where financial support has led to a high standard of living and a social safety net that provides free primary education and healthcare. French Polynesia's dependence on France, of course, brings a neocolonial political climate, but the economic and social safety net has meant that fishing is highly valued for cultural and recreational purposes rather than just for sustenance or economic livelihood. Indeed, eating fresh reef fish is central to the Polynesian sense of identity, and important events such as church gatherings, birthday parties, and *ma'a Tahiti* (large festive meals) invariably involve the consumption of locally caught reef fish.

Moorea also is well known as a center of coral reef scientific research. Moorea's coral reefs are some of the most studied tropical coral reef systems in the world. Two international research centers, one French and the other American, have hosted scientists specializing in coral reef research since the early 1970s, and dozens of scientific papers are published every year. Moorea's scientific community has compiled a detailed series tracking the island's ecosystem change through time. The American research facility (Gump Research Station) is administered by the University of California and is the field base for a National Science Foundation–funded Long Term Ecological Research site (Moorea Coral Reef LTER) that was established in 2004 to study the coral reef ecosystem. The French station, known as the Centre de Recherches Insulaires et Observatoire de l'Environnement (CRIOBE) has a similar Centre National de la Recherche Scientifique–funded monitoring program. The research conducted for this chapter involves the collaboration of social scientists, marine scientists from the Moorea Coral Reef LTER, and local fishers. Our project, titled "Recherche Collaborative Pour la Pêche à Moorea" (Collaborative Research for Moorea's Fishery), is focused on better understanding through collaborative science, the interrelations between fishing practices, livelihood strategies, and shifting dominance of coral and algae on reefs around the island.

Hungry Starfish and Coral Loss

From 2006 to 2007, Moorea's marine science research community began to notice a rapid spike in the abundance of crown-of-thorns starfish (*Acanthaster planci*), an organism that has attracted more attention from the scientific community than any other single species on coral reefs (Lison de

Loma et al. 2006; Lagouy 2007). Found across the tropical Indo-Pacific, this sea star is the world's largest. It can grow up to twenty legs, reach nearly half a meter in diameter, and can weigh up to six kilograms (Pratchett et al. 2014). It is covered with a starburst of thick, venomous, two- to three-centimeter-long thornlike spines, whose toxin can momentarily paralyze a swimmer and cause fits of vomiting. The spines, purportedly resembling the biblical crown of thorns, give the sea star its English common name.

Commonly referred to by the acronym COTS, the starfish has gained a notorious, and infamous, reputation among many marine scientists for its voracious appetite for coral polyps, which it consumes by extruding its stomach out of its body to digest the living tissue of the coral. Moreover, COTS have a propensity to undergo sudden population booms and emerge in large aggregations (Birkeland and Lucas 1990). During outbreaks, the organisms consume huge swaths of coral reef. First reported in Fiji in the 1930s, then in Japan in the 1950s, and later on the Great Barrier Reef in Australia in the 1960s, COTS outbreaks have now caused widespread damage on Indo-Pacific reefs (Zann, Brodie, and Vuki 1990; Brodie et al. 2005).

Despite this apparent increase, there is much debate in the marine science community as to whether COTS outbreaks are the result of recent anthropogenic drivers, such as increased nutrient delivery from land, or if they are a normal population dynamic that has been occurring for many thousands of years. Opinions about COTS outbreaks are so polarized that they have been dubbed a "Starfish War" (Raymond 1986), and only recently, due to the even greater existential threat of climate change, has attention on the starfish waned. Despite these divisions within the marine science community about COTS, some coral reef researchers consider the sea star outbreaks as a menace to coral reefs, and in some areas eradication programs have been established. It has been estimated that nearly seventeen million starfish have been killed or removed from reefs across the Indo-Pacific since the 1970s at the cost of nearly US$40 million (Pratchett et al. 2014). In fact, in 2012 the Australian government committed US$23 million to fund the Great Barrier Reef Marine Park Authority to implement a ten-year control program (Kwai 2018). France's applied research institute IRD (Institut de Recherche pour le Développement) has recently advocated for a citizen-based "lime juice fight" against COTS in New Caledonia and Vanuatu, in which lime juice injections are described as an effective control method (Moutardier et al. 2015; Dumas et al. 2015). Much research continues to investigate more radical techniques of population control, such as injecting the starfish with lethal toxins.

In our conversations with marine scientists working on Moorea, many of them expressed uncertainty about the course of action during outbreaks.

In 2018, for example, a technician from the CRIOBE station shared: "We know very little of the actual effects of human removal of COTS. Actually, removing them may have ecological consequences we are not aware of." During a COTS workshop convened by the Australian government in 2012, dozens of marine scientists—including from CRIOBE and the MCR LTER—agreed that research had failed to ask the important question of what causes a COTS outbreak to collapse (Schaffelke and Anthony 2015). Examples of proactive interventions appear to result in shifting the boom-and-bust dynamics to chronically recurring episodes. In other cases, interventions appear to effectively mitigate severe outbreaks, but only in circumscribed areas to protect a particularly valuable reef tract (e.g., for ecotourism). More broadly, scientists are beginning to focus their efforts to track the ecological effects of COTS interventions.

On Moorea, two COTS outbreaks have been documented by marine scientists, one in 1979 and another in 2009, although outbreaks had been reported on the island as early as 1969 (Trapon et al. 2011; Adam et al. 2011; Adjeroud et al. 2009; Rassweiler et al. 2020). In the more recent and better-monitored 2009 outbreak, underwater surveys revealed a dramatic increase in COTS densities on Moorea's outer reefs in 2007, eventually peaking in 2009 and then abruptly declining in 2010 (Kayal et al. 2012). In just one year, the density of starfish increased tenfold. In addition, a category four cyclone, Oli, battered the island in February of 2010. The combined effects of these disturbances reduced live coral cover from 40 percent to less than 5 percent, a 95 percent reduction.

This dramatic decline was a source of concern for some scientists who, at the time of the outbreak, were cautious about the possibility of recovery (Adjeroud et al. 2009). They advocated for "rapid intervention" (Lison de Loma et al. 2006: 13) in which eradication would be focused in the most infected sites and eventually consider "total eradication" around the island. Culling efforts from a previous outbreak in 1984 to ease the impact of COTS on reefs were noted. In addition, the local branch of an international NGO, Reef Check Polynésie, operating across French Polynesia and founded by a former director of CRIOBE, produced a report arguing how intervention was a necessary course of action (Lagouy 2007). Reef Check Polynésie, which received a 2007 grant from the French and French Polynesian governments, produced a flier outlining different eradication techniques and advocated for harvesting campaigns. We do note, however, that CRIOBE did not take part in or support these campaigns and that some scientists did suggest that recovery could occur without interventions. Despite varied opinion among scientists about the future recovery of corals from the COTS outbreaks, the French Polynesian government's Fisheries Service (presently DRM—Direction des Ressources

Marines—and formerly Service de la Pêche) encouraged COTS eradication during the outbreak. As a result, the local Moorea municipal government, schools, and NGOs organized sea star harvesting campaigns in 2009 during which some community members were paid on a per-kilogram basis to extract COTS from Moorea's reefs. Although some scientists were hesitant to predict recovery, subsequent studies since the 2009 outbreak have documented a rapid regrowth of coral cover (Holbrook et al. 2018; Adjeroud et al. 2018).

Local Knowledge about COTS

As part of an interdisciplinary research project funded by the National Science Foundation, we documented the local communities' perception of and response to the COTS outbreak. With several graduate students working alongside Moorean interpreters, we interviewed over 351 households as well as 15 key informants in 2014 and 2015 in three of Moorea's five administrative districts. Then in 2018–19 we conducted another series of interviews with more key informants on local perceptions of the past and present state of the lagoon and its marine resources (N=59) (refer to Rassweiler et al. 2020 for full details of the methods).

These interviews revealed that Mooreans are well aware of COTS, which they call *taramea,* and their coral destroying habits. Fishers talked about how in the past *taramea* were harvested, dried, ground up, and spread around garden plants as a pesticide, although this practice has disappeared. Generally, fishers were neutral about *taramea* and did not see them has a threat to the long-term health of the coral or reef fish. When asked if they changed their fishing practices because of the 2009 COTS outbreak, less than 25 percent of households stated that they changed their practices, and those that did change their practices were paid to extract COTS during the government-led eradication campaign. Most households did not feel it was necessary to kill *taramea* during the outbreak, and those fishers who did participate in eradication campaigns described the practice of removing *taramea* during outbreaks as a "new thing," and "the old-timers never mentioned anything like this." Many fishers discussed how they were instructed as children to leave the starfish alone and that disturbing the creatures might increase the intensity of an outbreak.

Local fishers who were involved in the culling activities described how an elderly woman was upset with the eradication efforts and publicly requested that they leave the *taramea* alone. Although there is no published information about the number of COTS removed from Moorea's reefs, community members recalled that COTS were piled on the beach, dried in the sun, and then burned. However, the idea that harvesting COTS is a necessary and effective course of action during outbreaks continues to pervade some local NGOs and fishing associations, who continue to seek both advice from the scientific community and grants to lead harvesting campaigns for future expected outbreaks.

To most fishers, though, *taramea* outbreaks are nonthreatening events, because it is well known that they occur every few decades on Moorea. They discussed how two sea snail species, *pu* (giant triton—*Charonia tritonis*) and *pu tara* or *pu pae ho'e* (giant spider conch—*Lambis truncata*), are predators of the COTS. They further cited traditional chants that describe starfish outbreaks and portray how swarms creep up from the outer reef ledges into shallower waters. Some fishers, as well as the head of the environmental department of the local municipality, talked about possible positive outcomes of *taramea* and explained their role in the ecosystem wherein they clean the reef of disease "like an antibiotic" or have a regenerative, reinvigorating effect. Indeed, marine scientists have also established that non-outbreak levels of COTS predation increase coral diversity because the starfish feed on the fastest-growing corals, such as plate and staghorn corals, enabling slower-growing species to become established (Great Barrier Reef Marine Park Authority 2017).

Everywhere across the Pacific, islanders have in-depth knowledge of COTS. Fishers interviewed in New Ireland, Solomon Islands, and Samoa all discussed a pattern of past outbreaks and were, in general, unconcerned about current COTS population booms. Another indication of Pacific Islanders' long-term relationship with COTS is evidenced by specific names assigned to the organism in many Pacific Island languages, such as *alamea* (Samoa), *rrusech* (Palau), and *bula* (Fiji) (Birkeland 1981).

COTS Effect on Fish Abundance

Our marine science colleagues documented not only shifts in coral cover after the COTS outbreak but also changes in the coral reef fish assemblage. Rapid and widespread coral loss, whether caused by COTS, coral bleaching, or strong storms, is generally assumed to shift coral reef fish species composition and overall fish biomass (Holbrook et al. 2018). An LTER-led analysis of 271,000 fish observed during an underwater census from 2009 to 2010 quantified the absolute and relative abundance of fish during the 2009–10 period. The biomass of the important food fish *Naso* (*ume* in Tahitian) fell from twenty-one to four kilograms per hectare, while the biomass of parrotfish from the family *Scarus*, another popular food fish known as *pa'ati* or *pahoro,* increased at roughly the same magnitude as *Naso* declined.

Our research team compared the marine science surveys of fishable biomass and species composition with the reef fish catch. To do this, we carried out an extensive roadside fish seller survey in 2014–15. Most reef fish on Moorea are sold along the island's coastal perimeter road. Makeshift metal racks are constructed, and fish are hung in *tui,* strings of freshly caught fish held together by passing a piece of twisted tree-bark string through the fish's stomach and mouth (figure 3.2).

To estimate the reef fish catch, our Tahitian collaborator drove the perimeter road every Sunday morning and interviewed every fish seller she encountered. In addition to a brief survey, she photographed the sellers' *tui,* making sure to place a scale bar near the hanging strings of fish, a technique that allowed us to record eighteen thousand fish. We then analyzed the photographs to identify the fish to the lowest taxonomic level possible (mostly to species) and measured the length of each fish in the photograph by comparison with the scale bar using established photo measurement techniques. Our surveys conducted in 2014 and 2015 were augmented by similar catch surveys conducted by French researchers from CRIOBE in 2007, 2008, and 2012. These combined datasets enabled us to analyze the

Figure 3.2. Strings of fish (*tui*) sold by roadside. © Terava Atger.

composition of the catch before, during, and after the environmental disturbances from the COTS outbreak and Cyclone Oli.

Our analysis showed that, overall, the roadside catch data mirrored the trends revealed through the underwater surveys (for a full description of the analysis see Rassweiler et al. 2020). *Naso* spp., for example, decreased from over a third of the catch prior to the disturbances to less than 10 percent after. In contrast, parrotfish increased from 56 to 66 percent. In fact, there was a rather tight correlation between biomass of the taxa measured during the underwater surveys and those documented in our roadside catch surveys.

However, when we asked fishers about their catch, the fish they ate, and the fish they purchased or sold, few reported any change. Just 1.5 percent of households stated that they changed the kinds of fish they ate, bought, or sold after the COTS outbreak, and just 13 percent indicated that they changed where, what, or how they fished. Of those that did change their behavior, some avoided the *taramea*-infested areas, others switched to new fishing grounds, while others participated in the municipal government's efforts to remove COTS from the reef.

Starfish Glut or Bloom?

To recap, marine scientists documented what they characterized as the most devastating and intense COTS outbreak ever recorded on Moorea's reefs. There were mixed views, however, about its significance to coral reef health, with some arguing that the outbreak posed a threat while others refrained from describing its impact as undermining reef health (Kayal et al. 2012). Yet, the French Polynesian press took a decidedly negative stance toward COTS. One local newspaper declared "Les coraux de Tahiti menacés par une étoile de mer" [Tahitian corals threatened by a starfish]. Indeed, major international media outlets have a long history of demonizing COTS. For example, during the first reports of COTS outbreaks on the Great Barrier Reef of Australia in the 1960s, *Time* magazine bluntly described the starfish as a "Plague at Sea" (Time 1969), while outbreaks in Micronesia led to this 1969 *New York Times* headline: "Scientists Say Coral-Eating Starfish Peril Pacific Islands" (Trumbull 1969). More recently, another *New York Times* headline declared, "Voracious Starfish Is Destroying the Great Barrier Reef" (Kwai 2018). These sentiments have led to a long-standing and widespread strategy to mitigate the effects of COTS outbreaks by destroying the organisms through eradication campaigns, a practice that was carried out on Moorea.

However, islanders' perceptions of the COTS outbreak and the decline of Moorea's coral contrasted with some in the marine science community and differed sharply with the government-led program to eradicate the sea stars. Moorea fishers, although well aware of the loss in coral and the coral-eating habits of COTS, did not find the change particularly important and did little to alter their behavior after the outbreak. Climate change–related shifts, such as sea temperature rise, were never mentioned as a possible cause of the outbreak. Moreover, changes in relative fish abundance documented both in the roadside catch surveys and during underwater diver surveys of the reef did not register as noteworthy among Moorea fishers. Few fishers noted a change in the fish they caught or ate. Similarly, a recent coral-bleaching event in Moorea (mid-2019), which has raised a great deal of concern among the local scientific community and local NGOs, does not seem to worry fishers. Coral bleaching, where thermal stress causes coral polyps to expel their symbiotic algae, turn white, and die, is one of the key climate change–induced disturbances affecting coral reefs worldwide. In contrast to the alarm coral bleaching has caused among coral reef scientists, a sixty-five-year-old Moorean fisher interviewed in 2019 mentioned that the ongoing bleaching event would help renew and strengthen the island's coral reefs.

Thus, we are presented with contrasting understandings of what constitutes a perturbation or change on Moorea's coral reefs. These differing standpoints of change in marine ecosystems may be due to the longer time horizon of the Polynesian fishers' knowledge base. While extensive time-series data have been collected in Moorea over the past forty years by both research stations, the temporal depth of scientific research is significantly shorter than the experience of fishers who draw on their own lifespan and the intergenerational transmitted knowledge of their parents and grandparents. The knowledge of fishers, as with all indigenous knowledge, develops and is sustained through a mixture of intergenerational transmitted knowledge, experience, regular interaction with the underwater environment, and the reception of other kinds of scientific and nonscientific knowledge. Fishers were able to evaluate the 2009 COTS outbreak in relation to others in their memory or the memory of previous fishers. This awareness of previous COTS outbreaks and their associated ecological outcomes could be the basis upon which fishers interpreted the 2009 population boom as a normal cyclical pattern rather than a unique and threatening change. The same may be said about the ongoing coral bleaching event: fishers noticed its particular intensity and geographical extent but do not find it alarming as they have witnessed past disturbances caused by bleaching and past recoveries of the reefs. However, intergenerational knowledge may not provide accurate guides to action involving current and future ecological changes associated with rising ocean temperatures and ocean acidification. These climate change–induced changes are unique to our Anthropocene epoch and may produce effects that lie outside the experience horizon of Polynesians' fifteen hundred years of accumulated place-based knowledge. Indeed, Western scientists face the same uncertainties and are not necessarily more prepared than fishers are to predict novel changes that have no historic analog.

We tentatively forward the notion that fishers and local community members may be more preoccupied by gradual and slow-paced changes than rapid, intense ecological changes. Indeed, rapid disturbances (such as COTS outbreaks, cyclones, or bleaching events), while erratic and unpredictable, are nonetheless expected and perceived to appear cyclically. When asking fishers what their main preoccupations are concerning their marine environment, the slow process of sedimentation is often mentioned. One reason may be the linearity of such changes that are perceived as irreversible.

Even more surprising is that the shifts we documented in both the roadside catch and the underwater reef surveys were unremarkable to fishers. The *Naso* species in particular dropped in both the reef counts and our counts of fish sold on the roadside. *Naso* spp. are a highly prized food fish on Moorea and widely sought after, yet fishers noted little change in their catch or in their diets. It could be that the shifts observed in the reef and roadside surveys represent large changes for each taxon but add up to relatively modest change when the suite of common food fishes are considered as a group. In our household surveys, fishers consistently reported *pa'ati* (*Scarus/Chlorurus* spp., terminal phase), *pahoro (Scarus/Chlorurus* spp., initial phase), *i'ihi* (*Myripristis* spp.), *tarao* (*Epinephelus* spp.), *pa'aihere* (*Caranx* spp.), and *ume* (*Naso* spp.) as the most common fish that they ate and caught. Because the decline in *Ume* was mostly offset by an increase in *pahoro* and *pa'ati*, the suite of fish remains consistent. As with the COTS outbreaks, shifts in the relative abundance of food fish may register as normal fluctuations within the local knowledge of fishers. If one of these fish were to completely disappear from the reef, it is possible that this would constitute a radical break from a "normal" catch.

In follow-up surveys, we asked fishers about the roadside and reef surveys, and they responded that their concerns focus more on fish behavior than on the abundance of fish on the reef. Fishers frequently commented how *Ume*, in areas where they are heavily fished, learn to be wary of fishers and evade them quickly by swimming to deeper water beyond the range of most free-diving spearfishers. Yet this does not necessarily result in fewer fish caught for skilled spearfishers. As one fisher noted, "A good spearfisher will find and catch the fish he desires." This suggests that Moorean fishers may grasp fish abundance as constituted in the *relationship* between fishers and their preferred targets rather than as an *attribute* of the fishery that is independent of the observer.

Discrepancies between fishers' and marine scientists' understanding of change may also be due to crucial knowledge gaps in tropical coral reef science. The forty-year running debate about the causes of COTS outbreaks is just one example. The same type and intensity of disturbance to a coral reef can vary greatly in the intensity, spatial scale, magnitude, and longevity of its impacts (Wilson et al. 2010). Marine scientists still struggle to predict coral reef decline associated with COTS, as well as thermal stress related coral bleaching. The fact that at least some of Moorea's coral reefs have rapidly returned to their pre-COTS levels of coral cover highlights the level of scientific uncertainty involved when predicting the effects of disturbance (Holbrook et al. 2018; Adjeroud et al. 2018).

Differences in Local Knowledges

The differences in how scientists and fishers understand the COTS outbreak and its noteworthiness raise a number of challenging questions for studies investigating local knowledge of climate change–induced ecological change. Much of the literature highlighting how local or indigenous people detect climate-related change relies, either explicitly or implicitly, on climate science to validate local knowledge claims about changing ecosystems. Crate (2008), for example, working among the Viliui Sakha of northeastern Siberia, discusses in detail how local elders lament the disappearance, due to warming temperatures, of winter, which they describe as a "white bull with blue spots, huge horns, and frosty breath" (Crate 2008: 570). The warming observed by the Sakha is assumed to be an outcome of "unprecedented global climate change" (Crate 2008: 570), yet many climate scientists fiercely deny the possibility that global climate change is locally visible (Rudiak-Gould 2013b). As discussed in detail by Rudiak-Gould (2013b), research such as Crate's accepts climate change science as a means to legitimize local ecological knowledge and emphasize that it is a viable and empirically sound body of knowledge. Moreover, by validating LEK, the voice of marginalized communities tends to gain more traction in decision-making.

This kind of commitment to bolster the legitimacy of LEK has been central to many indigenous knowledge studies and indigenous advocates for decades, but in many cases LEK is positioned not in concert with scientific, expert knowledge but *in opposition* to it (Brokensha, Warren, and Werner 1980; Agrawal 1995; Hobart 1993). A case in point is illustrated by "counter mapping" (Schofield 2016). This popular technique utilizes LEK to develop cartographic and other kinds of spatial data to represent the knowledge and interests of local and indigenous people that are overlooked in official cartographic representations. These techniques have emerged to reverse the long track record of international development schemes and conservation initiatives where expert knowledge tramples local adaptations and practices (Hobart 1993). In this body of research, science and expert knowledge are understood not as neutral forces for good but rather as hegemonic forces, tied with postcolonial power and, in many cases, oppression of the marginalized (Said 1979; Foucault 1990; Haraway 1988).

Indeed, on Moorea there are tensions between fishers and scientists. Like most Pacific Islands, life in French Polynesia has been subjected to countless impositions since the arrival of Europeans, including unequal trade, imposed religion, privatization of land tenure, monolingual French education, and broad cultural oppression (Thompson and Adloff 1971). Although overt colonial oppression has declined somewhat in French Polynesia, Moorean fishers have grown increasingly skeptical of the motives of the scientific community because fishers associate them with the implementation of a top-down and expert-led lagoon management plan that disproportionately restricts fishing activities compared to other kinds of uses, such as tourism and scientific activities (Walker 2001; Hunter et al. 2018). Known as the Plan de Gestion de l'Espace Maritime (PGEM), in 2004 it established eight no-take zones and other restrictions on harvesting marine life around the island. It is widely known on the island that the PGEM restrictions are often ignored by fishers. In fact, marine science evaluations conducted a decade after the establishment of the no-take zones have shown them to be ineffective in substantially increasing the biomass of fish inside of the reserve areas (Thiault et al. 2019).

In response to the PGEM, grassroots movements have emerged on Moorea, as well as on other islands in French Polynesia, that seek to increase local control over marine space. Many of these groups frame their community-led management as a form of a neotraditional management known as *rahui* (Bambridge 2016; Bambridge et al. 2019). One of the key elements of the emerging *rahui* groups who advocate for more community control is to have influence over management decisions in situations like the COTS outbreak. The fact that the French Polynesian government's Fisheries Service and Moorea's municipal government organized the COTS eradication campaign during the 2009 population boom suggests that they followed the lead of at least part of the scientific community about the COTS outbreak and maybe faced pressure from tourist operators, who feared that coral loss would harm tourism, rather than consulting fishers[1] and involving them in the decisions over marine management.

Moorea's *rahui* groups recognize that French Polynesian decision-makers base their management decisions, at least in part, on scientific knowledge, and as a result, some groups are now conducting their own quasi-scientific assessments of marine resources. One group, for example, carried out an underwater fish survey to assess areas for overfishing and reported their results in a detailed summary document. This same group, in overt conflict with both the local municipality and CRIOBE research teams, has adopted scientific claims about COTS and is advocating for their removal during outbreaks. However, rather than physically removing COTS, the members proudly claim possession of traditional knowledge that can influence ecological processes to mitigate COTS outbreaks. However, this and other such groups refuse to share this knowledge with scientists. Their strategy to withhold knowledge from the scientific community exemplifies how LEK and Western scientific knowledge may be pitted against one another in contexts of political struggles over management of environmental resources.

That local people such as Moorea's *rahui* organizations are adopting scientific methods and positioning LEK in ways to achieve political aims, however, raises red flags for many scientists who hold the view that science conducted by professionals is the most effective method to produce accurate knowledge (Carr and Heyman 2012). For example, Davis and Ruddle suggest that the most cited LEK literature lacks scientific validation and that this is problematic because of the "need for researchers to be held accountable to their knowledge claims" (Davis and Ruddle 2010: 893) through "systematic evaluation." For these researchers, Western science provides privileged access to phenomena, and, "like it or not, until replaced at some future time, Western science is the dominant paradigm that sets the prevailing standard" (Davis and Ruddle 2010: 881). The commitment is in some ways a reversion back to older pejorative understandings of non-Western knowledge that viewed local understanding as simply "tradition" or "belief" and that LEK only gains legitimacy when it is absorbed by peer-reviewed science.

Yet, much research from the field of science and technology studies has shown how scientific knowledge production, albeit extremely powerful and important, is never fully purified of its specific epistemic assumptions about social relationships, value, and behavior (Latour 1993; Turnbull 2000). Rather than a positionless view from nowhere, what Donna Haraway (1988) calls the "God trick," science, like all knowledge systems, is a situated practice that brings with it its own terms of validation (Lauer and Aswani 2009). More often than not aspects of LEK are disqualified in favor of science, and even when LEK is seen as a possible source of reliable information, bits of it are brought into the work of science only after being properly framed (Klenk et al. 2017).

Knowledge Spaces

That Moorean fishers appear to have been accurate in that the massive loss of coral documented by marine scientists did not constitute a persistent, irreversible change to the ecosystem suggests that peer-reviewed science is not always sufficient. We make note of this not to suggest that Moorea fisher knowledge is necessarily superior to that of the marine scientists but rather to highlight that all knowledge systems may produce useful insights and spur innovation if given the space to do so. Here, following Turnbull (1997) and Turnhout et al. (2012), we suggest that although there are critical differences between knowledges, we should seek ways to enable their coexistence. To open spaces for the coexistence of LEK and science, validity, accuracy, and verification are better conceived as *products of knowledge practices* rather than external criteria (Wynne 1996), and these critical criteria must be subjected to active debate, deliberation, and topics of inquiry rather than harnessed as resources and hidden behind claims that either science or LEK has privileged access to a "real" reality.

A symmetrical approach to knowledges should not be interpreted as a call that all knowledges are justified or that antiscientific thinking should be encouraged. Indeed, extreme relativism is what has led to the current political tactics in the United States, where large portions of the population can be swayed by powerful corporations who accuse scientists of being nothing more than lobbyists for their own interests. Likewise, the hostility toward science expressed by many indigenous rights activists is equally flawed since it relies on the spurious claim that indigenous or local knowledge invariably produces harmonious human/nonhuman relations and a socially just world. To circumvent the expert/nonexpert and LEK/science divides, it is vital to explicitly emphasize that the production of all knowledges is bound up with certain epistemic commitments, value-laden assumptions, and political positions. This approach not only encourages us to begin to critically examine, recognize, and research how knowledge claims are constructed but also serves as a means to rebuild the legitimacy of scientific knowledge while not denigrating nonexpert knowledge. As Sarewitz argues, "The social value of science itself is likely to increase if . . . value disputes have been brought out into the open, their implications for society explored, and suitable goals identified" (Sarewitz 2004: 399).

Moorea, we argue, provides a unique opportunity to implement these kinds of knowledge spaces and experimental partnerships since there is thriving local knowledge *and* a local science community operating and intertwining side by side. Recognizing this and the challenges we face, our research group is developing workshops to enable fishers and marine scientists to interrogate each other's methods, knowledge, and ways of knowing. Rather than just meeting and discussing, fishers and marine scientists will be asked to jointly conduct field-based assessments of fish abundance, examine each other's methods and knowledge claims for determining abundance, and discuss the issues at stake and the aims of generating knowledge along with the epistemic criteria it should be judged by. In this way, our hope is to produce knowledge about Moorea's coral reef system in a more open-ended and less dichotomized manner while also potentially redistributing recognized authority to and spurring action among fishers who have been marginalized from most formal knowledge- and decision-making processes.

Acknowledgments

Many thanks to Ms. Hinano Murphy for logistic support and Dr. Jean-Yves Meyer for assistance with permits. We gratefully acknowledge the support of the National Science Foundation (BSE 1714704, OCE 1325554, 325652) and the US NSF Moorea Coral Reef Long Term Ecological Research project. Permits for fieldwork were issued by the Haut-commissariat de la République en Polynésie Française (DRRT).

Matthew Lauer is a professor of environmental and ecological anthropology at San Diego State University. His research interests include environmental change, local knowledge, disasters, and conservation in Oceania. He is interested specifically in the production, spread, and consumption of local knowledge among Pacific Islanders experiencing rapid environmental change as well as how place-based scientific knowledge of coral reef environments is produced and intermingles with the local ecological knowledge of fishers and other stakeholders.

Terava Atger is an educator and outreach coordinator based in French Polynesia, with interests in the promotion of Polynesian languages and island ecosystems.

Sally J. Holbrook is a professor at the Department of Ecology, Evolution, and Marine Biology, University of California at Santa Barbara. Her interests focus on ecological processes, such as species interactions (interspecific competition, predation, herbivory) in coral reef ecosystems, with a goal of understanding how these processes act, along with environmental disturbances and anthropogenic impacts such as fishing, to shape the abundance, dynamics, and diversity of reef species. She is especially interested in the interactions and feedbacks between human activities and the reef ecosystem.

Andrew Rassweiler is an assistant professor at the Department of Biological Science, Florida State University. He studies coastal systems with an emphasis on two complementary topics: (1) the spatial management of marine resources, and (2) resilience and abrupt ecological state change. His research is motivated by the dual goals of advancing ecological theory and influencing management decisions, addressing these topics using modeling, analysis of existing datasets, observational fieldwork, and manipulative experiments.

Russell J. Schmitt is a professor at the Department of Ecology, Evolution, and Marine Biology, University of California at Santa Barbara. His research focuses on ecological responses to local and global perturbations and using that knowledge to better inform management and conservation of coastal marine ecosystems. Most of his work is concentrated on coral reefs, an extremely diverse and socially important ecosystem that is highly threatened by human activities and natural events. His work includes the dynamic coupling between ecological and social components of coral reef ecosystems.

Jean Wencélius is a postdoctoral researcher at the Department of Anthropology, San Diego State University. As a cultural anthropologist, he focuses his research on the interactions between humans, their environment, and the biological entities they use for their subsistence. Through interdisciplinary collaborations, he has sought to frame ecosystems as the integrated result of continuous feedbacks between human practices, sociocultural norms, and ecological dynamics. His most recent research focuses on the ethnography of reef fisheries in French Polynesia, investigating specifically fishing practices and strategies, local ichthyological knowledge, and the sociopolitical dynamics of marine management.

Note

1. We note that some local environmental and *rahui* groups on Moorea are now advocating for COTS removal during outbreaks.

References

Adam, Thomas C., Russell J. Schmitt, Sally J. Holbrook, Andrew J. Brooks, Peter J. Edmunds, Robert C. Carpenter, and Giacomo Bernardi. 2011. "Herbivory, Connectivity, and Ecosystem Resilience: Response of a Coral Reef to a Large-Scale Perturbation." *PLoS ONE* 6(8): e23717.

Adjeroud, Mehdi, Mohsen Kayal, Claudie Iborra-Cantonnet, Julie Vercelloni, Pauline Bosserelle, Vetea Liao, Yannick Chancerelle, Joachim Claudet, and Lucie Penin. 2018. "Recovery of Coral Assemblages Despite Acute and Recurrent Disturbances on a South Central Pacific Reef." *Scientific Reports* 8(1): 9680.

Adjeroud, Mehdi, François Michonneau, Peter J. Edmunds, Yannick Chancerelle, Thierry Lison de Loma, Lucie Penin, Loïc Thibaut, Jeremie Vidal-Dupiol, Bernard Salvat, and Rene Galzin . 2009. "Recurrent Disturbances, Recovery Trajectories, and Resilience of Coral Assemblages on a South Central Pacific Reef." *Coral Reefs* 28(3): 775–80.

Agrawal, Arun. 1995. "Dismantling the Divide between Indigenous and Scientific Knowledge." *Development and Change* 26: 413–39.

Aswani, Shankar, and Matthew Lauer. 2014. "Indigenous People's Detection of Rapid Ecological Change." *Conservation Biology* 28(3): 820–28.

Bambridge, Tamatoa, ed. 2016. *The Rahui: Legal Pluralism in Polynesian Traditional Management of Resources and Territories*. Acton: ANU Press.

Bambridge, Tamatoa, Francois Gaulme, Christian Montet, and Thierry Paulais. 2019. *Communs Et Océan: Le Rahui En Polynésie*. Papeete: Au Vent des Iles.

Berkes, Fikret. 2012. *Sacred Ecology: Traditional Ecological Knowledge and Resource Management*. 3rd ed. New York: Routledge.

Berkes, Fikret, Johan Colding, and Carl Folke. 2000. "Rediscovery of Traditional Ecological Knowledge as Adaptive Management." *Ecological Applications* 10(5): 1251–62.

Birkeland, Charles. 1981. "*Acanthaster* in the Cultures of High Islands." *Atoll Research Bulletin* 255: 55–58.

Birkeland, Charles, and John Lucas. 1990. *Acanthaster planci: Major Management Problem of Coral Reefs*. Boca Raton, FL: CRC Press.

Bohensky, Erin L., and Yiheyis Maru. 2011. "Indigenous Knowledge, Science, and Resilience: What Have We Learned from a Decade of International Literature on 'Integration'?" *Ecology and Society* 16(4): 6.

Brodie, Jon, Katharina Fabricius, Glenn De'ath, and Ken Okaji. 2005. "Are Increased Nutrient Inputs Responsible for More Outbreaks of Crown-of-Thorns Starfish? An Appraisal of the Evidence." *Marine Pollution Bulletin* 51(1): 266–78.

Brokensha, David, Dennis M. Warren, and Oswald Werner. 1980. *Indigenous Knowledge Systems and Development*. Washington, DC: University Press of America.

Brosius, J. Peter. 2006. "What Counts as Local Knowledge in Global Environmental Assessments and Conventions?" In *Bridging Scales and Knowledge Systems: Concepts and Applications in Ecosystem Assessment*, edited by W. V. Reid, F. Berkes, T. Wilbanks, and D. Capistran, 129–44. Washington, DC: Island Press.

Carey, Mark. 2010. *In the Shadow of Melting Glaciers: Climate Change and Andean Society*. Oxford: Oxford University Press.

Carr, Liam M., and William D. Heyman. 2012. "'It's about Seeing What's Actually out There': Quantifying Fishers' Ecological Knowledge and Biases in a Small-Scale Commercial Fishery as a Path toward Co-management." *Ocean & Coastal Management* 69: 118–32.

Conklin, Harold C. 1954. "The Relation of Hanunóo Culture to the Plant World." PhD diss., Yale University, New Haven, CT.
Connell, John. 2003. "Losing Ground? Tuvalu, the Greenhouse Effect and the Garbage Can." *Asia Pacific Viewpoint* 44(2): 89–107.
Crate, Susan A. 2008. "Gone the Bull of Winter? Grappling with the Cultural Implications of and Anthropology's Role(s) in Global Climate Change." *Current Anthropology* 49(4): 569–95.
———. 2011. "Climate and Culture: Anthropology in the Era of Contemporary Climate Change." *Annual Review of Anthropology* 40(1): 175–94.
Davis, Anthony, and Kenneth Ruddle. 2010. "Constructing Confidence: Rational Skepticism and Systematic Enquiry in Local Ecological Knowledge Research." *Ecological Applications* 20(3): 880–94.
Dumas, Pascal, Sompert Gereva, Grégoire Moutardier, Jayven Ham, and Rocky Kaku. 2015. "Collective Action and Lime Juice Fight Crown-of-Thorns Starfish Outbreaks in Vanuatu." *SPC Fisheries newsletter* 146: 47–52.
Foucault, Michel. 1990. *The History of Sexuality*. New York: Vintage Books.
Goldman, Mara. 2007. "Tracking Wildebeest, Locating Knowledge: Maasai and Conservation Biology Understandings of Wildebeest Behavior in Northern Tanzania." *Environment and Planning D: Society and Space* 25(2): 307–31.
Goldman, Mara, Paul Nadasdy, and Matthew D. Turner. 2011. *Knowing Nature: Conversations at the Intersection of Political Ecology and Science Studies*. Chicago: University of Chicago Press.
Great Barrier Reef Marine Park Authority. 2017. *Crown-of-Thorns Starfish Control Guidelines*. 2nd ed. Townsville: GBRMP.
Haraway, Donna. 1988. "Situated Knowledges: The Science Question in Feminism and the Privilege of Partial Perspective." *Feminist Studies* 14(3): 575–99.
———. 1991. *Simians, Cyborgs, and Women: The Reinvention of Nature*. New York: Routledge.
Hau'ofa, Epeli. 2000. "The Ocean in Us." *Contemporary Pacific* 10(2): 391–410.
Hobart, Mark. 1993. *An Anthropological Critique of Development: The Growth of Ignorance*. London: Routledge.
Holbrook, Sally J., Thomas C. Adam, Peter J. Edmunds, Russell J. Schmitt, Robert C. Carpenter, Andrew J. Brooks, Hunter S. Lenihan, and Cheryl J. Briggs. 2018. "Recruitment Drives Spatial Variation in Recovery Rates of Resilient Coral Reefs." *Scientific Reports* 8(1): 7338.
Hughes, Terry P., Andrew H. Baird, David R. Bellwood, Margaret Card, Sean R. Connolly, Carl Folke, Richard Grosberg, Ove Hoegh-Guldberg, Jeremy B. C. Jackson, Janice Kleypas, Janice M. Lough, Paul Marshall, Magnus Nystrom, Stephen R. Palumbi, John M. Pandolfi, Brian Rosen, and Joan Roughgarden. 2003. "Climate Change, Human Impacts, and the Resilience of Coral Reefs." *Science* 301(5635): 929–33.
Hunter, Chelsea E., Matthew Lauer, Arielle Levine, Sally Holbrook, and Andrew Rassweiler. 2018. "Maneuvering towards Adaptive Co-management in a Coral Reef Fishery." *Marine Policy* 98: 77–84.
Jasanoff, Sheila. 2004. *States of Knowledge: The Co-production of Science and the Social Order*. London: Routledge.
Johannes, Robert E. 1981. *Words of the Lagoon: Fishing and Marine Lore in the Palau District of Micronesia*. Berkeley: University of California Press.
Jupiter, Stacy D., Philippa J. Cohen, Rebecca Weeks, Alifereti Tawake, and Hugh Govan. 2014. "Locally-Managed Marine Areas: Multiple Objectives and Diverse Strategies." *Pacific Conservation Biology* 20(2): 165–79.

Kamya, Pamela Z., Maria Byrne, Benjamin Mos, Lauren Hall, and Symon A. Dworjanyn. 2017. "Indirect Effects of Ocean Acidification Drive Feeding and Growth of Juvenile Crown-of-Thorns Starfish, *Acanthaster Planci.*" *Proceedings of the Royal Society B: Biological Sciences* 284(1856): 20170778.

Kayal, Mohsen, Julie Vercelloni, Thierry Lison de Loma, Pauline Bosserelle, Yannick Chancerelle, Sylvie Geoffroy, Céline Stievenart, François Michonneau, Lucie Penin, Serge Planes, and Mehdi Adjeroud. 2012. "Predator Crown-of-Thorns Starfish (Acanthaster Planci) Outbreak, Mass Mortality of Corals, and Cascading Effects on Reef Fish and Benthic Communities." *PLoS ONE* 7(10): e47363.

Klenk, Nicole, Anna Fiume, Katie Meehan, and Cerian Gibbes. 2017. "Local Knowledge in Climate Adaptation Research: Moving Knowledge Frameworks from Extraction to Co-production." *Wiley Interdisciplinary Reviews: Climate Change* 8(5): e475.

Krupnik, Igor, and Dyanna Jolly, eds. 2002. *The Earth Is Faster Now: Indigenous Observations of Arctic Environment Change*. Fairbanks, AK: Arctic Research Consortium of the United States.

Kwai, Isabelle. 2018. "A Voracious Starfish Is Destroying the Great Barrier Reef." *New York Times*, 6 January, A9.

Lagouy, Elodie. 2007. *Etat Des Lieux Des Étoiles De Mer Épineuses Acanthaster Planci, Taramea, En Polynésie Française*. Papeete, French Polynesia: Service de la Pêche.

Latour, Bruno. 1993. *We Have Never Been Modern*. Cambridge, MA: Harvard University Press.

———. 1999. *Pandora's Hope: Essays on the Reality of Science Studies*. Cambridge, MA: Harvard University Press.

Lauer, Matthew. 2017. "Changing Understandings of Local Knowledge in Island Environments." *Environmental Conservation* 44(4): 336–47.

Lauer, Matthew, and Shankar Aswani. 2009. "Indigenous Ecological Knowledge as Situated Practices: Understanding Fishers' Knowledge in the Western Solomon Islands." *American Anthropologist* 111(3): 317–29.

———. 2010. "Indigenous Knowledge and Long-Term Ecological Change: Detection, Interpretation, and Responses to Changing Ecological Conditions in Pacific Island Communities." *Environmental Management* 45(5): 985–97.

Lauer, Matthew, and Jaime Matera. 2016. "Who Detects Ecological Change after Catastrophic Events? Indigenous Knowledge, Social Networks, and Situated Practices." *Human Ecology* 44(1): 33–46.

Lazrus, Heather. 2012. "Sea Change: Island Communities and Climate Change." *Annual Review of Anthropology* 41: 285–301.

Leenhardt, Pierre, Matthew Lauer, Rakamaly Madi Moussa, Sally J. Holbrook, Andrew Rassweiler, Russell J. Schmitt, and Joachim Claudet. 2016. "Complexities and Uncertainties in Transitioning Small-Scale Coral Reef Fisheries." *Frontiers in Marine Science* 3(70): 1–9.

Lison de Loma, Thierry, Yannick Chancerelle, and Franck Lerouvreur. 2006. "Evaluation Des Densités D'acanthaster Planci Sur L'île De Moorea." *Rapport CRIOBE UMS 2978 CNRS-EPHE,-RA149*. Retrieved 24 July 2019 from http://ifrecor-doc.fr/items/show/1515.

McMillen, Heather L., Tamara Ticktin, Alan Friedlander, Stacy D. Jupiter, Randolph Thaman, John Campbell, Joeli Veitayaki, Thomas Giambelluca, Salesa Nihmei, and Etika Rupeni. 2014. "Small Islands, Valuable Insights: Systems of Customary Resource Use and Resilience to Climate Change in the Pacific." *Ecology and Society* 19(4): 44.

Mimura, Nobuo, Leonard Nurse, Roger McLean, John Agard, Lino Briguglio, Penehuro Lefale, Rolph Payet, and Graham Sem. 2007. "Small Islands." In *Climate Change*

2007; Impacts, Adaptation and Vulnerability; Contribution of Working Group Ii to the Fourth Assessment Report of the Intergovernmental Panel on Climate Change, edited by M. L. Parry, O. F. Canziani, J. P. Palutikof, P. J. van der Linden and C. E. Hanson, 687–716. Cambridge: Cambridge University Press.

Moutardier, Grégoire, Sompert Gereva, Suzanne C. Mills, Mehdi Adjeroud, Ricardo Beldade, Jayven Ham, Rocky Kaku, and Pascal Dumas. 2015. "Lime Juice and Vinegar Injections as a Cheap and Natural Alternative to Control Cots Outbreaks." *PLoS ONE* 10(9): e0137605.

Oreskes, Naomi, and Erik M. Conway. 2011. *Merchants of Doubt: How a Handful of Scientists Obscured the Truth on Issues from Tobacco Smoke to Global Warming*. New York: Bloomsbury Press.

Pratchett, Morgan, Ciemon F. Caballes, Jairo Rivera-Posada, and Hugh P. A. Sweatman. 2014. "Limits to Understanding and Managing Outbreaks of Crown-of-Thorns Starfish (*Acanthaster* Spp.)." *Oceanography and Marine Biology: An Annual Review* 52: 133–200.

Pratchett, Morgan S., Ciemon F. Caballes, Jennifer C. Wilmes, Samuel Matthews, Camille Mellin, Hugh P. A. Sweatman, Lauren E. Nadler, Jon Brodie, Cassandra A. Thompson, Jessica Hoey, Arthur R. Bos, Maria Byrne, Vanessa Messmer, Sofia A. V. Fortunato, Carla C. M. Chen, Alexander C. E. Buck, Russell C. Babcock, and Sven Uthicke. 2017. "Thirty Years of Research on Crown-of-Thorns Starfish (1986–2016): Scientific Advances and Emerging Opportunities." *Diversity* 9(4): 41.

Rassweiler, Andrew, Matthew Lauer, Sarah E. Lester, Sally J. Holbrook, Russell J. Schmitt, Rakamaly Madi Moussa, Katrina S. Munsterman, Hunter S. Lenihan, Andrew J. Brooks, Jean Wencélius, and Joachim Claudet. 2020. "Perceptions and Responses of Pacific Island Fishers to Changing Coral Reefs." *Ambio* 49(1): 130–43.

Raymond, Robert. 1986. *Starfish Wars: Coral Death and the Crown-of-Thorns*. South Melbourne: Macmillan Co. of Australia.

Rudiak-Gould, Peter. 2013a. *Climate Change and Tradition in a Small Island State: The Rising Tide*. New York: Routledge.

———. 2013b. "'We Have Seen It with Our Own Eyes': Why We Disagree about Climate Change Visibility." *Weather, Climate, and Society* 5(2): 120–32.

Said, Edward. 1979. *Orientalism*. New York: Vintage Books.

Sarewitz, Daniel. 2004. "How Science Makes Environmental Controversies Worse." *Environmental Science & Policy* 7(5): 385–403.

Schaffelke, Britta, and Ken Anthony. 2015. "Crown of Thorns Starfish Research Strategy." *AIMS: Australian Institute of Marine Science*.

Schofield, John, ed. 2016. *Who Needs Experts? Counter-mapping Cultural Heritage*. New York: Routledge.

South, G. Robin, Denis Goulet, Seremaia Tuqiri, and Marguerite Church, eds. 1994. *Traditional Marine Tenure and Sustainable Management of Marine Resources in Asia and the Pacific*. Suva: International Ocean Institute-South Pacific.

Thiault, Lauric, Laëtitia Kernaléguen, Craig. W. Osenberg, Thierry Lison de Loma, Yannick Chancerelle, G. Siu, and Joachim Claudet. 2019. "Ecological Evaluation of a Marine Protected Area Network: A Progressive-Change Bacips Approach." *Ecosphere* 10(2): e02576.

Thompson, Virginia, and Richard Adloff. 1971. *The French Pacific Islands: French Polynesia and New Caledonia*. Berkeley: University of California Press.

Time. 1969. "Plague in the Sea." *Time* 94(11): 61.

Trapon, Mélanie L., Morgan S. Pratchett, and Lucie Penin. 2011. "Comparative Effects of Different Disturbances in Coral Reef Habitats in Moorea, French Polynesia." *Journal of Marine Biology*: 1–11.

Trumbull, Robert. 1969. "Scientists Say Coral-Eating Starfish Peril Pacific Islands." *New York Times*, 21 July, C35.

Turnbull, David. 1997. "Reframing Science and Other Local Knowledge Traditions." *Futures* 29(6): 551–62.

———. 2000. *Masons, Tricksters and Cartographers: Comparative Studies in the Sociology of Scientific and Indigenous Knowledge*. Amsterdam: Harwood Academic.

Turnhout, Esther, Bob Bloomfield, Mike Hulme, Johannes Vogel, and Brian Wynne. 2012. "Listen to the Voices of Experience." *Nature* 488(7412): 454–55.

Uthicke, S., M. Logan, M. Liddy, D. Francis, N. Hardy, and M. Lamare. 2015. "Climate Change as an Unexpected Co-factor Promoting Coral Eating Seastar (*Acanthaster planci*) Outbreaks." *Scientific Reports* 5(1): 8402.

Walker, Barbara Louise Endemaño. 2001. "Mapping Moorea's Lagoons: Conflicts over Marine Protected Areas in French Polynesia." *Proceedings of the Inaugural Pacific Regional Meeting of the International Association for the Study of Common Property, Brisbane, Australia (September)*: 2–24.

Watson-Verran, Helen, and David Turnbull. 1995. "Science and Other Indigenous Knowledge Systems." In *Handbook of Science and Technology Studies*, edited by S. Jasanoff, G. E. Markle, J. C. Peterson and T Pinch, 114–19. Thousand Oaks, CA: Sage.

Wilson, S. K., M. Adjeroud, D. R. Bellwood, M. L. Berumen, D. Booth, Y.-Marie Bozec, P. Chabanet, A. Cheal, J. Cinner, M. Depczynski, D. A. Feary, M. Gagliano, N. A. J. Graham, A. R. Halford, B. S. Halpern, A. R. Harborne, A. S. Hoey, S. J. Holbrook, G. P. Jones, M. Kulbiki, Y. Letourneur, T. L. De Loma, T. McClanahan, M. I. McCormick, M. G. Meekan, P. J. Mumby, P. L. Munday, M. C. Öhman, M. S. Pratchett, B. Riegl, M. Sano, R. J. Schmitt, and C. Syms. 2010. "Crucial Knowledge Gaps in Current Understanding of Climate Change Impacts on Coral Reef Fishes." *Journal of Experimental Biology* 213(6): 894–900.

Wynne, Brian. 1996. "May the Sheep Safely Graze? A Reflexive View of the Expert-Lay Knowledge Divide." In *Risk, Environment and Modernity: Towards a New Ecology*, edited by Scott Lash, Bronislaw Szerszynski, and Brian Wynne, 44–79. London: Sage.

Zann, Leon, Jon Brodie, and Veikila Vuki. 1990. "History and Dynamics of the Crown-of-Thorns Starfish *Acanthaster planci* (L.) in the Suva Area, Fiji." *Coral Reefs* 9(3): 135–44.

Chapter 4

Urban Transformations in the Hydric Landscapes of Belém, Brazil

Environmental Memories and Urban Floods

Pedro Paulo de Miranda Araújo Soares

Introduction

In the late nineteenth century, the famous English naturalist Henry Walter Bates used to walk through the *várzea* forests of the Una Basin and sail through the streams that connected the Guajará Bay to the vicinity of what was then downtown Belém, Brazil. Bates described the Una Basin as his "favorite spot" and a "paradise for naturalists" (Bates 1944: 83). The contemporary urban imagery tells a different story, however. The status of paradise changed from the mid-twentieth century onward as the Una Basin endured the impacts of what Belém's policymakers envisioned as modernization (Costa et al. 2018).[1]

In the name of progress, modernity, and ultimately development, both nationally and internationally funded economic projects attracted a massive influx of investments and migrants to Belém between the 1960s and the late 1980s (Loureiro 1992). Among these immigrants, the poorer occupied the lowlands next to the streams that surrounded the city, giving rise to a land occupation form known as *baixadas*.[2] As the population density intensively amassed, it was common for inhabitants to build their houses not only on the riverbanks but also on the riverbeds in wooden suspended constructions called *palafitas*.[3]

The already extant hazards surrounding this population increased when factories installed facilities to make it easier to discharge industrial

Notes for this chapter begin on page 104.

waste into the Guajará Bay next to Una's basin watercourses. The banks of the Una River and its tributaries became dotted with various factories producing paper, vegetable oil, screws, packaging, and soap (Penteado 1968). Industrialization unfolded with the growing population density, resulting in increased environmental degradation of the Una Basin. It is not uncommon to find old residents in the Una Basin who recall the catastrophic image of fish floating on the surface of the Una River and other streams when the paper factory was opened in the mid-1960s (Costa et al. 2018). However, that was only the beginning of problematic water issues.

In an attempt to build drainage facilities, impose basic sanitation, and "advanced" water supplies the government of the State of Pará in which Belem sits, in a partnership with the municipal administration along with the federal government and an international financial institution, performed several rectification and canalization works along the city's watercourses that were connected to the Guajará Bay.[4] The project was known as Una Basin Macro-drainage. This chapter is centered on the impacts of these sanitation policies[5] and climate change, mainly on the floods in the city of Belém,[6] which were caused by the overflow of the drainage systems deployed since the 1990s, and on the Una Basin. For some inhabitants, Belém's urbanization implied an increasingly waterlogged reality, reflecting the disastrous combination between urban planning shortcomings and climate change effects.

The study is based on ethnography conducted in an impoverished, flood-prone neighborhood. Using the concept of environmental memory (Devos et al. 2010), it explores changes in the ways that people understand and relate to daily encounters with river waters and to flooding events, especially following a massive urban readjustment program. During fieldwork conducted between 2013 and 2016, I noted that some technicians and engineers involved in the Una Watershed Project stood against its paradigm as public policy. "The river has died after the project," they said. This assumption is examined throughout this chapter. The research supports that the policies actually generated critical changes in ways people relate to urban nature and to the hydric landscapes, among other social and environmental issues in the urban context.

This work argues that when socionatural disasters—that is disasters that can be seen as hybrids between nature and culture—happen, the institutionalized thresholds between nature and culture are disorganized. The regulation of relations with nature via public policies does not eradicate the environmental memory of the inhabitants from Una Basin, as will be seen throughout the chapter. This means that even if the works of macrodrainage neglect the river or the *igarapés*[7] that once existed, the old watercourses continue to exist in the memory of the inhabitants, especially when disasters in the form of floods return the channels to their once natural, chaotic, and destructive condition, resembling the rivers they have been in the past.

The chapter is organized as follows: To begin, it offers its contribution to the debate of climate change. It then addresses the Una Watershed Project as an example of climate change mitigation and urban readjustment that has taken place in the city of Belém. Afterward it presents an overview of the theoretical background regarding disaster studies, nature and culture dynamics, and environmental memory. Lastly, it discusses some ethnographic findings about how people react to floods and their consequences in scrambling the ontological realms of nature and culture employed in the everyday life of Una watershed residents. The conclusion reflects the implications of the material for discussing public policies on flooding and urban planning.

Climate Change in the Urban Amazon

With regard to the theme of this volume, the contribution of my chapter is to bring the urban Amazon to a central discussion about climate changes. Recent reports about climate changes in the Amazon (Nobre 2014; Marengo 2018) call attention to problems such as deforestation, biodiversity loss, erosion, and river siltation, as well as savanization and desertification risks. Although the reports recommend public policies on the part of local government and civilian society engagement in socioenvironmental issues, there is yet no mention of the climate change impacts in the cities of the Amazon. It reinforces Eduardo Brondizio's proclamation about the problem: to the academy and to the NGOs, Amazonian cities are invisible (Brondizio 2016). Consequently, the environmental issues and the climate change impacts in the cities are ignored, as if they are bound to the forest and to the Amazonian rural world.

However, research accomplished in the last decade show how Amazonian cities need to be included in the considerations about climate change. Farias's work (2012), for instance, discusses the relationship between climate changes and urban floods in Belém, with a focus on adaptation and vulnerability, both social and environmental, of the populations affected by these disasters. Although the definition of climate change includes human action's effects over the environment, when Farias refers to the urban floods in Belém, she supports that these are results of the sea level's increase and intense precipitation in seaside towns, as well as the urbanization process. The urbanization process, in turn, is treated as a synonym of population increase, and the occupation of areas by the poorest population next to rivers and swamps is considered inappropriate by the public authorities.

A further study adds more layers of comprehension to the issue. On the one hand, Mansur et al. (2017) credits a great responsibility for urban floods in Belém to climate change and to the uncontrolled population growth since the 1970s. On the other hand, the work recognizes that these factors must be related to the cumulative result of politic decisions taken over the years. In other words, the effect of the climate change over cities such as Belém have been intensified through factors such as the inappropriate public policies of drainage and sanitation, lack of awareness by engineers and technicians about the local reality, and the neglect of the local population to participate in decision-making processes concerning the urban policies that have direct impacts on their lives (Mansur et al. 2017).

Both Farias's and Mansur et al.'s works merit placing the Amazonian cities at the center of discussions about environment, refusing the contradiction between the "built" urban and the "natural" environmental, which serves as an excuse for the invisibility of Amazonian cities in the context of climate change. Unlike former studies, they bring together environmental analysis and urban studies, reinforcing the urban and social features of what is considered the environmental. Cities as dynamic centers of global capitalism tend to compress in their territory the effects of the social and economic system, including the severe environmental impacts, which generally affect the poorest urban populations more frequently and intensely.

Cities such as Belém, a metropolis in the Global South, offer exemplars to understand how social problems are intensified by ecological factors. In recent years, events like floods are happening more often in Belém, in part due to demographic factors and inadequate storm drainage policies, but also on account of sea level rise[8] and higher incidence of rains.[9] Those social and environmental changes increase the social vulnerability of the poorest populations who suffer from the immediate impacts of climate change, especially due to the lack of public policies addressing the mitigation of hazards such as urban floods. However, as will be shown, the main problem in Belém is not necessarily the lack of public policies.

The Una Watershed Project

In Belém, official discourses and the public sentiment stress that the city's process of urbanization has been historically grounded in the antithesis between water and civilization (Cardoso and Ventura Neto 2013). Otherwise put, the foundational myth of Belém relies upon the idea that the city was formed by conquering supposedly wild territories that originally belonged to water: swamps, creeks, and floodplains.[10] Thus, Belém—or at least its ideal of urbanization—is built in opposition to nature (Soares 2016).

On the pretext of overcoming the restraints of nature to achieve its goal of urbanization and modernity, Belém has gone through several slum-cleaning processes through public drainage works that have substantially transformed the city's landscape. Among those initiatives is the Una Watershed Project, which generated massive impacts on the riparian landscapes of the *baixadas* in the Una watershed. In spite of not being the main subject of this work, one must keep in mind that the Una Watershed Project is an expression of the conflicting relationship between the city of Belém and its waters. Taking place from 1993 to 2004, the Una Watershed Project received resources from the Interamerican Development Bank (IDB) and was responsible for draining wetlands, forming soil embankments, and implementing a basic sanitation system over a large amount of the city (Pará 2006).

On the books, the Una Watershed Project presents outstanding numbers regarding its accomplishments. A report by the Sewage Company of Pará (COSANPA in Portuguese) lists that the Una Watershed Project built 25,731 individual septic tanks, 91 collective cesspits, 307 kilometers of sewage network, 2,164 inspection wells, 3,887 cleaning terminals, and a drying bed of septic tanks (Pará 2006: 11). The strategy of creating concrete channel systems to drain water was replicated in other parts of the Una Basin, resulting in more than 24 kilometers of storm drainage channels. The Una Watershed Project cost, after all, over US$300 million (Pará 2006). Besides sanitation, the main goal of the project was to promote a socioeconomic transformation in the city, improving the well-being of about 600,000 people, representing approximately 120,000 families (Pará 2006).

The Una Watershed Project was completed with the installation of floodgates at the mouth of the Una and Jacaré river channels in order to avoid floods caused by the entrance of Guajará Bay waters into the city (Pará 2006). From then on, the old rivers and *igarapés* became drainage channels, and their flow toward Guajará Bay was partially blocked by the floodgates.

In reality, although the Una Watershed Project contributed to a very ambitious urban readjustment project that aimed to solve habitation problems and flooding in the city of Belém, the hydrologic landscapes of the Una Basin were instead converted into an immense flood-control system with no other purpose than the effective removal of the extreme amounts of tropical rainwater from the city (see figure 4.1). The transformations carried out by the Una Drainage Project resulted in an altered environment that required massive investments and extensive labor to maintain. The equipment and machines used in the maintenance of the public drainage works were diverted, sold, and improperly appropriated by groups related to the office of the newly elected mayor when he began his tenure in 2005 (Costa et al. 2018). Many areas of the watershed were deliberately excluded from the benefits of the Una Drainage Project. Also, the municipality did not carry out a set of complementary works of drainage, which were required in the contract with the IDB (Costa et al. 2018).

The political and ecological failure of the Una Watershed Project to maintain the public works of drainage and sanitation has, once more, made the population of the watershed vulnerable to environmental hazards and flooding (Soares and Cruz 2019). As a case of structural violence (Farmer 2005), there is a chain of complicity that goes from co-opted grassroots organizations in the Una watershed neighborhoods to local public administration and the IDB. This has resulted in changes in the ways through which Belém's inhabitants relate to the environment. Currently, these relationships are mediated, in part, through public policy related to drainage, sewer, and water management systems, which are embedded in social processes of inequality and marginalization manifest through urban flooding.

Environmental Memory and Disasters

In this chapter I approach water not just as a natural element but also as a set of processes, practices, and meanings that are essential to understanding urban policymaking and people's responses to it. What is considered urban infrastructure stems from negotiations between political actors, ideological struggles, and power relations that result in the uneven distribution of natural resources and environmental hazards related to the water.[11]

The flood season combined with intense tropical rains has always tended to provoke the overflow of watercourses. In the Amazon, a great amount of lands called *várzea* remain submerged during months of big tides and rains. This hydrologic dynamic does not necessarily imply the occurrence of a disaster. In rural areas, the population who live on riverbanks develop migration strategies involving areas that are not flooded during this time of year (Coelho 2013). Urbanization and sedentarization in large cities change completely the relationship between people with these hydroclimatic extremes. In these contexts, the urban populations—mostly the poorest—become the most vulnerable to floods. The literature on anthropology of disasters particularly focuses on these questions, since it addresses not only the natural phenomena itself but also both social and physical elements that aggravate or reduce the impacts of a natural hazard (Oliver-Smith 2002: 27).

In the case of Belém, the literature on disasters sheds light upon two points that might be emphasized. The first one is that, as it was proposed by Oliver-Smith (2002), the distribution of disasters' effects is always unequal. If environments such as cities are socially built, then the effects of disasters tend to be distributed among urban populations as a reflection of the social relations that take place in the city. In following, the poorest population of Belém is the most vulnerable because the occupation process of the city's territory allowed the wealthiest to occupy the higher and, therefore, more valuable lands and forced the poorest in consequence to inhabit the wetlands next to the rivers, subjecting them to their floods (Trindade Junior 1997). The second point is that there are vulnerability patterns to disasters that are historically built. This means that a disaster can be the result of cumulative political decisions made throughout time (Nelson and Finan 2009), which in Belém entails the technological solutions embraced to manage urban waters.

Another aspect that should be highlighted is the specificity of the disasters[12] that occur in Belém. It concerns the cyclical disasters, as named by Nelson and Finan (2009) in their work about the drought in the Brazilian northeast. In fact, there is a parallel between the lack of water in the northeast of Brazil and the overflowing of watercourses in Belém, both of which are historical, seasonal, and predictable. The reasons are ecological but also political. They also both present themselves as "slow-onset phenomena" instead of "rapid-onset phenomena such as earthquakes and nuclear accidents" (Oliver-Smith 2002: 25). In the case of Belém, cyclical disasters like floods are intensified by urbanization and the social inequalities created and reproduced in the process mentioned above, resulting in disasters that combine natural and social aspects.

The environmental memory perspective (Devos et al. 2010) enables the researcher to address issues such as flooding not just as a physically disruptive, one-time event. Rather, it is possible to see it as a long-term process, which further reflects the considerations of Hoffman and Oliver-Smith (2002) on what a disaster is. There is a wider dynamic of class, politics, and production of urban space that precedes the flooding, as well as a recovery process that includes political struggles for rights over sanitation and adequate housing. In addition, there are also struggles over the social production of meaning on flooding, which contrasts with the framing of disasters offered by the media and other social institutions (Button 2002).

The media and public opinion often treat flooding as a natural and isolated event in which the poor are blamed for their situation because their homes are built in inappropriate areas and they dump garbage and waste into drainage canals (Soares and Cruz 2019). However, flooding is a symptom of broader social issues that are manifested through sanitation and other infrastructure shortcomings in Belém. The focus on flooding being an isolated event masks more complex and systematic issues that lead to the current situation. In other words, the media, as well as the conventional wisdom, treat floods in Belém only as one-time environmental hazards, while they should be interpreted as social and cyclic disasters (Nelson and Finan 2009).

Disasters, Nature, and Culture

Besides the discussion about disasters, I also draw on Philippe Descola's (1998, 2003, 2011) and Bruno Latour's (2013) works concerning the overlap between the ontological domains of "nature"[13] and "culture."[14] Descola argues that naturalism[15] is specifically a Western ontology, not finding parallel in other societies. Among Amerindians, for instance, even when their cosmologies identify beings belonging to nature, they do it so these domains interpenetrate and dialogue, far from being hermetic as they are in Western societies. Still, Descola claims that even in the colonized Amazonian society, the relationship between culture and nature also occurs in line with more dialogical and porous ontological borders, bearing in mind the examples from the relations between human beings and animals in Amazonian contexts (1998). In another work, Descola (2011) reflects on how the separation between nature and culture in Western societies works only in the discursive plan, given that in their everyday lives, Europeans tend to ignore the apparent splitting between the natural and the cultural realms in the face of what the author calls "ontological slides." Included are the humanization of domestic animals, genetic manipulation, the actions of human beings over the environment, and the findings about chimpanzees' fabrication of tools and transmission of its techniques.

Descola's criticism to the unsustainable splitting between nature and culture in the Western world resonates with Bruno Latour's (2013), who cites modernity as a greater expression of the Westernized thought that separates society (culture, speech, and politics) from technique and whose object is the nature conceived as external and independent in relation to social and cultural processes. Latour identifies two processes operating dialectically through which modernity expresses itself via science. The first process is what Latour calls purification, which means the compartmentalization of scientific knowledge, creating nature and culture as independent objects. Paradoxically, purification generates a second process he calls hybridization, which is the producing of quasi-objects that are not entirely composed from either nature or culture but are both at the same time.

An example that engages both these processes is the implementation of public policies that were destined to promote urbanization, mitigate environmental hazards and climate change impacts, and modify inhabitants' relationships with the local environmental, mostly with water. Large-scale projects like the Una Watershed Project institutionalize boundaries between ontological domains of nature and culture in the river-channel margins of the city. As a process of purification, when the river becomes a channel or simply a drainage ditch, a "natural" marker is transformed into a built or "cultural" form of environment, although this study questions how permanent this transformation really is given that the policy itself results in processes of hybridization. The public works of macro-drainage existing in Belém are examples of hybrids or quasi-objects, since they combine politics with natural actors like tropical rainstorms, tides, soil types, vegetation, and the preexisting natural drainage system in the city territory. Also, these natural and political actors compose networks that can be addressed in several scales of analysis, to begin with all management levels through which the project is produced: municipal, state, and federal, without neglecting the role of financial institutions such as the IDB in the formulation and implementation of the policies.

Besides, although changes in the city environment occur, in everyday life people continue to live near waterways and still try to make sense of the transformations carried out by the Una Watershed Project, especially when channels overflow flooding streets and households. Simply put, everyday life practices and events may keep the river alive in the memories of local communities, even while public policies seek to domesticate and sterilize urban hydric landscapes, revealing other ways of producing hybrids between city and nature.

Some questions arise: How do people relate to water after the conclusion of the Una Watershed Project? How do disasters, such as urban floods, affect the distinction between the ontological realms of nature and culture as well as the relationship between humans and nonhumans in the Una watershed?

Urban Flooding and the Thresholds between Nature and Culture

The following discussion draws on the idea of urban flooding in Belém as a cyclical disaster disrupting the ways through which urban planning has institutionalized environment and human relations.[16] As Virginia Acosta pointed out in the recent Critical Issues on Risk and Disasters panel at the IAUES World Congress, disasters are *ventanas criticas* (critical windows) that permit us to see cultural dynamics. The work of Susanna Hoffman (2002) is an inspirational model to help us think about how disasters disrupt people's ways of organizing the world as they know it. It also emphasizes how disasters call into question the "promethean project of modernity"[17] (Kaika 2005: 5), in which technology and technocratic policies aiming to control nature are seen as key components to building cities.

Hoffman's (2002) analysis of the Oakland, California, fire in 1991 points to the symbolism of disasters and their role in collapsing the ontological dualism (nature/culture). The ethnographic examples presented by the author show how the fire made relations between humans and nonhumans collapse. The Oakland middle-class neighborhood was nearly destroyed by the fire. In the process, domestic animals ran into the woods, and the inhabitants feared that their pets had become wild animals. Also, the efforts to reconstruct houses and yards symbolized the inhabitants' desire to bring back the neighborhood to culture from a chaotic state of nature. The savage landscape produced by the fire was to be re-domesticated and reintroduced to the logic of culture. If the fire was the disrupting element in Oakland, water created the same effect in Belém.

In Belém, large-scale infrastructure public works projects such as the Una Watershed Project failed to promote an urban citizenship in relation to water, resulting in recurrent and repeated floods caused by the overflowing of the canal network redesigned by the drainage project. Just like the fire in Oakland, these floods had the same impact on disrupting the dualism between nature and culture. The following are some examples.

Example 1

Every year, rainfall and tides overfill the current drainage system. Heavily contaminated waters overflow the canals and cesspools, invading households and bringing pollution into direct contact with people. When I first entered Mr. Costa's house after a flood, I was quite shocked. He lives on his own in a two-story wooden house near the Galo Channel, the same home as his childhood. The flood not only destroyed furniture and domestic utensils but it also left a thick layer of mud on the floor and other places that had been in contact with water. Symbolically, the effects of the destruction caused by flooding resemble those caused by the fire in Oakland. In both instances, landscapes that humans had domesticated were taken by the chaotic agency of nature. Besides the confusion of nature and culture generated by the flood, water also scrambled domestic and public features. On the one hand, water brought into the domestic space what should belong to the public sphere: garbage, sewage, and dead animals from the canal. On the other hand, domestic objects, documents, and photo albums were taken away by the overflowing waters.

Example 2

The Una Watershed Project transformed the former hydrographic basin into a network of canals and galleries in which the flow of water is controlled by floodgates located at the mouth of the river. During most of the year, when there is little rain, the canals resemble ditches. Nevertheless, the rainy season provokes an increase in water levels, which transforms the canals' appearance. This shift also changes the inhabitants' attitudes toward the canals. When the Galo Channel is close to overflowing, some dwellers see an opportunity to swim there, just as if it were a river again. They do so despite knowing that the water is considerably polluted. In those occasions, the bleak appearance of the canals transforms into the semblance of a thriving hydric landscape like the living waterways of the past. Rainwater feeds the channel, diluting dirt and sewage. The watercourse regains its volume, and its coloring changes: from pale gray, which reeks of sewage and filth, its waters become bulky and muddy just as many other Amazonian rivers. The perception of the residents on the channel is altered by the rain. A day after the flooding, residents will say, "All of it appeared just like a river."

Example 3

Residents often report encounters with wild animals, especially reptiles, during rainy seasons. It seems that such animals come from the rivers and forests surrounding the city when they get stuck in the drainage network. Or, possibly, they simply find the network a habitable environment where they can easily find food and shelter. When the system floods due to intense rainfalls, those animals approach the nearby streets and households, especially the low-lying terrains and wetlands where residents build their homes. Seeming to combine fantasy and facts, a local resident, Mrs. Barroso, once told me that an anaconda (*Eunectes murinus*) used to live beneath a woman's house near the São Joaquim Channel. Mrs. Barroso said that the fragile household used to shake when the serpent moved in the flooded ground. The family began to worry when they realized the serpent was growing exponentially, and it eventually captured and ate a cat. Fearing for her children, she called the fire department, who apprehended the creature.

The final story above is an extreme example of interaction between humans and dangerous animals in the Una Basin. However, the presence of animals brought by the rains is not unusual. After a flood in 2015, Mr. Costa told me his neighbor found a group of small turtles in her backyard. They could not say if the turtles had come from another household in the neighborhood or if they actually lived on the Galo Channel and were brought to the backyard by the flood.

These cases are also reported on the local news. In 2011, a newspaper published a story revealing the presence of hundreds of tilapia (*Oreochromis niloticus*) in a channel in the Una Basin (Portal Orm 2011). At the same time, dead fish started to appear over the surface of Visconde de Inhaúma Channel, calling the attention of the local inhabitants and the public health authorities. The story reported that experts recommend the inhabitants not eat those fish. However, residents stated that they had the habit of fishing the tilapia, and those fish were already part of the local people's diet. A researcher interviewed by a newspaper suggested that the fish belonged to a particular pool or cistern in the neighborhood and that they were taken into the channel during the last flood. The same researcher predicted that the tilapia would adapt themselves to the channel's waters, remaining there as part of the aquatic fauna.

Visiting the research locus during fieldwork in 2016, I saw a shoal of fish in the waters of this channel. They were diverse in size, but their gray color and red scales indicated they were tilapia. Mr. Alfredo, an eighty-five-year-old inhabitant who was walking me around the neighborhood that day, confirmed and explained: those fish were a remnant from the old *igarapé* that existed a long time ago in the exact place where the channel was built. He said that this channel was different from the other ones because its water is clean, which means that the same water from the old *igarapé* continues to pour into the midstream of the channel and into the backyards of the neighborhood. Just as Mr. Alfredo and other residents in some areas of the Una Basin still can identify, the waters from springs next to the old *igarapés* still flow even after they were transformed into channels. This tale shows the awareness of the in situ residents of the constant everyday life-relationship with the city's hydric landscape, a feature that was neglected by the technocratic paradigm that guided the Macro-drainage project of the Una Basin.

The examples above show how everyday practices and cyclic flooding disrupt the institutionalized thresholds between nature and culture. They call attention to the uncertainty of the attempts to control nature via technology, which again constitutes what Maria Kaika (2005) calls the "promethean project of modernity." In these episodes, natural forces people believed to be tamed regain their agency, revealing that they actually have remained aggressive and unmanageable. In Belém, for instance, an ordinary rainstorm readily overcomes the recently implemented drainage and sanitation technologies. Through floods, water invades neighborhoods, destroys households, and puts humans in contact not only with waste, sewage, and gray mud but also with dangerous and unwelcome animals.

The *igarapé* (natural entity) may be dead when transformed into a channel (cultural entity), the latter built and understood as resulting from the extinction of the first. It is not unusual to hear some inhabitants refer to the channels as "trenches" or "ditches," mostly on occasions that bring up the problems of sewage and risk of contamination through water contact. In this context, the relation between risk, blame, and technology becomes evident. Technology, as it was pointed by Mary Douglas and Aaron Wildavsky (1992), was once responsible for overcoming the entailments of moral and danger, when the lack of scientific knowledge about natural dangers was seen as a moral problem. In Belém, the very use of science and technique has resulted in an anachronistic macro-drainage system, which receives sewage and waste but is unable to bear the city's pluviometric demand. In this case, technology applied to public policies has turned into a risk factor for flooding. Technocratic solutions did not solve the problem of floods. Instead, these events have become even more frequent and intense over the years.[18]

Generally associated with disease, garbage, and floods, the drainage channel is understood by some local inhabitants and experts as a river that died so public policies of urbanization and sanitation could live, disregarding the complexities of watercourses and relationships between people, water, plants, and animals in the Amazonian context. In the absence of a satisfactory synthesis between the city and its rivers, the channel system represents a technical solution for nature's instrumentation with high social and environmental costs.

Figure 4.1. A drainage channel in the Una Basin, Belém, 2017. © Pedro Paulo de Miranda Araújo Soares.

Nonetheless, the data related in this chapter implies that the "death" of the watercourses in Belém is not permanent and the ideas of *igarapé,* river, and channel are negotiated, situational categories that vary in accordance with the perspective of who is describing the landscape and their relationship with it. Also, these points of view stem from the beholder's position in the power struggles in Belém's socioenvironmental arena. These categories and values adhere to a vast set of images about water and its symbolism in the Amazon that evokes a hydric psyche (Bachelard 2008) but also points to an anthropological problem when referring to the relations between humans and nonhumans (Descola 1998, 2003; Latour 2013; Viveiros de Castro 2002).

Conclusions

Where once Henry Walter Bates saw and wrote of a vibrant and lush paradise now exists an environment overrun with sewage and disease. Within this context, climate change effects such as sea level rise and increased rainfall are combined with infrastructural deficiencies, generating environmental hazards that affect specially the urban poor. In Bates's favorite spot, what remains today is a complex social-ecological landscape, where flooded and eroded urban settings strain the quality of life and livelihood of impoverished and increasingly stratified social classes.

The Una Watershed Project was an expression of urban planning in Belém consisting of a vertical, unilateral, and universal sanitation and drainage policy dominated by a technical and scientific "rationality" on what constitutes "modern" cities (Scott 1998). This sort of project involves the replication of public policy models and an articulation between city and nature that disregards local tactics and strategies for adaptation to climatic extremes and urban infrastructure deficits. It seems that the scientific and technical language used in the Una Watershed Project insulated policymakers from local knowledge and overlooked the Amazonian livelihoods based on direct contact with water.

Within this context, the unending occurrence of floods suggests that the river, which has been transformed by technocratic public works, is seen as a river once again by the population, even after it was turned into a mere drainage canal and a sewer disposal. Everyday practices, as well as flooding episodes, can shatter the institutionalized ways of managing urban waters.

In a context where water is a constant feature in people's everyday lives and the thresholds of nature and culture remain unclear, flooding can be seen as an indicator of the technological failure of the city, a city that is constantly reclaimed by nature through floods. In an Amazonian city such as Belém, which has been through ambitious urban readjustment programs, the persistence of floods is a sign that taming nature may not be a beneficial operation. This brings up the question about how flooding and its consequent ontological disruptions may help to reflect upon the effectiveness of public policies on drainage and sanitation in the Amazonian scenario.

Pedro Paulo de Miranda Araújo Soares holds a PhD in anthropology and is a PNPD/CAPES scholar and visiting professor at Federal University of Pará (UFPA, Brazil). He is also a member of the Assisting Program to Urban Reform (PARU) and the Research Program on Urban Policy and Social Movements in the Globalized Amazon (GPPUMA) of the Graduate Program in Social Work at the Federal University of Pará (PPGSS-UFPA).

Notes

1. Parts of this text were abridged from Costa, Soares, and Dias 2018.
2. *Baixadas* are known as a precarious type of habitation and land use, which is characterized by high population density, lack of urban infrastructure, and occupation of the soil on floodplains or lowlands near watercourses (Trindade Júnior 1997).
3. *Palafita* is a local name for precarious stilt houses in the *baixadas* of Belém.
4. This practice can be observed in many areas of the city throughout the twentieth century and, more recently, in the Una Basin itself. There are exceptions, of course, such as the Tucunduba Basin Project, which consisted of public work projects that aimed at maintaining the natural attributes of rivers in an area comprised of five districts (Barbosa 2003). The Tucunduba River is still navigable and local communities still use the watercourse for swimming and fishing, despite its sewage contaminated waters.
5. In Brazil, a federal law defines basic sanitation as a set of public works and equipment divided into four components: sewage, water management, solid waste management, and urban drainage (Brasil 2007). In this chapter, special attention is given to urban drainage.
6. The city was founded in 1616 in an area inhabited by Tupinambá tribes before the coming of Spanish, Dutch, English, and Portuguese settlers. Belém is currently the second largest city in the Amazon Delta regions (1.4 million inhabitants).
7. *Igarapé* is a local expression that means a watercourse of varying dimensions that flows into the main river.
8. Vinagre et al. (2017) analyzed the functionality and the drainage system behavior of the Una Basin using the software SWMM (Storm Water Management Model). The assessment estimated tides of 3.15 meters maximum within 3 hours of severe rainfall, simulating a rain that occured on 9 May 2011. In 2020, watermark levels in Belém reached 3.70 meters during rainy season, which indicates how sea level rise is a growing factor in shifting flood patterns.
9. Santos et al. (2019) confirmed not only that the annual average rainfall in Belém has increased between 1961 and 2010 but also that wet seasons have been increasing in intensity in the last thirty years, mostly under the influence of the El Niña phenomena.
10. In 1973, Augusto Meira Filho wrote that Belém's vocation for "progress" and "opulence" would only be accomplished by "conquering lowlands, forests, and swamps" (Meira Filho 1973: 44). These words express an ideology of progress and modernity based on suppressing and mastering nature so the city can thrive. The

very same ideology oriented public policies on sanitation implemented in Belém throughout the twentieth century and in the first two decades of the 2000s.

11. Both the excess and the absence of water in the Global South are not only coincidental but also complementary. Flooding and water shortages are connected in cities like Belém. Matthew Gandy states that the constitution of the "hydrological subject" should be a measure of urban citizenship in different social-environmental contexts (2014).
12. Among disaster agents there are natural hazards (atmospheric, hydrological, geological, biological), technological hazards (dangerous material, destructive processes, mechanical and productive), and even social hazards such as war, terrorism, civil conflict, use of hazardous material processes or technologies (Oliver-Smith 2002: 25).
13. "Nature" here is understood as the natural elements along with all beings that dwell in it. The commonsense notion of nature includes biochemical phenomena, physical and metabolic processes involving those natural elements usually depicted as prehumen and external to society (Heynen, Kaika, and Swyngedowu 2006)
14. In this work, "culture" appears as a synonym of human action over the nature. This action—the work—is characterized by Marx (2009) as the essence of human condition and establishes a dialectic relationship between subject and object. This dialectic is reassessed later by Lévi-Strauss (2009) in his distinction between nature and culture. The Lévi-Straussian scheme suggests that humans act over nature, domesticating it while imposing order and particular rules upon it, which can be perceived by human beings as universal and chaotic (Lévi-Strauss 2009).
15. Dualist ontology assumes that, whereby nature is the common ground to all living things around the world, only humans are capable of developing cultural faculties. Yet, human beings participate simultaneously in both cultural and natural domains, because they do not cease to exist as a biological species.
16. According to Latour (2013) and Lemos (2003), technicians and experts tend to insulate themselves from politics and decision-making processes, depicting sociocultural features in terms of technical solutions addressing only nature.
17. "The historical geographical process that started with industrialization and urbanization and aimed at taming nature through technology, human labor, and capital investment" (Kaika 2005: 5).
18. In light of floods that occurred in March 2020, both municipal and state governments decreed a state of emergency for the first time in history in Belém's metropolitan area and in parts of the countryside.

References

Bachelard, Gaston. 2008. *A água e os sonhos*. São Paulo: Martins Fontes.

Barbosa, Maria José de S. 2003. *Estudo de caso: Tucunduba*. Belém: UFPA.

Bates, Henry Walter. 1944. *O naturalista no Rio Amazonas*. São Paulo: Brasiliana.

Brasil. 2007. *Lei nº 11.445, de 5 de janeiro de 2007: Estabelece diretrizes nacionais para o saneamento básico e dá outras providências*. Retrieved 8 March 2020 from http://www.planalto.gov.br/ccivil_03/_ato2007-2010/2007/lei/11445.htm.

Button, Gregory V. 2002. "Popular Media Reframing of Manmade Diasters: A Cautionary Tale." In *Catastrophe and Culture: The Anthropology of Disaster*, edited by Susanna Hoffman and Anthony Oliver-Smith, 143–58. Santa Fe: School of American Research Press.

Brondizio, Eduardo S. 2016. "The Elephant in the Room: Amazonian Cities Deserve More Attention in Climate Change and Sustainability Discussions." Nature of Cities. Retrieved 15 August 2016 from https://www.thenatureofcities.com/2016/02/02/the-elephant-in-the-room-amazonian-cities-deserve-more-attention-in-climate-change-and-sustainability-discussions/.

Cardoso, Ana C. D., and Raul da S. Ventura Neto. 2013. "A evolução urbana de Belém: trajetória de ambiguidades e conflitos socioambientais." *Cadernos Metrópole* 15(29): 55–75.

Coelho, Roberta F. C. de. 2013. *Ribeirinhos urbanos: Vida e modos de vida no Puraquequara*. Manaus: Edua.

Costa, Jose A. J., Pedro P. M. A. Soares, Vitor M. Dias. 2018. "Regaining Paradise Lost: Global Investments, Mega-Projects, and Seeds of Local Resistance to Polluted Floods in Belém." Nature of Cities. Retrieved 3 March 2019 from https://www.thenatureofcities.com/2018/12/12/regaining-paradise-lost-global-investments-mega-projects-seeds-local-resistance-polluted-floods-belem/.

Descola, Phillipe. 1998. "Estrutura ou sentimento: A relação com o animal na Amazônia." *Mana* 4(1): 23–45.

———. 2003. *Par-delá nature et culture*. Paris: Gallimard.

———. 2011. *Más allá de la naturaleza y de la cultura*. Bogotá: Jardín Botânico de Bogotá Celestino Mutis.

Devos, Rafael V., Ana. P. M. Soares, Ana L. C. da Rocha. 2010. "Habitantes do Arroio: Memória ambiental águas urbanas." *Desenvolvimento e Meio Ambiente* 22: 51–64.

Douglas, Mary and Aaron Wildavsky. 1992. *Risk and Blame: Essays on Cultural Theory*. London: Routledge.

Farias, Glorgia B. de L. 2012. "Cidades, Vulnerabilidade e Adaptação às Mudanças Climáticas: Um estudo na Região Metropolitana de Belém." Master's thesis, Universidade Federal do Pará, Belém.

Farmer, Paul. 2005. "On Suffering and Structural Violence." In *Pathologies of Power: Health, Human Rights and the New War on the Poor*, edited by Paul Farmer, 27–50. Berkeley: University of California Press.

Gandy, Matthew. 2014. *Fabric of Space: Water, Modernity, and the Urban Imagination*. Cambridge, MA: MIT Press.

Heynen, Nik, Maria Kaika, and Erik Swyngedowu, eds. 2006. *The Nature of Cities*. New York: Routledge.

Hoffman, Susanna. 2002. "The Monster and the Mother: The Symbolism of Disaster." In *Catastrophe and Culture: The Anthropology of Disasters*, edited by Susanna Hoffman and Anthony Oliver-Smith, 113–42. Santa Fe: School of American Research Press.

Hoffman, Susanna, and Anthony Oliver-Smith, eds. 2002. *Catastrophe and Culture: The Anthropology of Disasters*. Santa Fe: School of American Research Press.

Kaika, Maria. 2005. *City of Flows: Water, Modernity, and the City*. New York: Routledge.

Latour, Bruno. 2013. *Nunca fomos modernos*. São Paulo: Editora 34.

Lemos, Maria C. de M. 2003. "A Tale of Two Policies: The Politics of Climate Forecasting and Drought Relief in Ceará, Brazil." *Policy Sciences* 36: 101–23.

Lévi-Strauss, Claude. 2009. "Natureza e Cultura." *Antropos* 3(2): 17–26.

Loureiro, Violeta R. 1992. *Amazônia: Estado, homem, natureza*. Belém: Cejup.

Mansur, Andressa V., Eduardo S. Brondizio, Samapriya Roy, Pedro P. M. A. Soares, and Alice Newton. 2017. "Adapting to Urban Challenges in the Amazon: Flood Risk and Infrastructure Deficiencies in Belém, Brazil." *Regional Environmental Change* 17: 1–16.

Marengo, José A., and Carlos Souza Jr. 2018. *Mudanças climáticas: Impactos e cenários para a Amazônia.* São Paulo: ALANA/Articulação dos Povos Indígenas do Brasil/Artigo 19/Conectas Direitos Humanos/Engajamundo/Greenpeace/Instituto Socioambiental/Instituto de Energia e Meio Ambiente/Programa de Pós-Graduação em Ciência Ambiental Universidade de São Paulo/Instituto Nacional de Ciência e Tecnologia para Mudanças Climáticas Fase 2.

Marx, Karl. 2009. *A ideologia alemã.* São Paulo: Expressão Popular.

Meira Filho, A. 1973. *Contribuição à História de Belém.* Belém: Imprensa Oficial do Estado.

Nelson, Donald R., and Timothy J. Finan. 2009. "Week Winters: Dynamic Decision-Making in the Face of Extended Drought in Ceará, Northeast Brazil." In *Political Economy of Hazards and Disasters,* edited by Eric C. Jones and Arthur D. Murphy, 107–32. Lanham, MD: AltaMira Press.

Nobre, Antônio D. 2014. *O futuro climático da Amazônia: Relatório de avaliação científica.* Articulación Regional Amazônica.

Oliver-Smith, Anthony. 2002. "Theorizing Disasters: Nature, Power, and Culture." In *Catastrophe and Culture: The Anthropology of Disasters,* edited by Susanna Hoffman and Anthony Oliver-Smith, 23–48. Santa Fe: School of American Research Press.

Oliver-Smith, Anthony, and Susanna M. Hoffman. "Introduction: Why Anthropologists Should Study Disasters." In *Catastrophe and Culture: The Anthropology of Disasters,* edited by Susanna Hoffman and Anthony Oliver-Smith, 3–22. Santa Fe: School of American Research Press.

Pará. 2006. *Informações gerais sobre o Projeto Una.* Companhia de Saneamento do Pará. Belém: Governo do Estado.

Penteado, Antônio. 1968. *Belém—Estudo de Geografia Urbana.* Vol. 2. Belém: UFPA.

Portal Orm. 2011. *Morte misteriosa de peixes.* Retrieved 20 March 2020 from http://noticias.orm.com.br/noticia.asp?id=520996&percent7Cmorte+misteriosa+de+peixes#.VWYwgc-4TIU.

Santos, Josiane S. dos, Edson J. P. da Rocha, José A. de Souza Jr., Jaqueline S. dos Santos, Flávio A. A. dos Santos. 2019. "Climatologia da Amazônia Oriental: Uso de prognósticos climático como ferramenta de prevenção de ameaças naturais." *Revista Brasileira de Geografia Física* 12(5): 1853–71.

Scott, James C. 1998. *Seeing like a State: How Certain Schemes to Improve the Human Condition Have Failed.* New Haven, CT: Yale University Press.

Soares, Pedro P. M. A. 2016. "Memória Ambiental da Bacia do Una: Estudo antropológico sobre transformações urbanas e políticas públicas de saneamento em Belém (PA)." PhD diss., Universidade Federal do Rio Grande do Sul, Porto Alegre, Brazil.

Soares, Pedro P. M. A., and Sandra H. R. Cruz. 2019. "A Ecologia Política das inundações na Bacia do Una em Belém (PA)." *Emancipação* 19(1): 1–15.

Trindade, Saint-Clair, Jr. 1997. *Produção do espaço e uso do solo urbano em Belém.* Belém: UFPA/NAEA.

Vinagre, Marco V. de A., Leonardo A. L. Bello, Andréia do S. C. de S. Cardoso, Paulo G. P. de O. Folha Neto, Vitor G. Rabêlo. 2017. "Modelo de gestão de drenagem urbana aplicado à Bacia do Una em Belém-PA." *Revista da Universidade do Vale do Rio Verde* 15(1): 253–67.

Viveiros de Castro, Eduardo. 2002. "Perspectivismo e multinaturalismo na América indígena." In *A Inconstância da alma selvagem e outros ensaios de Antropologia,* edited by Eduardo Viveiros de Castro, 347–99. São Paulo: Cosac Naify.

Part II

Situations and Decisions

"For it is only criminals who presum e to damage other people nowadays without the aid of philosophy." This sentence from Robert Musil's famous novel *The Man without Qualities* would apply perfectly to current corporate and government decision-making. Just substitute "philosophy" with "vested interest," and among those, chiefly by economic interests. From there the permutations could proceed endlessly. The chapters in part II provide worthy examples of how, in different places and situations, decisions are made across the globe with one main common goal: economy first.

This section of *Cooling Down* explores different issues facing a variety of places, including migration and housing in crowded cities of Bangladesh; a slowly vanishing coastline in the United States; disputed urban planning in New Zealand; environmental crisis in the East African countries of Kenya, Somalia, and Ethiopia; and finally to melting glaciers and fake snow in Austria.

Tasneem Siddiqui, Mohammad Jalal Uddin Sikder, and Mohammad Rashed Alam Bhuiyan examine migrant movement and living conditions in one of the most vulnerable countries of the world, Bangladesh. The Anthropocene as described by the authors is both overwhelming and particularly informative for the understanding of how climate change relates to other socioeconomic variables, in particular life-threatening impoverishment. As the research team writes, "Climate change does not displace people directly but, rather, exacerbates various forms of vulnerability," which contribute to their displacement. Climate-driven migrations may create new roofs and jobs, but they do so by forcing already deprived people to "environmentally hazardous regions, coupled with inadequate facilities for food, shelter, sanitation, and healthcare."

The chapter authored by Brian Orland, Meredith Welch-Devine, and Micah Taylor reveals that issues of social inequality occur everywhere, but not in the same manner. Their research details how people, compelled into forced displacement directly following increasingly intense hurricanes, on the coast of Georgia, United States, perceive the transpiring climatic phenomena, envision the near future, and calculate forms of adaptation. What in Bangladesh is a matter of survival instead entails concerns of well-being and freedom of choice in Georgia. The decision on whether to move or not involves not only the effects of climate change but also distinctions of income, gender, education, and age.

Based on very long fieldwork, Paul Schneider and Bruce Glavovic offer a brimming description of how residents of the Coromandel Peninsula, Aotearoa New Zealand, deal with ongoing climate change manifestations. Coromandel is inhabited by people of European and Maori descent. It is a place, therefore, in which different worldviews are ingrained. In addition, some of the villages on the peninsula swell by some 600 percent during summer. Contested interests on land ownership, urban planning, culturally diverse landscape features, coastal erosion, erratic local authorities and policies, sea level rise, and disputed access to the seafront and beaches contribute for continuously deferred actions and decision-making concerning the evolving effects of global warming. Here the present and looming climate conundrum is the outcome, not merely in terms of inequalities in status and income, but also in social and cultural complexities, native and colonial, resident and tourist, leisure and labor, and tottering contestation.

Africa is notable in its exposure to climate change. Severe droughts and concomitant infestations of locusts, deforestation, and a growing number of overpopulated cities contribute to this situation. Meanwhile, as in Bangladesh, poverty and all forms of power privation tremendously magnifies virtually all imaginable climate change consequences. A. Peter Castro brings to us "tales of conflict and displacement" taking place in Kenya, Somalia, and Ethiopia. In Kenya, green policies such as those related to carbon sequestrating are justifying unjust initiatives and indubitably harkening to colonial pasts. In Somalia and Ethiopia, where both droughts and floods are occurring more frequently, millions have recently been displaced, and famine lingers. At the same time, global "environmental refugees" actually propelled by such dire circumstances are yet devoid of legal status and are yet being portrayed by the media as political or economic refugees or, worse, human invaders.

As temperatures in the Alps continue to climb, causing glacier melt, the government of Austria is trying to keep winter tourists coming, a major asset in their economy, by creating manufactured snow. Herta Nöbauer's chapter about "the cryosphere environment of the Austrian Alps" is an account of how climate change, national environmental policies, local entrepreneurs, fake snow, jobs, landscapes as national identity symbols, and tourism shape an anthropological picture of the Anthropocene. Nöbauer takes us to Pitztal Valley, with its surrounding mountain peaks, ski slopes, and glaciers, through an ethnography of snow and "vertical globalization." In the valley, the melting glacier is the preeminent symbol of the rapid process of global warming. Shifting meteorological conditions are affecting daily lives, relied-on landscape, and the identities of hundreds of ski resort workers. The rising temperature is at the same time changing what used to be a hospitable and predictable future with long-assured economic security into one that is no longer guaranteed.

Chapter 5

Climate Change and Mitigation in Bangladesh

Vulnerability in Urban Locations

Tasneem Siddiqui, Mohammad Jalal Uddin Sikder, and Mohammad Rashed Alam Bhuiyan

Introduction

The 2014 report on the Intergovernmental Panel on Climate Change (IPCC) acknowledges that, during the twenty-first century, climate change is projected to increase the displacement of people from both rural and urban areas, particularly in developing countries and among low-income groups (IPCC 2014b: 20). Households with a poor resource base situated in rural areas are more vulnerable to the shocks, stresses, and negative impacts of climate-related events (Warner et al. 2012; Siddiqui and Billah 2014; Islam and Siddiqui 2016). Bangladesh is one of the most vulnerable countries facing the adverse effects of climate change. The International Displacement Monitoring Centre (IDMC) data show that, from 2009 to 2014, around four million Bangladeshis were displaced due to different hazards (IDMC 2015). Based on a comparison of the census results regarding population growth in different districts between 2001 and 2011, Black, Kinveton, and Schmidt-Verkerk (2013) projected that, from 2011 to 2050, as many as sixteen to twenty-six million Bangladeshis will migrate from their place of origin due to flooding, storm surges, river bank erosion, and rising sea levels. This includes both the displaced population as well as labor migrants.

Notes for this chapter begin on page 127.

The aim of this chapter is to develop an understanding of those who have moved to urban locations in the context of climate change and assess their gains and vulnerabilities in urban settings. The major research questions pursued in this study include, have the migrants or their households, through migration, been able to resolve some of the challenges that led them to decide to move in the first place? Can their migration experience be termed a successful adaptation practice?[1] How integrated are the new migrants in the urban locations? Do the migrants face new forms of vulnerabilities in their urban setting? What are the sources of those vulnerabilities? How do they resolve these?

This chapter is divided into six sections. The first section presents the aim of the chapter, outlines the major research questions and conceptual framework, and also describes the methodology employed in the research. The second section reveals various types of migration experience of the respondents. It also explains the situation under which the migrants or their households took the migration decision. The third section presents an assessment of the job and income situation of the migrants in the city. The fourth section evaluates their living arrangements in different informal settlements. The fifth section examines other human security concerns of the migrants. The sixth section draws the major conclusions and also provides some modest recommendations.

Conceptual Framework

This chapter is based on the understanding that climate change does not displace people directly but, rather, exacerbates various forms of vulnerability that contribute to displacement (Kolmannskog 2012). It also draws on the study by Jayawardhan (2017) that demonstrates that anthropogenic climate change affects most of the inhabitants of a community, yet socio-economic inequalities make marginalized groups more vulnerable to it. When the vulnerability of affected persons reaches the threshold where life and livelihood in the areas of origin become unsustainable, then the individuals concerned are forced to leave their habitual residence (De Compos et al. 2019). This study does not explore the climate change–induced migration experiences of households from a relatively better-off economic background but concentrates on relatively poorer households who have been residing in various informal urban settlements in three cities. To understand the achievements and vulnerabilities, the analysis is based on the studies of Craig (2015), Uddin (2018) and Adger et al. (2020). This group of scholars finds that migration responses to climate change provide a new roof over the head of migrants while also offering new jobs to many migrants, but at the same time they also expose migrants to new forms of vulnerabilities[2] in the destination areas. These studies show that migrants living in the urban areas tend to be located in low-lying, environmentally hazardous regions, coupled with inadequate facilities in terms of food, shelter, sanitation, and healthcare. The extent of this vulnerability differs among the migrants across locations, gender, age, ethnicity, region of origin, and income/occupation groups (Jones, Mahbub, and Haq 2016).

Craig (2015), Uddin (2018), te Lintelo et al. (2018), Rahman, Haughton, and Jonas (2010), Siddiqui et al. (2021), and Williams et al. (2016), who have studied informal settlements in Mumbai, Shanghai, Dhaka, Chittagong, and Durban, demonstrate that, in those areas, the residents face severe flooding, poor water quality, the prevalence of mosquitos, health vulnerability such as vector-borne diseases, exposure to communicable diseases, damp housing conditions, inadequate sanitation facilities, and so on. Rahman, Haughton, and Jonas (2010) and Chen, Chen, and Landry (2013) find a poor physical environment with a nonexistent solid waste disposal system in most of the settlements, resulting in the high prevalence of (water-borne) diseases among the children. Foresight (2011), Akter (2012), Sinthia (2013), and Randall (2013) reveal that households opted for migration to reduce their immediate social, financial, and environmental pressures and improve their capacity to manage shocks and stresses, and so they are likely to be trapped in their destinations and exposed to other forms of vulnerabilities. Recently, Adger et al. (2020) find that migrants think that they are better off and happier in their urban setting. Black et al. (2013) argue that, with or without the influence of climate change, some people will move to the urban areas for job opportunities. They can enjoy a better life when the urban infrastructures are developed and climate change adaptation funds can provide a source of finance for materializing this.

Methodology

We draw on data generated by the Refugee and Migratory Movements Research Unit (RMMRU) for a study on "Adaptation Strategies of Poor Urban Migrants in the Context of Climate Change: A Case Study of Informal Settlements," supported by the Deutsche Gesellschaftfür Internationale Zusammenarbeit (GIZ) GmbH. The research adopts a mixed-method approach. A particular interest of the study was to understand the nature of internal migration from rural areas to nearby cities. It deliberately avoided megacities, as a few large-scale studies have been conducted on conditions of migrants working in the megacities Dhaka and Chattogram. The three selected sites are situated in the northwestern region of Bangladesh.[3] The Millennium Development Goal (MDG): Bangladesh Progress Report 2008 notes that the northwestern region of Bangladesh carries a higher burden of poverty due to naturally triggered disasters (especially riverbank erosion and flooding) and is subject to a local phenomenon called *Monga*, interplay between a lack of jobs and food entitlement. The report identified that these two key factors contribute significantly toward creating a major rural urban migration flow from that region (GoB 2009).

Through conducting key informant interviews (KIIs) with members of the municipal corporations, government functionaries,[4] and inhabitants of the three cities, a total of twenty-three informal settlements were identified.[5] The research team conducted a quick survey of around twenty-six households in each informal settlement. The survey started from the left side of each settlement and enlisted every fourth house, stopping once twenty-six households had been listed. In total, the survey covered six hundred migrant households. By applying stratified random sampling, sixty-eight migrants were selected for in-depth interviews from the larger sample (six hundred). The fieldwork of this study took place over a three-month period (July–October 2014). Each of the selected households was interviewed twice. The initial interview provided specific information on rural-to-urban migration. The subsequent interview provided the householders with an opportunity to reflect further on their gains, risks, and vulnerabilities within the urban settlement.

Why Migrate?

The survey results of six hundred dwellers of informal settlements in Rajshahi, Natore, and Sirajganj reveal that almost all of the migrants in urban locations come from a very poor economic background. They hardly have any formal education, and almost all of them are married. Their current average family size is 4.26 members, 90 percent of them initially migrated at a very young age (22 years old), and at least half of them are now settled in the city and have their families with them. Some of them are temporary migrants, and a section of these are cyclical migrants. Their family remains behind in the rural areas when they migrate every month. Another section stays in an urban location, but these return home during the harvest season. The members of a section of these families engage in seasonal migration at a particular time to work in the brick kilns for six months. For a number of these, this is not their first destination. They previously moved to another urban location before arriving at their current one. During the interviews it was revealed that, on average, they have around sixteen years of migration experience.

For the majority, migration was not a sudden decision, and almost half of the migrants or their families had been considering migration as an option for a long time. The study finds that climatic stress is intertwined with the economic and social motivations underlying migration. One-third of the respondents had lost their homestead due to climate stresses (flood, drought, and riverbank erosion) as well as poverty. Eighty percent of them do not possess any agricultural land. The adult members of the households are desperate to earn an income, especially in the case of female-headed households. Abdul Latif's case sheds some light on how highly complex micro-, macro-, and mesolevel factors interact to produce the migration decision. Latif is originally from Chapainawabganj (in the northwestern district of Bangladesh), and he migrated, more or less permanently, to the city of Rajshahi with his wife and two sons seven years ago. He did not inherit any agricultural land from his father, and the only land he owns is homestead land. His village lies next to the India/Bangladesh border. In the village, he was unemployed. He, along with a few other villagers, used to work as an agricultural laborer on the other side of the border. He would cross the border in the morning, work all day, and then return across the border in the evening. Local people have always done this, long before the partition of India. Fifteen or sixteen years ago, the border security was tightened, and he and other workers faced difficulty in continuing to work. Then, he and his cousin joined a local *thikadar*,[6] who was himself a worker, and went to Rajshahi to engage in road work. For several years, he continued to work there for months at a time before returning home. In the meantime, he lost his homestead land to river erosion. For a further three years or so, he, his wife, and his children lived on land belonging to one of his relatives while he migrated back and forth in order to work in Rajshahi. The river eventually also submerged his relative's land, so, along with a few other families, the family moved to Rajshahi permanently. A few of his neighbors from the village accompanied him. For the last seven years, he has lived in Rajshahi in the Bhodra lakeside settlement.

In drought-affected areas, those with agricultural land faced difficulties in continuing with cultivation due to the downward trend in the groundwater level. Rich households could afford to continue to cultivate the land by paying a higher price for irrigation water. Families also were prompted to migrate if they had extended family or neighbors living in urban locations who passed on information about available jobs or offered an initial place to stay. In Rajshahi, like Latif, several other respondents also initially migrated with the support of *thikadars*. When asked why they chose this particular city, the main reasons inevitably included its proximity to their place of origin and access to information about the city through their social network.

Fear of the unknown also deterred some from moving longer distances to cities. Those who still had family members close by preferred not to travel too far in order to safeguard their house and family who remained in the rural areas. This explains why the majority of the migrants (64 percent) are from the same or neighboring districts. The possibility of having a roof over their head and access to a livelihood plus the availability of jobs in the urban locations are among the major driving forces behind their migration decisions. Nonetheless, this economic desperation is influenced or, to some extent, caused by climate change, disasters, and social as well as demographic circumstances. Moreover, the pull factors such as rapid urbanization and the consequent need for labor force play an equally important role to the push factors in the areas of origin. Mediating factors, such as access to social networks at the destination, also play a role. The following sections examine the situation of migrants in urban locations with the aim of understanding how many of the targets (that motivated them to migrate) have been achieved. We will analyze this on five grounds, namely: access to jobs, access to housing, access to drinking water, access to hygienic sanitation, and access to electricity and personal security.

Access to Livelihoods at the Urban Locations

Generally, through migration, most migrants manage to find work. A large number of them work in the service sector. Their jobs include: rickshaw, auto rickshaw, and van puller; bus and auto-tempo driver or conductor; hotel/restaurant boy; garage/workshop laborer; cook; cleaner; security guard; carpenter; scrap collector; private tutor; etc. In Chala *upazila*, a subdistrict of Sirajganj, some migrants work in the loom industry, while others are employed in the rice-processing mills, flour mills, jute mills, etc. Members of some of the households of the Haiadarpara railway colony settlement of Sirajganj seasonally migrate to other districts, traveling as far as Chattogram, Rangamati, and Khagrachari (southeastern districts) to work in the brick kilns for a period of six months at a stretch under a contractual arrangement. They also participate in earthwork. During the harvest period, the members of some of these households migrate to Tangail, Manikganj, Jamalpur, Sylhet, and Sunamgonj. A small percentage of the members of these households have migrated permanently and are employed in the readymade garment and other manufacturing and textile industries in Dhaka, Savar, Konabari, Gazipur, Narayangonj, and Narshingdi. The income of households with one or more seasonal outmigrant is comparatively higher and more secure compared with other types of migrant households. Although the migrants find jobs in the urban areas, the majority of these positions, apart from a few, are informal sector jobs. The level of income earned from these jobs is very low. These migrants' average monthly household income is only Bangladeshi Taka (BDT) 7,382 (US$96). With an average family size of 4.26, most of these households live below the poverty line, so the question is whether they migrated only for a job or whether they wanted a decent job with a reasonable salary that would allow them to live at least above the poverty line.

Now let us try to understand the nature of the jobs performed by some of these migrants. Getting a job is not a spontaneous event. Cyclic migrants do sometimes face problems finding work. It is not always assured. Some of them wait a few days, even up to a week, before securing employment. During our interview with Rafiq in Shatopolli Adarsha Gram of Natore, he stated that he had been waiting as long as twenty days. In some instances, the migrants even return home without finding work. Consequently, the migrants suffer financial losses while seeking employment. They have to bear their food and accommodation expenses. Most of those who have remained in the city permanently are not engaged in stable jobs. They face termination without notice. If, for any reason, a migrant cannot find work, he or she cannot earn anything that day.

Some of the tasks performed by the migrants are unhygienic and injurious to health. The risks vary according to the type of job performed. Those who work in the tanneries of Sirajganj are extremely vulnerable to breathing problems, asthma, and skin disease. Those who work as scrap and metal collectors, particularly in Rajshahi, gather waste items such as, torn sandals, dust, hair, polythene, broken glass, plastic bottles, etc. The sharp edges of these materials make them prone to suffering accidents, and they are also exposed to broken glass, syringes, and other toxic materials. These families also involve their children in their scrap collection work.

Now we will discuss the conditions of the migrants' residence in the urban areas and attempt to understand the risks associated with the urban living conditions.

A Roof over My Head

Forty years ago, a few migrants started one of the first informal settlements named Uttar Patuapara settlement in Natore. It is located by the side of a long canal that passes through the municipality. The canal is connected to the Padma River. Over the years, the canal has become narrow. This is due both to the loss of water flow caused by Farakka Barrage[7] on the Bangladesh-India border and also to the unauthorized land grabbing by some locally influential people. The inhabitants of this settlement have never experienced eviction. Although temporary, the settlement has electricity, taps with drinking water, and sanitation facilities.

The migrants who have been living in informal settlements on railway land have experienced multiple evictions. With the help of influential politicians, subsequently, they have returned to the same area and been able to rebuild their shanties. The aim of local politicians in helping this low-income population is to ensure their support at political events such as processions and rallies. Permanent migrants also try to maintain good relationships with politicians as they try to secure a national voter identity card (ID) from the location where they are currently residing. The local politicians assist them with this endeavor in order to use them as a voter bank.

Migrants are keen to obtain an ID in their current place of residence in order to access the services provided by the government's social safety net measures. This is necessary also in order to obtain loans from the local banks or nongovernmental organizations (NGOs). One of the sites of Natore is a good example of a government initiative for ensuring safe, sustainable shelter for migrants. The government allocates land, and the migrants build houses on it. These houses are well built. The NGOs, under a microcredit program, provide loans to these migrants. Some also borrow from moneylenders. Those who live on government-allocated land do not face the fear of eviction. In the words of a fortunate migrant, they have not only been able to ensure a stable house but have also increased their position in society. A female head of a migrant household in Natore stated, "Please don't call us slums-dwellers . . . we have prestige. . . . My elder son is now studying at BUET [Bangladesh University of Engineering and Technology, one of the top universities in Bangladesh]. We arranged a good marriage for my daughter. . . . They are now living a happy life . . ."

In Rajshahi, the main settlements are located beside the Padma River, adjacent to the railway lines. Recently, the city government invested in the renovation of the city, and many migrants have been displaced because of these development initiatives. The city mayor states that he plans to build houses for those who have been displaced. During our interviews with the migrants, they stated that since eviction they have been living by the roadside and on public and private vacant land. In Sirajganj, the migrants mostly settle on the banks of the Jamuna River, on the side of the embankment, and on unused railway property. The local residents also build shanties on unused government land, and migrants pay rent to live in these shanties. Migrants who live on the embankment and by the roadside feel insecure because of the location of their houses. Another migrant, Nadia, also from the Bongram settlement in Rajshahi, stated, "I'm scared that my young son might have a road accident. . . . When I hear the heavy trucks moving at night, I wake up scared that they might crash into our shelter and kill us." These comments reflect the migrants' different types of vulnerabilities.

Two kinds of shelter exist in all locations. The relatively better ones are made of tin and have a concrete floor. The poorer ones are constructed of mud, thatch, or bamboo. All of the houses are very small, being 266.34 square feet on average. Over 70 percent of the households do not own the house in which they currently live. As the migrants are aware that they might be evicted at any time, they are not willing to spend large sums of money building their houses, or on or long-lasting repairs. The floating migrants build their shanties out of cheap, low-quality materials like bamboo and plastic. Every year, the residents need to spend money to repair their shanties because these cheap materials decay so quickly. They also have to repair their shanties particularly after storms. Nonetheless, some of them consider themselves lucky to have a roof over their head. Nikheel Kanti resides in a very poor-quality shanty but thinks that, in the future, his family will be able to move into a better one.

Environmental concerns are also significant in the urban locations. In all three areas under consideration, migrants settle at the riverside, which is prone to flooding and river erosion. In these areas, due to the low quality of housing materials, these migrant households suffer badly during extreme weather conditions. One migrant Abdul, from the informal settlement located beside the Rajshahi Railway Station, stated ironically, "Weather cuts both ways for us. During the winter we suffer from cold spells and during summer from extreme heat." Migrants also face human-made environmental stresses in their dwellings. Rojina, the wife of a migrant who resides in the Padma residential area of Rajshahi, stated that "in the last twenty years, first, I lived with my parents and now with my husband's. We never saw strong thunderstorms like we had last year that caused great damage to our dwelling. We had to find extra money to do the repair work. Now, we must always be prepared to rebuild our shanty again in cases of excessive rain and storms." In Natore's Shatopolli Adarsha Gram, those living at the side of the canals of Turag River are exposed to industrial waste produced by the Carew chemical company. Due to chemical reactions, their tin sheds become rusty very quickly. The dwellers have lost their ability to catch fish due to water pollution.

Some of the migrants identified other forms of insecurity in their settlements. Children's exposure to drugs is one of these. Lipi of Shohidgonj settlement of Sirajganj has two small children, and she stated that she would like to move to a more secure place where her children would not be exposed to drugs. In order to obtain protection, migrants need to have a good relationship with the local politicians. These politicians, in exchange, want them to participate in political events. Taking part in these events can be risky. A few women migrants in Sirajganj and Natore, on the other hand, are concerned about their physical and sexual security. Women also expressed the insecurity of leaving their young children indoors when they are working outside. Lata leaves her daughter unattended while she goes to work in a nearby house. She stated: "My employer told me there are bad people who kidnap children." Sultan, the wife of a rickshaw puller, lives with her family in Dhulipara settlement of Sirajganj. She stated: "I'm scared during the night when my husband's busy pulling a rickshaw. My two daughters are unsafe, as an unknown man with a bad intention may enter our home. . . . One day, I was terrified when two gangs were fighting each other and suddenly damaged my house." Compared to male migrants, women migrants as well as male migrants' wives expressed more concern about the physical and sexual security of their children.

Access to Drinking Water, Sanitation, and Electricity

Tube wells remain the predominant source of drinking water for migrant households in all of the locations. Pipe water is available in a few locations. In Sirajganj, those who live on the riverbank find it difficult to obtain safe drinking water during the monsoon floods. Some of the tube wells become submerged by water, forcing migrants to travel a long distance to obtain drinking water and spend extra money for a rickshaw or boat to transport the water. Sadeq, a migrant from Zelepara settlement of Sirajganj, stated: "There are houses that have tube wells, but the owners are unwilling to allow us to use them as this may reduce the longevity of the pump. They lock up their tube well with iron chains."

In Rajshahi, it is also difficult to access safe drinking water. However, those living in the city center have access to the city corporation's water supply. Sometimes, they also visit houses in the nearby residential areas to obtain water that people kindly provide. But during the summer it is very difficult to obtain water because the water supply also falls. One migrant from Borda lakeside settlement of Rajshahi has access to a tube well. He revealed a new problem, as it seemed to him like the groundwater level had fallen, so for the last five to six years, during some seasons, it became difficult to obtain water through the tube well. Recently, the Natore municipality has begun to supply pure drinking water in the summer. The local NGO officer and key informants state that Natore does not suffer from a shortage of water as the area has one of the largest water bodies of the country named "Chalan Beel."[8] However, at the three study sites, the conditions of the tube wells are very poor, and some are out of service. People there are waiting for influential political leaders or government officials to set up new tube wells. Again, the problems associated with accessing safe drinking water are worse for those living at the riverside than those living in the city centers.

The sanitation system has been identified as one of the biggest problems in the settlement areas. The houses that are built or rented by the migrants on the government-allocated land in Natore have proper water-sealed toilets. The rest of the settlement dwellers face a shortage of toilets. More than half of the respondents use a common (community) latrine. In Sirajganj, those living at the riverside have only a few community toilets for a large number of users. For those living on the embankment, it is very hard to find space to set up new latrines. In some cases, they use the riverside as their toilet. Many households have created a temporary toilet beside their kitchen. These toilets have grave consequence for hygiene, as some of these areas are flooded very easily. In Rajshahi, all of those who live by the roadside use open spaces, and others use private and community latrines. Migrants who have received a voter ID are now lobbying the influential political leaders to set up latrines for them. Not only is the number of toilets insufficient but their quality is also poor. In all three locations, half of the latrines are not water sealed. Lima, the wife of a migrant from the Dadapur road settlement of Natore, stated that "my daughter is two years old. She suffers off and on from an upset stomach. When we visit our village, she suffers less. One of my neighbors told me not to allow her to play in the common open space, as it is too close to the toilets. I think she's right. The community toilets here are open latrines."

More than 80 percent of the respondents have electricity in their houses, while 20 percent use kerosene for lighting. Those who live on the riverbank in Sirajganj cannot access electricity, while those who live on the embankment do so via illegal connections and must regularly bribe the local power supplier. In Rajshahi, those who live in the city areas have electricity, but those who live at the riverside do not. In Natore, most houses can access electricity. When the migrants were asked to reflect on what they liked most about urban living, access to electricity is appreciated the most by those who have it, as, according to them, it helps their children to study; however, power cuts are common. Nafisa from the Narodnodi settlement of Natore complained that power cuts occur at odd times, usually when the children need to study.

Summary and Conclusions

This chapter has attempted to shed light on the lives and livelihood of climate-affected poor internal migrants living in informal urban settlements. The aim of this chapter was to understand their perceptions of their gains and losses as a result of migrating. In other words, it has attempted to assess their evaluation of the outcome of their migration decision: did they think that they would be able to use migration as a successful adaptation tool to overcome some of the climate-related and other stresses they faced in their areas of origin that led them to migrate in the first place? In retrospect, was this the best choice? What are the challenges and vulnerabilities that they face in their new urban life? More importantly, do they feel that those vulnerabilities can be reduced?

The chapter began by exploring the role of climate change–related stresses in inducing migration. It became obvious from the statements of the migrants and their life partners that their migration decision has been a culmination of the interplay of many complex factors, including macrolevel realities such as the hurried, unplanned border settlement between India and Bangladesh by the British rulers during the partition of India, where the agricultural land is located on one side and the villages where people live are located on the other side; individuals' experience of the loss of their homestead land to river erosion or the financial inability of an individual to continue farming due to the increased cost of irrigation, against a backdrop of the slow-onset climate change of increased drought and its effect in lowering the groundwater level; exposure to information about job opportunities in urban areas; and access to help provided by the *thikadars*. The study finds that migration to urban locations has provided better job opportunities for the migrants—they can earn an income by pursuing various trades, although these generate modest sums for their subsistence. The migrants perceive this as fulfillment of the most important target of their migration, which is income adaptation. Those who lost their homestead land can now have a roof over their head. Some of their children attend school, while others have even managed to send their children to the best universities in the country. To them, these are great achievements compared to what would have happened if they had remained in their villages. Given all of their options, the majority was of the opinion that moving to the urban location was a better choice than staying put in their villages. But do they think that it is enough for them? Perhaps not—they aspire to many other things.

With some exceptions, these migrants are predominantly employed in the informal sectors, and their average income is around US$96. The average family size is 4.26 people. A simple headcount calculation would show that the majority of them live below the poverty line. Some jobs are hazardous, while others expose them to respiratory, skin, and many other types of diseases. The jobs that they perform are essential for the city, but their working conditions are poor. They live with a dream of formalizing their services to the government. They also hope for the passage of a minimum wage policy, as they wish to enjoy basic labor rights.

On average, the size of the current homestead of these migrants is only 270 square feet. A large number of migrants live with the constant fear of eviction. One-third have gradually moved into improved housing, and the rest also aspire to move out of their current shanties. Many new migrants end up living in environmentally unsustainable areas. Those who live in the riverside settlements regularly experience flooding, polluted water, bad odors from garbage, etc. Dwellers in some of the settlements currently cannot access safe drinking water year round. In the established settlements, the migrants have access to hygienic toilets, but, for many others, the insufficient number of toilets and unhygienic positioning of those that do exist have exposed many to health problems. Migrants who have access to electricity deeply appreciate it and stress its importance for their children's education. Unfortunately, many of them fear that the electricity lines might be disconnected at any time, as they are illegally installed.

The study also reveals that the experience of living in the informal settlements varies based on the location and gender of the respondents. In some locations, access to drinking water is a huge problem, while in others it is not; similarly, the situation regarding access to electricity is also location specific. Major differences have been observed among men and women when asked to prioritize the problems in their urban lives. Men stress job-related problems, while the women focus more on problems linked to personal safety and security, sexual harassment, a lack of childcare facilities, etc.

An important finding of the research is that the migrants do not always passively take whatever is available in the settlements but are willing to accept challenges to improve their situation. On various occasions, certain migrants have demonstrated their agency. Through influencing the local politicians or city mayor, they have been able to introduce electricity into their dwellings, or negotiate water-sealed toilets, and also lobby for the installation of new tube wells.

This leads to the conclusion that, based on the macro- and microlevel individual realities, influenced by the environmental, demographic, social-economic, and political surroundings operational in both the origin and destination locations, and mediated by intervening agencies and networks, a percentage of the villagers would stand by their decision to migrate. More importantly, migration can be one of the many adaptation choices for people faced with climatic stresses. To make migration choices more effective, the traditional policies of regarding rural-to-urban migration as a threat or problem of urbanization should be replaced with a more accommodative policy framework. Instead of explaining rural-to-urban migration only from the push perspective, the urban pull factors should also be taken into consideration. The growth hubs of Bangladesh are located in the urban areas. These growth centers require a workforce. That is one of the reasons why the country is rapidly becoming urbanized. Therefore, urban planners need to think in terms of creating space for service providers. The study has particularly explored alternative cities instead of the two megacities of Bangladesh. This has allowed us to argue for the decentralization of the growth centers to avoid overcrowding in a few cities. If the country is committed to achieving the Sustainable Development Goals (SDGs) target 11, related to creating safe, sustainable, and inclusive cities for all, then policies must be in place to eliminate precarious work and the living conditions of new migrants in the urban locations.

Acknowledgments

We acknowledge the contributions of the Deutsche Gesellschaftfür Internationale Zusammenarbeit (GIZ), Bangladesh, for supporting the Refugee and Migratory Movements Research Unit (RMMRU) to conduct the research on "Adaptation Strategies of Poor Urban Migrants in the Context of Climate Change: A Case Study of Informal Settlement in Natore, Sirajgang and Rajshahi." We are also thankful to the RMMRU for providing us opportunities to carry out the research, on which this chapter is based.

Dr. Tasneem Siddiqui is professor of political science, University of Dhaka, and founding chair of the Refugee and Migratory Movements Research Unit (RMMRU), Bangladesh. She advises research projects of the RMMRU. She is known nationally, regionally, and internationally for her insightful works. She has vast experience in migration, gender, public-migration policy, administrative and policy reform in the migration sector, and other relevant issues. She has written pioneering works on the migration–climate change nexus, migration-gender nexus, return migration–skill development nexus, etc. She is deft on advocacy issues.

Dr. Mohammad Jalal Uddin Sikder serves as an associate professor of development studies at the Daffodil International University (DIU), Bangladesh. He also engages as an adjunct senior research fellow at the RMMRU.

Sikder has also been involved in a number of studies under the Migration out of Poverty RPC, University of Sussex, United Kingdom. He was a recipient of the NTS-Asia Research Fellowship from Nanyang Technological University, Singapore. His research interests include labor migration recruitment processes and returnee reintegration; remittances, development, dependency, and inequality; irregular cross-border migration, human smuggling, and trafficking; migration and TVET skills for employment; forced migration and Rohingya (Myanmar) refugees' situation; climate change–induced migration; and urban resettlement and livelihood. Sikder has coauthored a book titled *Remittance Income and Social Resilience of Migrant Households in Rural Bangladesh* (Palgrave MacMillan, 2017).

Mohammad Rashed Alam Bhuiyan is a PhD student at the Centre for Trust, Peace and Social Relations (CTPSR), Coventry University, United Kingdom. He will be studying "The Role of Intermediaries in Understanding the Relations of Migration (In)equality and Development" under the UKRI-GCRF-funded Migration For Development and Equality (MIDEQ) project (www.mideq.org). He is also a faculty member of the Department of Political Science at the University of Dhaka and adjunct research fellow of RMMRU. Recently he did an MSc in Sustainable Development from the University of Exeter with Commonwealth Shared Scholarship. His expertise lies in studying various forms of internal and international migration, adaptation to climate change, sustainable and inclusive urban development, and migrants' rights and well-being. Previously, he served on the RMMRU as a research associate. He has worked across several research projects in collaboration with the Universities of Exeter, Sussex, Southampton, and other research units, primarily focused on international labor migration, internal migration (climate-induced migration and displacement), adaptation to climate change, sustainable cities, and inclusive urban development.

Notes

1. In this chapter, we follow the definition of adaptation employed by Working Groups II and III in the IPCC (The Intergovernmental Panel on Climate Change) *Fifth Assessment Report*. They define the term as "the process of adjustment to actual or expected climate and its effects." IPCC is the United Nations body entrusted with periodical evaluation of the scientific premise of climate change, its associated impacts and risks ahead, and solutions for adaptation and mitigation. Since its inception in 1988 it has been producing state of climate change–related

assessment reports periodically. IPCC *Fifth Assessment Report* (AR5) published in 2014 presented the climate change–related scientific knowledge, new results, and updates since the publication of the IPCC *Fourth Assessment Report* (AR4) in 2007. The IPCC Working Group II (WGII) assesses the impacts, adaptations, and vulnerabilities related to climate change. The IPCC Working Group III (WG III) mainly deals with climate change mitigation, or assessing methods for reducing greenhouse gas emissions in the atmosphere. For details see, IPCC 2014a.

2. The term "vulnerability" describes the degree of exposure to risk, stress, and shock as a result of the adverse effects and consequences of the livelihood and socioecological systems (Masten 1994: 7; Adger 2000: 348).
3. Bangladesh is divided into seven divisions: Dhaka, Chittagong, Rajshahi, Sylhet, Barisal, Khulna and Rangpur, each of which is itself subdivided into 64 districts or *zila*. These districts are further subdivided into 493 sub-districts or *upazila*, each with its own police station (except for those in the metropolitan areas), and are further divided into several *unions* made up of multiple villages.
4. Interviews were held with the mayors and ward commissioners of Natore Municipality and Sirajganj Municipality and the chief executive officer and ward commissioners of Rajshahi City Corporations.
5. In Natore Municipality area: Shatopolli Adarsha Gram, Uttar Patuapara, Jhautola, Kandivita, Moddhopara, Dadapur road, Station Rail, Mirpara Narod Nodi, and Chalan Beel. In Sirajganj Municipality area: Haidarpara Railway Colony, Kamrapara, Dhulipara, Chowdhuripara, Zamtola, Notun Vangabari, Shohidgonj, KhademerPul, Zele Para, and Rishi Para. In Rajshahi corporation area: Bhodra lakeside, Bongram, Padma Residential area, and Rajshahi railway station.
6. A contractor or middleman.
7. Farakka Barrage was installed across the Ganga River located (Murshidabad district Indian state of West Bengal), roughly 18 kilometres far from the Shibganj border of Bangladesh.
8. Chalan Beel is a wetland in Bangladesh and located across Singra and Gurudaspur *upazilas* of Natore District as well as its outspreads in some *upazilas* of Pabna and Sirajganj Districts.

References

Adger, William Neil. 2000. "Social and Ecological Resilience: Are They Related?" *Progress in Human Geography* 24(3): 347–67.

Adger, William Neil, Ricardo Safra de Campos, Tasneem Siddiqui, and Lucy Szaboova. 2020. "Commentary: Inequality, Precarity and Sustainable Ecosystems as Elements of Urban Resilience." *Urban Studies* 57(7): 1588–95.

Akter, Sonia. 2012. "The Role of Microinsurance as a Safety Net against Environmental Risks in Bangladesh." *Journal of Environment & Development* 21(2): 263–80.

Black, Richard, Dominic Kniveton, and Kerstin Schmidt-Verkerk. 2013. "Migration and Climate Change: Toward an Integrated Assessment of Sensitivity." In *Disentangling Migration and Climate Change*, edited by Thomas Faist and Jeanette Schade, 29–53. Berlin: Springer.

Chen, Juan, Shuo Chen, and Pierre F. Landry. 2013. "Migration, Environmental Hazards, and Health Outcomes in China." *Social Science & Medicine* 80: 85–95.

Craig, Gary. 2015. "Migration and Integration: A Local and Experiential Perspective." Birmingham: Institute for Research into Superdiversity, University of Birmingham.

De Campos, Ricardo Safra, Samuel Nii Ardey Codjoe, W. Neil Adger, Colette Mortreux, Sugata Hazra, Tasneem Siddiqui, Shouvik Das, D. Yaw Atiglo, Mohammad Rashed Alam Bhuiyan, Mahmudol Hasan Rocky, and Mumuni Abu. 2019. "Where People Live and Move in Deltas." In *Deltas in the Anthropocene*, edited by Robert J. Nicholls, W. Neil Adger, Craig W. Hutton, and Susan E. Hanson, 153–71. Cham: Palgrave Macmillan.

Foresight. 2011. "Migration and Global Environmental Change: Final Project Report." London: The Government Office for Science.

GoB (Government of Bangladesh). 2009. *Millennium Development Goals: Bangladesh Progress Report 2008*. Dhaka: General Economics Division, Planning Commission, Government of Bangladesh.

Hunter, Lori M., Jessie K. Luna, and Rachel M. Norton. 2015. "Environmental Dimensions of Migration." *Annual Review of Sociology* 41(1): 377–97.

IDMC (International Displacement Monitoring Centre). 2015. *Global Estimates, 2015: People Displaced by Disasters*. Geneva: International Displacement Monitoring Centre (IDMC).

IPCC (Intergovernmental Panel on Climate Change). 2014a. *Climate Change 2014: Synthesis Report. Contribution of Working Groups I, II and III to the Fifth Assessment Report of the Intergovernmental Panel on Climate Change*. Geneva: Intergovernmental Panel on Climate Change (IPCC).

———. 2014b. *Planned Relocation as an Adaptation Strategy*. Geneva: Working Group II, the Intergovernmental Panel on Climate Change (IPCC).

Islam, Mohammad Towheedul, and Tasneem Siddiqui. 2016. "Migratory Flows in Bangladesh in the Age of Climate Change: Sensibility, Patterns and Challenges." In *Refugees and Migration in Asia and Europe*, ed. Beatrice Gorawantschy, Megha Sarmah, and Patrick Rueppel, 49–66. Singapore: Konrad-Adenauer-Stiftung.

Jayawardhan, Shweta. 2017. "Vulnerability and Climate Change Induced Human Displacement in Consilience." *Consilience: The Journal of Sustainable Development* 17(1): 103–42.

Jones, Gavin, A. Q. M. Mahbub, and Md Izazul Haq. 2016. "Urbanization and Migration in Bangladesh." Dhaka: United Nations Population Fund (UNFPA).

Kolmannskog, Vikram. 2012. "Climate Change, Environmental Displacement and International Law." *Journal of International Development* 24(8): 1071–81.

Masten, Ann S. 1994. "Resilience in Individual Development: Successful Adaptation Despite Risk and Adversity." In *Risk and Resilience in Inner City America: Challenges and Prospects*, edited by Margaret C Wang and Edmund W. Gordon, 3–25. Hillsdale, NJ: Erlbaum.

McLeman, Robert, and Berend Smit. 2006. "Migration as an Adaptation to Climate Change." *Climatic Change* 76(1–2): 31–53.

Rahman, M. Maksudur, Graham Haughton, and Andrew E. G. Jonas. 2010. "The Challenges of Local Environmental Problems Facing the Urban Poor in Chittagong, Bangladesh: A Scale-Sensitive Analysis." *Environment and Urbanization* 22(2): 561–78.

Randall, Alex. 2013. *Climate Change Driving Migration into China's Vulnerable Cities*. London: Chinadialouge.

Siddiqui, Tasneem, Lucy Szaboova, William Neil Adger, Ricardo Safra de Campos, Mohammad Rashed Alam Bhuiyan, and Tamim Billah. 2021. "Policy Opportunities and Constraints for Addressing Urban Precarity of Migrant Populations." *Global Policy*. Special Issue: *Precarity, Mobility and the City* 12(S2): 91–105.

Siddiqui, Tasneem, Mohammad Rashed Alam Bhuiyan, Dominick Kniveton, Richard Black, Mohammad Towheedul Islam, and Maxmilan Martin. 2018. "Situating Migration in Planned and Autonomous Adaptation Practices to Climate Change in Bangladesh." In *Climate Change Vulnerability and Migration*, edited by S. Irudaya Rajan and R. B. Bhagat, 119–46. London: Routledge.

Siddiqui, Tasneem, and Motasim Billah. 2014. "Adaptation to Climate Change in Bangladesh: Migration the Missing Link." In *Adaptation to Climate Change in Asia*, edited by Sushil Vachani and Jawed Usmani, 117–41. Cheltenham: Edward Elgar.

Sinthia, S. Ahmed. 2013. "Sustainable Urban Development of Slum Prone Area of Dhaka City." *World Academy of Science, Engineering and Technology* 7(3): 328–35.

Tacoli, Cecilia. 2009. "Crisis or Adaptation? Migration and Climate Change in a Context of High Mobility." *Environment and Urbanization* 21(2): 513–25.

Te Lintelo, Dolf J. H., Jaideep Gupte, J. Allister Mcgregor, Rajith Lakshman, and Ferdous Jahan. 2018. "Wellbeing and Urban Governance: Who Fails, Survives or Thrives in Informal Settlements in Bangladeshi Cities?" *Cities* 72 (Part B): 391–402.

Uddin, Nasir. 2018. "Assessing Urban Sustainability of Slum Settlements in Bangladesh: Evidence from Chittagong City." *Journal of Urban Management* 7(1): 32–42.

Warner, Koko, Tamer Afifi, Kevin Henry, Tonya Rawe, Christopher, Smith, and Alex De Sherbinin. 2012. "Where the Rain Falls: Climate Change, Food and Livelihood Security and Migration; An 8-Country Study to Understand Rainfall, Food Security and Human Development." Barcelona: Care France, United Nations University Institute for Environment and Human Security (UNU-EHS), Center for International Earth Science Information Network at the Earth Institute of Columbia University.

Williams, Liana J., Sharmin Afroz, Peter R. Brown, Lytoua Chialue, Clemens M. Grünbühel, Tanya Jakimow, Iqbal Khan, Mao Minea, V. Ratna Reddy, Silinthone Sacklokham, Emmanuel Santoyo Rio, Mak Soeun, Chiranjeevi Tallapragada, Say Tom, and Christian H. Roth. 2016. "Household Types as a Tool to Understand Adaptive Capacity: Case Studies from Cambodia, Lao Pdr, Bangladesh and India." *Climate and Development* 8(5): 423–34.

Chapter 6

Localizing Climate Change

Confronting Oversimplification of Local Responses

Brian Orland, Meredith Welch-Devine, and Micah Taylor

Introduction

People like to live near the coast. They feel happier and healthier there, and have thus historically migrated in the direction of the ocean (Wheeler et al. 2012). Over half of the US population lives along the coast, and this number is growing (Burger and Gochfeld 2017). Meanwhile, there is ample evidence that sea level rise is increasing the frequency of nuisance flooding in coastal communities and heightening the vulnerability of coastal settlements to storm-driven tidal surge and wind damage (Bilskie et al. 2016). The last several years have seen those effects combined in the destruction and disruption wrought by Hurricanes Katrina, Sandy, Matthew, Irma, Harvey, Maria, Florence, and Michael. So, are people fleeing coastal areas at risk of damage and flooding as a result of projected sea level rise and increased storm frequency and severity? The evidence is that after those storms wreaking the most disruption—e.g., Hurricanes Katrina, Sandy, Harvey—some people relocate and fail to return (Fussell, Sastry, and Vanlandingham 2010). However, people's responses to "near misses" are much less clear. For "wake-up call" storms that do not bring widespread devastation and loss of life, like Hurricane Matthew in the southeastern United States, there is little evidence that exposure to the immediate impact of a storm persuades people to consider migration. Many reasons have been advanced for this reluctance to move, among them economics, family ties, and sense of place. Policymakers rely on understanding those triggers, or the incentives that they can use to persuade people to take action, so it is critical to understand the individual values that could be engaged by those policies.

In this chapter, we focus on the state of Georgia in the southeastern United States. As many as 10 percent of Georgia's coastal residents are projected to be displaced by sea level rise by the year 2100 (Hauer et al. 2016), and those who remain will most likely be affected by extreme weather events of increasing severity (Gutmann et al. 2018). We use a mixed-methods approach, employing surveys and interviews directly following Hurricane Matthew (2016), to investigate how the complexity of residents' attitudes, perceptions, and beliefs about climate change contribute to their subsequent adaptation decisions in a population recently impacted by a damaging storm. Using events such as this as touch points allows researchers to discuss future scenarios with residents in a grounded way. For a short period of time, the scope and nature of anticipated climate-related changes were made clearer to residents and were isolated from the myriad other sociodemographic changes that will affect these communities in the coming years.

Residents of coastal Georgia live on a low-elevation littoral plain fringed by extensive marshes. Sea level rise as a consequence of climate change is already evident in people's daily lives. The causeway connecting one community, Tybee Island, to the mainland is more frequently flooded—twenty-three times in 2015—as a consequence of high tides, onshore winds, and locally heavy rain. In contrast to much of the eastern seaboard of the United States, the Georgia coastline is not heavily developed with accompanying coastal fortifications. It is a self-evidently vulnerable area, exposed to storms originating in the Atlantic Ocean. Much of the damage caused by Hurricane Matthew was from wind and rain rather than storm surge; however, sea level rise inexorably raises the likelihood of sustaining damage in such events as it raises the base level over which the other events unfold.

Working against any plans or intentions to migrate are people's ties to place and the attractive features of coastal living. The coastal city of Savannah, Georgia, is the oldest in the state, established in 1733, and is an acclaimed example of city planning. Originally the center of a British colony in the southeast United States, the city has evolved from a trading past based on agriculture to a thriving center of commerce, industry, and beach and cultural tourism. The smaller coastal cities of Brunswick, Darien, and St. Marys also have their roots in the eighteenth century, settled originally to occupy the coast and later becoming fishing and trading centers. The economy of the region was shaped by slave labor. After the U.S. Civil War, former slaves remained in the area, giving the region a rich mix of cultures with deep historical roots (Morgan, 2010). In more recent years, though, the entire southeast coast of the United States has boomed as a location for

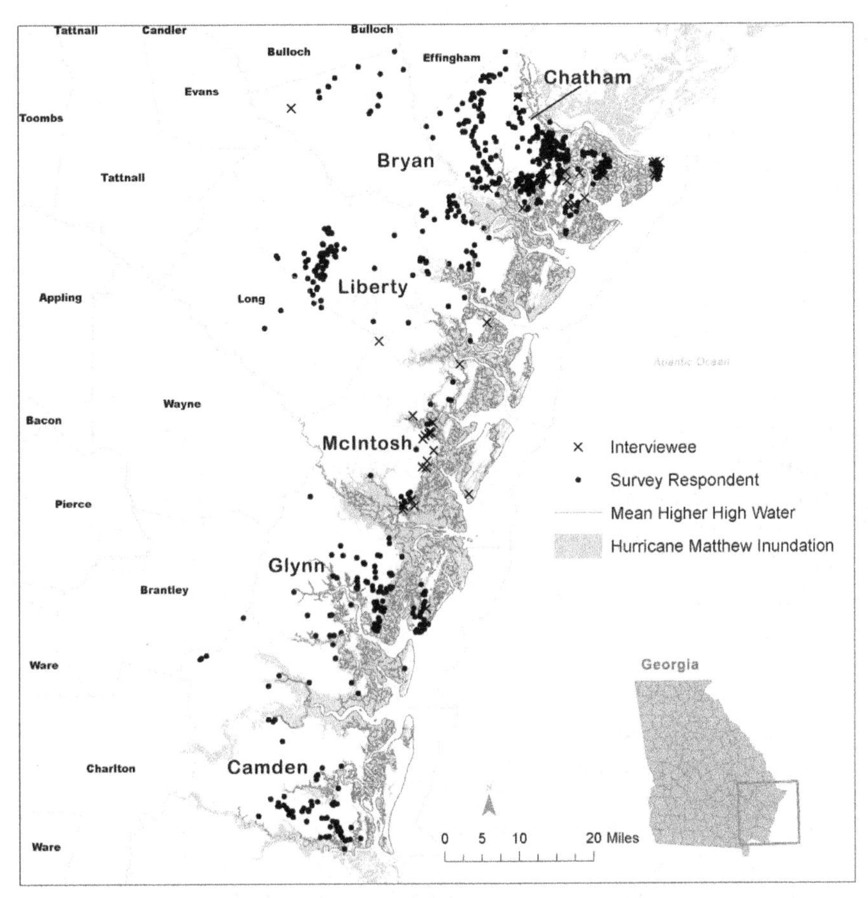

Figure 6.1. Coastal Georgia showing the normal Mean Higher High Water line, the estimated surge extent of Hurricane Matthew, and survey and interview respondent locations. Map by Micah Taylor, USA 31.489946° N, -81.499712° W. Esri, HERE, Garmin, © OpenStreetMap contributors, and the GIS user community. Accessed December 2018.

a range of industries and as a destination for retirement and tourism, with concomitant growth in permanent homes, second homes, and resorts. It is in the midst of these conflicting pressures, attracting and repelling, that residents of the coast and their communities are increasingly faced with important decisions about their plans for the future.

Background

Climate change is expected to cause a slow, incremental, but unavoidable rise in sea level, and it will spur increasingly violent storms originating over warmer coastal waters. A substantial literature has arisen that discusses the possible scope and severity of the anticipated changes together with their significant economic and human consequences (Rahmstorf 2017; Hauer et al. 2016). Hauer (2017) projects the displacement of 2.5 million people from the Miami/Fort Lauderdale/West Palm Beach area alone. However, it is not clear how people whose homes are at risk of rising sea levels or increasing storm damage will migrate or what incentives or triggers will cause them to move.

In October 2016, Hurricane Matthew passed northward off the coast of Georgia, resulting in new record tide levels at Fort Pulaski, adjacent to the city of Savannah. In the course of this event, the area experienced all of the anticipation and trauma of evacuation, return, and cleanup, but the resulting damage affected relatively few people, homes, and businesses by comparison with Hurricanes Katrina, Sandy, and Harvey. As a result, we realized in the aftermath of Hurricane Matthew the possibility of interviewing and surveying a population sensitive to and knowledgeable about the impacts of severe storms on their lives yet largely able to return to their homes and, thus, be available for our questioning.

Adaptation to Climate-Related Change

There are two main adaptation paths in response to climate-related change. For some people in some situations, it may be possible to protect or directly adapt their home to withstand the changing sea level or storm frequency and severity. After Hurricane Sandy destroyed homes on Staten Island, New York, some residents chose to rebuild their homes by raising the living areas substantially above flood level and using more storm-resistant construction (Bukvic, Smith, and Zhang 2015; Bukvic and Owen 2016). For others, the response, again dictated by personal circumstances, would be to relocate to an area safe from current and anticipated risks.

Butler, Deyle, and Mutnansky (2016) argue that responses to sea level change ought to be highly amenable to thoughtful planning because they take place incrementally and slowly over decades. While their work focuses on the behaviors of Florida municipalities and their plans for the future, their findings indicate that even planning agencies often have a wait-and-see attitude and focus on short-term and economic physical armoring of shorelines versus avoidance and retreat. It may not be surprising that individuals, then, display similar behaviors and are reluctant to make the serious and life-changing commitment to relocate. Moreover, in some case, residents do not perceive the burden of adaptation as being individual; instead, they see it as a governmental response, meaning that they expect external assistance either in managing the threat or in subsidizing any migration actions. Hoogendoorn, Fitchett, and Grant (2016) reported that because of the uncertainty of flood events in coastal South Africa, residents expected governmental agencies to find mechanisms to either protect or compensate people.

Residents' perceptions of risk are closely linked, and in part are a precursor, to their adaptation strategies. There is only sparse literature on how perceptions of risk shape migration response, and much is associated with developing populations facing threats of famine and loss of livelihoods. Evacuation from disaster-impacted areas has received much more attention. McCaffrey, Wilson, and Konar (2018) studied wildfire evacuation decisions and identified two distinct population types, those who evacuate and those who stay in place. In both groups, there was a substantial subset of "wait and see" respondents.

While governments will inevitably be centrally involved in planning and facilitating large-scale adaptation measures, their actions will be strongly shaped by the perceptions and decisions of residents in their jurisdictions. Policy implementation will only be successful if it accords with individual inclinations and provides them appropriate support (Song and Peng 2017). Song and Peng's study examined the likelihood of residents in Panama Beach, Florida, relocating away from low-lying areas in response to rising sea level. They examined people's perceptions of risk, their experiences of past hazards, their ability to cope with change, and their relocation destinations. Residents' characteristics affect their relocation responses, but not necessarily in ways that are easy to interpret. While well-educated people might be expected to understand the risks, they were more reluctant to relocate; the attitudes of friends and family might be more influential in decisions to move than the scientific information available. Hazard awareness is positively associated with willingness to relocate, but direct past experience of hazards had little effect. The emerging lesson from these studies is that the range of adaptation strategies available is vast, as is the range of individual responses, and is shaped by individual characteristics, coping abilities, and social influences. While surveys of attitudes and behaviors reveal the broad range of responses to changing conditions, they are limited in their ability to tell us why people make the choices and decisions that they do.

Displacement and Migration

Much of the literature examining migration patterns after severe weather events focuses on two issues: first, the sudden and dramatic dispersal of people after events like Hurricanes Katrina in New Orleans and Sandy in the New York/New Jersey region; and second, the planned-for but wrenching removal of entire populations from mid-ocean islands. The lessons of these events may or may not be transferable to the issues of slowly rising risk for populations that, on the face of it, do have places to move to and avoid catastrophe.

Hurricane Katrina scattered one million evacuees across the United States (Grier 2005). The population of New Orleans has not returned to pre-storm levels, but studies looking at returning displaced residents have revealed a series of possible, and plausible, reasons for not returning. Landry et al. (2007) found that rates of return could be affected by a range of demographic factors, but the results were highly variable. For example, they found higher proportions of middle-income households planning to return, but it was challenging to separate the influence of factors such as home ownership, the economic resilience of individuals, and that some neighborhoods of historically low-income populations were the most affected by flooding and subsequently uninhabitable. Groen and Polivka's 2010 analysis of people returning after Katrina revealed similar patterns: a decrease in the percentage of blacks in the population, i.e., more whites and Hispanics, and a decrease in the number of lower-income/education families, i.e., more residents with higher income/education. This observation was reinforced by Fussell, Sastry, and Vanlandingham (2010), who found that black residents returned at a lower rate than white residents did but that the disparity disappeared when they controlled for housing damage. Blacks tended to live in areas more affected by flooding. Studies of post–Hurricane Sandy relocation reveal the same kinds of patterns. Older residents were more willing to consider relocation, as were homeowners facing extensive repairs. Most would prefer to stay, but the cost of flood proofing by raising homes is prohibitive for many (Bukvic, Smith, and Zhang 2015; Bukvic and Owen 2016). Indeed, financial considerations can overturn plans to return and rebuild (Bukvic, Smith, and Zhang 2015; Bukvic and Owen 2016).

In the aftermath of a major disaster like Hurricanes Katrina, Sandy, and Harvey, relocation, albeit temporary, is unavoidable, and much of the shaping of the response is in the hands of agencies at all levels of government, from local to federal. By contrast, response to slower-onset changes such as sea level rise or increased storm frequency are subject to individual perceptions and interpretations. Stojanov and colleagues (2017) explored residents' perceptions of recent and future climate change impacts in the Maldives, low-elevation islands in the Indian Ocean, and their willingness to consider moving away. More than 50 percent of respondents perceived the threats as serious and accepted that migration might be an option. However, to individuals, the risks of sea level rise were not as serious as other important cultural, economic, and social challenges. Willox, Harper, and Edge (2012) examined communities' responses to climate warming in Labrador, Canada, where increasing temperatures are disrupting hunting, fishing, travel, and the look and feel of the landscape. They point to the impacts of those changes on place attachment and emotional health and well-being and to the challenges of developing adaptations and mediations. What is most evident, however, from these readings of the literature is that, for the numerous displacement studies, we have extensive social science survey-based evidence for a variety of behaviors but limited direct knowledge about the individual decisions that led to the summary observed behaviors. Conversely, for those studies that do explore individual responses to the threat of displacement, the expressions of cultural, spiritual, and place values appear central to decision-making but lack the generalizability and comparability that a policymaker would need to be able to propose a response. Our concurrent quantitative and qualitative studies sought to connect the synoptic and granular insights that these approaches offer.

Attachment to Place

Adger et al. (2012), in a broad review of the cultural dimensions of adaptation to climate change, argue that scholars have not paid enough attention to the role of place and identity in understanding individuals' decisions to remain in place or relocate. Groen and Polivka (2010) examined determinants of return migration after Katrina. Despite alluding to "sense of place" potentially being a factor in decisions about migration or return, their findings focused on demographic and economic factors, such as the cost of damage recovery, and failed to examine the "attractor" values of place, familiarity, neighbors, or even jobs. In examining the attitudes of coastal residents of Panama Beach, Florida, Song and Peng (2017) suggest that social ties and emotional attachments are hindrances to relocation, but their study focuses on attitudes toward planned retreat from coastal threats; they do not report on the values respondents sought by continuing to live on the coast.

In other studies that specifically address sense of place, the conception of sense of place that emerges is more one of attachment to knowledge systems (that cannot be readily addressed by policy) than to physical location (that can be regulated, etc., by policy and investment) (Hoffman 2017). Willox, Harper, and Edge (2012), in their examination of Inuit ties to the land in Canada, where traditional lifestyles were threatened by disrupted hunting, fishing, and traveling, people's responses were not so much about attachment to particular physical places as they were to the traditional practices carried out in that landscape. While it is understandable that most studies are focused on future behaviors, it is not clear that the questions asked in prior surveys fully address the range of factors that will affect decisions and actions to evacuate or migrate. Moreover, it is not clear how sense of place might interact with demographic or spatial location issues to shape migration responses.

Effects of Location

One factor that may have received less attention yet seems central to any consideration of migration is the role of proximity to a source of flooding or storm damage, both physical and perceptual, in decisions to relocate. Milfont and colleagues (2014) studied the relationship between New Zealanders' belief in the reality of climate change and their proximity to the shoreline. The model they developed controlled for height above sea level, regional poverty, and individual differences in gender, age, education, and wealth, indicating a connection between physical proximity and the psychological acceptance of climate change. Conversely, Bukvic and colleagues (2018), surveying residents in the aftermath of Hurricane Sandy, found only minor effects of proximity to shoreline on willingness to relocate. They suggested that factors such as residents' confidence in being able to adapt or retreat may play a bigger role than physical location in migration decisions, although those exposed to repeated flooding and offered buyouts were more likely to consider relocation.

A second issue with respect to spatial location is whether people have directly experienced or observed impacts of past storms. The effect of direct experience on the development of environmental attitudes and behavior has been noted by numerous authors (e.g., Duerden and Witt 2010) and undoubtedly contributes to the credibility of environmental projections (Dong et al. 2018). However, in considering the spatial extent of rising sea level, and, hence, its impact on populations (e.g., Hauer 2017), planners and policymakers have usually treated the phenomenon as an orderly "bathtub" rise of level to a new projected shoreline or flood zone boundary. Instead, the dynamics of storm surge driven by onshore winds and shaped by coastal geomorphology can dramatically extend the influence and evidence of flooding (figure 6.2). Bilskie et al. (2016) is an example from a growing literature indicating the inland extent of potential flooding under future storm conditions. Musser, Watson, and Gotvald (2017) collected high water mark data (Koenig et al. 2016) to illustrate the extension of post–Hurricane Matthew flooding beyond mapped flood lines. While many past studies have looked at the influence of proximity to shorelines or flood zones as possible influences on migration behavior

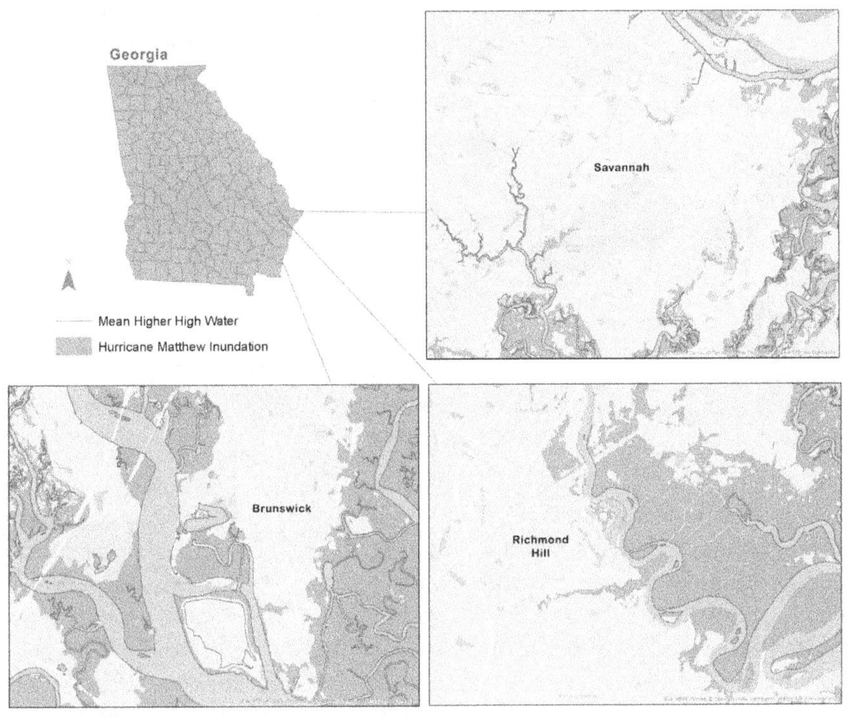

Figure 6.2. Estimated surge extent of Hurricane Matthew in Savannah, Brunswick, and Richmond Hill. Map by Micah Taylor, USA 31.489946° N, -81.499712° W. Esri, HERE, Garmin, © OpenStreetMap contributors, and the GIS user community. Accessed December 2018.

(e.g., Milfont et al. 2014; Stojanov et al. 2017), and others have looked at past experience of storm-related damage on future behavior (e.g., Bukvic and Owen 2016), we have not found any that consider the effects of residents' proximity to, or direct observation of, past flooding or other damage on their attitudes or behavior.

Study Design

Our study was designed to address some of the gaps we have described. We used a mixed-methods approach that would enable us to examine both population-level effects through an online survey and individual-level perceptions and expected behaviors through ethnographic interviews. Our approach follows the model below (figure 6.3).

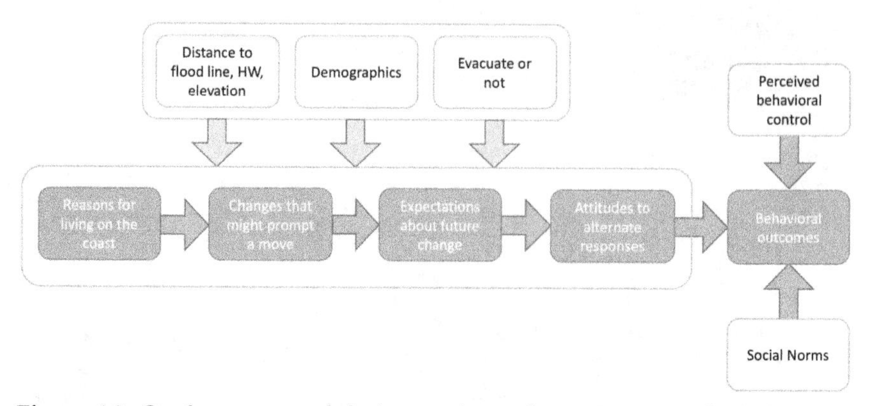

Figure 6.3. Study conceptual design: main study components shaded and external factors in open boxes. © Brian Orland.

First, we examine why people live on the coast, what changes might prompt them to consider moving away, and their expectations for future climate-related change. We then look at their attitudes toward future change and their expected response behaviors as shaped by the social norms expressed by those around them and their perceived level of personal control over outcomes. At each step, we investigate effects of distance from shoreline and observed high water marks, of demographic variables, and of past compliance with evacuation notices.

A survey of residents regarding their intentions to migrate addressed all six coastal counties of Georgia (figure 6.1). In-depth interviews were focused in Chatham County, the location of Savannah and growing in population, and in McIntosh County, rural and declining in population. While the coastal counties of Georgia play an increasingly important role in the economy and lifestyle of the state, except for Chatham County they are sparsely populated. Those with strong economic development, Chatham and Glynn, are growing. Liberty and McIntosh Counties are experiencing decline (table 6.1).

Surveys: A 139-item survey of residents of the six Georgia coastal counties was conducted via Qualtrics Panels in May 2017, seven months after Hurricane Matthew in October 2016. There were 2,509 surveys distributed via email. After removal of out-of-area responses, duplicates, and other low-quality responses, we have an analysis set of 991 responses. Survey questions were modeled on a number of prior studies. While there is an extensive literature associated with evacuation-related behavior (e.g., Pham et al. 2020), there have been fewer studies of intentions to migrate away from the coast in the face of climate-related change. A study of public understanding and intended behavior in the face of sea level change in

Table 6.1. County populations and populations at risk of coastal Georgia, sorted north to south. US Census, 2017. Source: American Community Survey, reported by Headwaters Economics, headwaterseconomics.org/par. Table by Brian Orland.

	Total Population	Percent change 4/1/10–7/1/16	Children under 5 as percent of total	Elderly over 65 as percent of total	White as percent of total	Black/African American as percent of total	Individuals in poverty as percent of total	No high school degree as percent of total
Georgia	**10,310,371**	**6.4%**	**6.5%**	**12.3%**	**59.8%**	**31.2%**	**17.8%**	**14.2%**
Coastal total	**539,319**	**7.4%**	**7.1%**	**13.3%**	**58.7%**	**33.8%**	**17.3%**	**10.9%**
Chatham	289,082	9.0%	6.7%	13.8%	53.3%	39.8%	18.0%	10.7%
Bryan	36,230	19.9%	7.3%	10.0%	78.4%	15.2%	13.3%	9.6%
Liberty	62,570	-1.4%	10.0%	7.5%	46.8%	41.3%	16.9%	10.3%
McIntosh	13,927	-2.8%	4.7%	20.1%	61.8%	35.1%	20.1%	17.6%
Glynn	84,502	6.1%	6.1%	17.5%	66.9%	25.4%	18.7%	12.4%
Camden	53,008	4.9%	7.7%	11.2%	74.5%	18.2%	14.0%	9.0%

the United Kingdom (Thomas et al. 2015) was used to guide our survey instrument. The Climate Change Attitude Survey (Christensen and Knezek 2015) and an attitudes and migration survey (Wilmot 2009) were used to guide the wording of individual questions.

Interviews: In March 2017, we conducted sixty-six interviews with seventy-two respondents. Interviews were designed to examine how individuals frame the problems they face and what their strategies are for responding to extreme weather events. We asked about rationales and motivations for their attitudes and behaviors during and after the storm. We selected interviewees to capture diversity in components of social vulnerability (e.g., age, ethnicity) and to approximate the demographic makeup of the coastal region. Fifty-six percent of respondents were female, and 44 percent were male. The majority, 75 percent, identified as White, with 22 percent identifying as Black or African American and 3 percent as mixed race. They ranged in age from twenty-five to ninety-one, with 53 percent above sixty-five. Median household income was $50,000 and ranged from $0 to $400,000 (table 6.2). Interviews covered migration histories, political economic contexts, storm experiences and attitudes, and adaptation possibilities. All interviews were recorded and transcribed for coding.

Spatial location: Respondents to our surveys were asked to provide their street address so that we could assess their proximity to shoreline and flood zones as well as elevation above sea level; 742 respondents provided that data.

Table 6.2. Interview and survey respondents versus census demographic characteristics. US Census, 2017. Source: American Community Survey, reported by Headwaters Economics, headwaterseconomics.org/par. Table by Brian Orland.

	Total Population	Elderly over 65 as percent of total	Gender, Female as percent of total	White as percent of total	Black/ African American as percent of total	Individuals in poverty as percent of total	No high school degree as percent of total	Graduate degree as percent of total
Georgia	10,310,371	12.3%	51.0%	59.8%	31.2%	17.8%	14.2%	30.4%
GA Coastal	539,319	13.3%	51.1%	58.7%	33.8%	17.3%	9.7%	23.9%
Interviews	72	53.0%	56.0%	75.0%	22.0%	–	–	–
Surveys	991	10.2%	69.1%	67.6%	20.0%	21.7%	3.4%	14.4%

Study Method

The aim of the study, as stated above, was to explore the relationships between perceptions, attitudes, and outcomes as shaped by demographic, social, and environmental factors. These are diagrammed in figure 6.3, the main study components shaded and the external factors in open boxes. Our general strategy for reporting results is to use the survey data to identify and evaluate responses to underlying constructs in each of the major components, then show how those respond to external factors such as demographic variables, past evacuation behavior (as a proxy measure of risk tolerance), and proximity to impacted areas. We then explore reasons for the responses by reference to the interview data that provides much richer means of revealing motivations of our respondent populations.

Survey data: Demographic variables examined included Age Class (18–24, 25–44, 45–64, 65+), Gender, Household Income, Educational Achievement, and Ethnicity (table 6.2). Expecting that evacuation behavior might provide insights into anticipated migration behavior, we asked whether survey respondents evacuated or not during the Hurricane Matthew event; 436 (62.8 percent) reported that they evacuated, 258 (37.2 percent) stayed in place. In addition to the survey data, we calculated four additional variables expressing proximity to the ocean and other water bodies: (1) the distance between each participant address location and Mean Higher High Water (MHHW) as determined by the National Oceanic and Atmospheric Administration (NOAA); (2) elevation of the address above Mean Higher High Water (MHHW); (3) the distance between each participant address location and the edge of the storm inundation zone taken from the surge forecast data and validated with US Geological Survey (USGS) High Water Mark data. Variables with multiple ordinal levels (e.g., Age category, Household Income) were examined using One-Way Analysis of Variance; Categorical variables, in this case all with two classes, were examined using unequal population t-tests. Because the aim of this chapter is to explore the relationships between perceptions, attitudes, and outcomes, we used data reduction for each component (figure 6.3) of our survey via Principal Components Analysis (PCA), assuming that Likert scale data had interval-quality characteristics and using varimax rotation to preserve statistical independence of the derived constructs for use in the succeeding components of the study.

Interview data: We use the Principal Components from the quantitative phase of our study to guide analysis and discussion of the qualitative data. For instance, since the PCA of questions regarding people's reasons for living on the coast revealed three factors that we titled Lifestyle, Family

Ties, and Job/Financial, discussion of interview data related to the same question addresses the same three factors.

Attachment to Place

Factors That Encourage People to Live on the Coast

People are attached to their homes on the coast for a variety of lifestyle, family, and economic reasons. The area's beauty, pace, affordability, and suitability as a retirement destination are reported as important by two-thirds or more of respondents. We asked respondents, "Why have you chosen to live in coastal Georgia?" Six out of ten responses mentioned that the factors reported were important or very important to them (table 6.3). Principal Components Analysis of responses to the eleven questions revealed three factors achieving Eigen values >1.0—Quality of Life, Family Ties, and Pragmatic (a combination of raising kids, job-related considerations, and financial investment factors)—that accounted for 60.6 percent of total variance. Cronbach's alpha for the six items in Lifestyle was 0.84, generally regarded as Good, and 0.69 for Family Ties, which is Acceptable, indicating strong internal consistency in responses to these factors. The 0.36 figure for Pragmatic indicates the individual variables were weakly related (Pairwise correlations 0.08, 0.13, 0.27).

Table 6.3. Reasons for living in coastal Georgia. ^{Q} = Quality of Life, ^{F} = Family Ties, ^{P} = Pragmatic. The superscript letter indicates the factor grouping. Data by Brian Orland.

	Slightly / not at all important	Important / very important
I enjoy the area's natural beauty ^{Q}	16.4%	78.8%
I like the pace of life ^{Q}	19.1%	74.5%
It's an affordable place to live ^{Q}	21.2%	71.0%
It's a good place to retire to ^{Q}	21.4%	66.3%
I feel a strong connection to the coast ^{Q}	26.5%	65.8%
I enjoy the recreational opportunities ^{Q}	26.4%	65.0%
I have family and friends in the area ^{F}	18.8%	63.8%
I grew up in the area ^{F}	21.8%	39.3%
It's a good place to raise kids ^{P}	21.1%	59.3%
I moved for job-related reasons ^{P}	22.2%	45.7%
It's a good financial investment ^{P}	36.5%	42.0%

There was a small effect of length of coastal residence on rated importance for Lifestyle factors as reasons for living on the coast at the $p=0.01$ level ($F [1, 740] = 6.74$) and a larger effect for Family Ties, $p=<0.0001$ ($F [1, 740] = 163.84$). For many respondents, time spent living on the coast equates to time spent building friendships and raising families, so that stronger relationship is not surprising. In contrast, people of all ages and length of residence are attracted by the quality of life factors contributing to Lifestyle. Previous experience of hurricanes did not affect the importance of Lifestyle, but those with no previous experience in an area prone to hurricanes expressed greater importance for Family Ties ($M=0.19$, $SD=0.98$) than those with experience ($M=-0.25$, $SD=0.97$); $t(740)=6.00$, $p<0.0001$. While this seems contradictory, that those with less experience of hurricanes would be more concerned about ties to family and friends, the Georgia coast prior to Hurricane Matthew had not sustained any direct hurricane damage since 1898, so long-term residents had no previous experience to draw upon.

Lifestyle was more important for older respondents, $p<0.0001$ ($F [3, 738] = 10.33$) but Family Ties were less important, $p=0.0003$ ($F [3, 738] = 6.25$). There were small positive effects of Household Income, $p=0.038$ ($F [6, 735] = 2.24$) and of Level of Education, $p=0.025$ ($F [6, 735] = 2.42$), on the importance of Lifestyle. Higher Household Income, $p=0.002$ ($F [6, 735] = 3.47$) and Education, $p<0.0001$ ($F [6, 735] = 5.44$), had negative effects on the importance of Family Ties. Although age, household income, and level of education were not highly intercorrelated, each has logical connections to the importance of lifestyle factors and less importance of family ties. Older people choose the coast as a place to retire to, wealthier people may have chosen the area as the location of a second home, and education might be associated with either age or income.

Lifestyle was more important for whites ($M=0.11$, $SD=0.99$) than for African Americans ($M=-0.28$, $SD=0.96$); $t(661)=-4.40$, $p<0.0001$. Blacks and African-Americans, however, found Family Ties more important ($M=0.35$,

Table 6.4. Correlation table, major demographic characteristics. Data by Brian Orland.
** Correlation at the $p<0.0001$ level, * Correlation at the $p=0.02$ level.

	Time lived on the coast	Age of respondent	Household income	Educational attainment
Time lived on the coast	–			
Age of respondent	0.225**	–		
Household income	0.086*	0.217**	–	
Educational attainment	0.063	0.291**	0.459**	–

SD=0.90) than whites (M=-0.07, SD=1.02); t(661)=4.82, p<0.0001. Gender had no effect on importance of Lifestyle. Females, however, attached more importance (M=0.07, SD=1.02) to Family Ties than males did (M=-0.16, SD=0.93); t(738)=-3.06, p=0.002. Evacuation behavior was not related to reasons for living on the coast.

Distance from Mean Higher High Water (MHHW), from the high-water mark associated with Hurricane Matthew (HWM) and with Elevation above MHHW, each were negatively related to Lifestyle factors as reasons to live at the coast—p<0.0001 (F [1, 740] = 15.80), p=0.002 (F [1, 740] = 9.76), p<0.0001 (F [1, 740) = 19.06) respectively, suggesting that location close to the shore is closely related to the importance of Lifestyle. Distance factors were not related to Family and Friend Ties.

Change Factors That Might Prompt Considering Migration

In the survey, respondents were asked, "Imagine your life in coastal Georgia changed. For each of the following changes, please indicate whether it might make you consider moving away from your current home." Four (of fifteen) items were cited as reasons for considering leaving the coast by 60 percent or more respondents after Matthew—Sea-level rise threatens your home, Increase in crime, No longer feels like a relaxed area, and Storm damage becomes more frequent—demonstrating the mix of economic and social considerations that impact decisions. PCA of the fifteen questions in table 6.5 revealed three factors that would likely induce people to move: Loss of Quality of Life, Increased Costs, and Job/Family Losses. Together, these accounted for 53.3 percent of total variance. Cronbach's Alpha for the six items in Loss of Quality of Life was 0.83, generally regarded as Good, and 0.79 for Increased Costs, also Good, but the 0.54 for Job/Family Loss is low, indicating that the individual variables were weakly related (Pairwise correlations 0.21, 0.28, 0.36).

While in general the importance of factors that might cause people to move away matched, in reverse—the importance of reasons for living at the coast—the responses to the loss of values were not as strongly stated. There was a small negative effect of length of coastal residence on rated importance for Loss of Quality of Life as a reason for leaving the coast at the p=0.006 level (F [5, 736] = 3.27), but no effect for Increased Costs as a rationale. There was a small negative effect of higher Household Income, p=0.042 (F [6, 735] = 2.19) on the importance of Loss of Quality of Life. African Americans attached more importance (M=0.01, SD=1.04) to Loss of Quality of Life than whites did (M=-0.06, SD=0.98); t(661)=6.00, p<0.0001. No other demographic or distance variables had effects on the importance of Loss of Quality of Life. Those who evacuated found Increased Costs

Table 6.5. Considerations for moving away. Factor groupings: ^{Q} = Loss of Quality of Life, ^{C} = Increased Costs, $^{J/F}$ = Job/Family Loss. Data by Brian Orland.

	Probably / definitely not move	Probably / definitely move
No longer feels like a relaxed area ^{Q}	32.4%	63.3%
Environmental pollution increases ^{Q}	36.4%	59.3%
Sense of community declines ^{Q}	45.0%	49.6%
Loss of area's natural beauty ^{Q}	45.8%	48.9%
Reduced access to public lands for recreation ^{Q}	52.7%	40.5%
Cultural and historical aspects decline ^{Q}	56.8%	37.6%
Sea level rise threatens your home ^{C}	18.7%	75.7%
Increase in crime ^{C}	26.2%	71.0%
Storm damage becomes more frequent ^{C}	36.1%	60.5%
Increased cost of living ^{C}	38.6%	58.1%
Large profit from selling property ^{C}	31.1%	57.0%
Increased property taxes ^{C}	42.7%	48.8%
Loss of my employment $^{J/F}$	35.8%	47.4%
Family and friends move away $^{J/F}$	51.2%	39.3%
Children grow up and move away $^{J/F}$	46.1%	28.3%

more important (M=0.06, SD=0.96) as a reason to move than those who stayed in place (M=-0.14, SD=1.06); t(642)=2.46, p=0.01. Older people attached slightly less importance to Increased Costs, p=0.0002 (F [3, 738] = 6.62). African Americans attached more importance (M=0.18, SD=0.98) to Increased Costs than whites did (M=-0.07, SD=0.99); t(661)=2.71, p=0.007. There were no other effects of demographic or distance variables.

Expectations about Future Change

Survey respondents were asked to report on their concerns for both their region and for themselves personally with respect to sea-level rise and damaging storms. Concern for both impacts is similar at the level of the region. However, for sea-level rise personal concern is lower (Table 6.6). Coastal residents are closely aware of where their homes sit with respect to sea level and distance from areas affected by storm surge. Those outside the threatened areas would expect no lasting damage.

Table 6.6. Regional and personal expectations regarding sea level rise and damaging storms. Data by Brian Orland.

	Not at all concerned	Slightly unconcerned	Neutral	Slightly concerned	Extremely concerned
To what extent are you concerned about sea level rise affecting the region?	13.34%	10.11%	18.19%	45.42%	12.94%
To what extent are you concerned about sea level rise affecting you personally?	16.58%	19.81%	36.93%	12.53%	14.15%
To what extent are you concerned about damaging storms affecting the region?	5.39%	7.28%	14.96%	49.87%	22.51%
To what extent are you concerned about damaging storms affecting you personally?	7.28%	8.63%	14.15%	47.31%	22.64%

None of the demographic, evacuation, or distance factors exhibited effects on survey respondents' expectations regarding sea level rise, that is, length of residence on the coast, gender, or distance between home and the ocean. There are small effects of age and household income on expectations of damaging storms, with those over sixty-five reporting somewhat lower expectations, but the effects were not linear with respect to either factor.

Attitudes to Adaptation Responses

The Theory of Reasoned Action (Fishbein and Ajzen 1975) posits that intentions to perform behavior, in this case adaptation measures, will be determined by an individual's positive attitude toward the behavior and their belief that others want them to perform the behavior. The Theory of Planned Behavior (Ajzen 1985) adds the element of self-efficacy, namely the individual's conviction that they can undertake the adaptation action. While the theories have been extensively applied in health fields to study intentions to exercise or undergo therapy, they have broad applicability to areas where people have volitional control over outcomes and are sufficiently informed to form a confident intention. They have been used to investigate inconsistencies in people's expressed intentions and actual migration behavior (Lu 1998) and with respect to international migration (Bilgili and Siegel 2015; Groenewold, Bruijn, and Bilsborrow 2012; Klabunde and Willekens 2016). With few exceptions (e.g., Lu 1998), most examine international migration for economic and security reasons. We have found no instances of application of these models to the issue of migration in the face of coastal climate-related hazards.

If a projected 10 percent of Georgia coastal residents are displaced by sea level rise (Hauer et al. 2016), then it will be critical to know what intentions and behaviors arise from their emerging beliefs in order for migration to be planned and managed. From the responses to the questions above, it seems clear that there are few policymaking opportunities in our examinations of the effects of demographic and distance-to-shore variables on the reasons people live at the coast or what might cause them to consider moving. Interactional models of reasoned action describe the relationships between beliefs and attitudes and behavioral outcomes. We examined how attitudes regarding sea level rise and increased storm severity might result in intentions to migrate, using both survey and ethnographic methods.

A group of questions addressed attitudes toward sea-level rise and increased storm damage: "What are your attitudes toward different possible responses to sea level change and increased storm frequency and severity?" The single strongest response was, "I will take the necessary measures to stay in my home," with 61.6 percent in agreement with the statement (table 6.7). Two other statements achieved more agreement than disagreement, that both local and federal government should do more to protect homes. All statements about attitudes to moving elsewhere skewed toward disagreement. PCA of the ten questions in table 6.6 revealed three factors with respondents' agreement with statements about attitudes: Government Entitlement, Personal Responsibility, and Government Appreciation. Together, these accounted for 63.1 percent of total variance. Cronbach's Alpha for the four items in Government Entitlement was 0.84, generally regarded as Good. A 0.63 rating for Government Appreciation is low, but there are only two variables contributing to the factor. A 0.41 rating for Personal Responsibility is Poor, and the individual variables are only weakly related (Pairwise correlations ranging from -0.17 to 0.51). Nevertheless, the groupings do represent three important attitudes: "I'm entitled to having the government protect me, but not by moving me," "I have to take personal responsibility for responding," and "I'm grateful for government's role in protecting me and my community."

There was no effect of length of coastal residence on the value of Government Entitlement. Agreement on the value of entitlement declined with increasing Age, $p=0.0003$ ($F[3, 738] = 6.25$), and Household Income, $p=0.0001$ ($F[6, 735] = 4.56$). African Americans were more strongly in

Table 6.7. Attitudes to adaptation responses. Factor groupings: E = Government Entitlement, P = Personal Responsibility, G = Government Appreciation. Data by Brian Orland.

	Mean on -2, 0, +2 scale	Disagree / Strongly disagree	Agree / Strongly agree
LOCAL government should do more to protect my home [E]	0.156	23.3%	37.5%
The FEDERAL government should do more to protect my home [E]	0.113	27.1%	37.2%
The FEDERAL government should help me move somewhere safer [E]	-0.338	46.2%	23.6%
LOCAL government should help me move somewhere safer [E]	-0.363	46.9%	22.4%
LOCAL government is doing a good job to protect my home [G]	-0.007	22.7%	23.7%
The FEDERAL government is doing a good job to protect my home [G]	-0.186	28.4%	14.8%
I would take the necessary measures to stay in my home [P]	0.534	11.8%	61.6%
I could not recover from losses or damage to my home [P]	-0.156	43.8%	31.5%
I would like to relocate elsewhere [P]	-0.302	45.4%	26.1%
I think about moving to another part of my community to avoid future losses or damage [P]	-0.367	48.2%	24.1%

agreement ($M=0.42$, $SD=1.02$) about the value of Government Entitlement than whites were ($M=-0.15$, $SD=0.93$); $t(661)=6.23$, $p<0.0001$. No other demographic or distance variables had effects on levels of support for Government Entitlement. There were no effects of demographic or distance variables on respondents' agreement that Local and Federal Government were doing a good job. In fact, 53.5 percent and 56.7 percent, respectively, responded in the neutral category, indicating ambivalence to government efforts.

Actions That Might Be Taken in Response to Climate-Related Change

We asked respondents the extent to which they agreed with statements about possible responses in the face of change: "The statements below reflect the actions you might take in the future in response to sea level change and increased storm frequency and severity." Only two statements received strong expressions of agreement: "I will storm- and flood-proof my home" and "I will move to be closer to family and friends if my home is threatened." "I intend to move back to where I moved from" received strong disagreement. Intentions to move to safer locations and within a five-year time frame received neutral or negative responses. PCA of responses to the eleven questions in table 6.7 revealed only two factors achieving Eigen values >1.0—Intention to Move and Stay in Place—that accounted for 51.9 percent of total variance. Cronbach's Alpha for the seven items in Intention to Move was 0.82, generally regarded as Good, and 0.49 for Stay in Place indicates the individual variables were weakly related (Pairwise correlations 0.21 to 0.28).

Comparing the mean values of the variables that comprise Intention to Move shows that there is net *disagreement* on intentions to move and that agreement *declined* further with increasing time lived on the coast, $p=0.007$ ($F [5, 736] = 3.19$), increasing age, $p<0.0001$ ($F [3, 738] = 29.94$),

Table 6.8. Actions that might be taken. Factor groupings: ^{M} = Intention to Move, ^{S} = Stay in Place. Data by Brian Orland.

	Mean on -2, 0, +2 scale	Disagree / Strongly disagree	Agree / Strongly agree
I will move to be closer to family and friends if my home is threatened ^{M}	0.185	26.9%	42.8%
I will move to somewhere I can get flood insurance ^{M}	0.059	25.9%	34.4%
I intend to move to another home in the next five years ^{M}	0.007	37.5%	39.4%
I intend to move somewhere safer but still close to my current home ^{M}	-0.214	38.5%	25.8%
I will move in the next five years to be closer to friends and family ^{M}	-0.315	46.6%	22.8%
I intend to move within five years to somewhere hurricane risk is lower ^{M}	-0.322	47.4%	24.2%
I intend to move back to where I moved from ^{M}	-0.549	52.7%	18.2%
I will storm- and flood-proof my home ^{S}	0.507	11.9%	53.8%
I will stay where I am, whatever happens ^{S}	-0.097	35.4%	28.5%
I intend to stay here as long as I get government assistance for repairs ^{S}	-0.156	35.7%	26.8%

household income, $p=0.002$ ($F [6, 735] = 3.46$), and educational attainment, $p=0.04$ ($F [6, 735] = 2.23$). Those who had evacuated ($M=0.06$, $SD=0.95$) were more likely to consider moving than those who stayed in place ($M=-0.13$, $SD=1.06$); $t(642)=2.35$, $p=0.02$. African Americans were more positive toward moving away ($M=0.29$, $SD=1.01$) than whites were ($M=-0.12$, $SD=0.98$); $t(661)=4.38$, $p<0.0001$. Increasing distance from the Hurricane Matthew High Water Mark (USGS Seed and Stain data) was weakly related to expressed intentions to move away, $p=0.01$ ($F [1, 740] = 6.05$), as was increased elevation about MHHW, $p=0.006$ ($F [1, 740] = 7.65$). As for people's reasons for living at the coast, these are counterintuitive findings suggesting that living close to the ocean is a more powerful "pull" factor than a reason for moving.

The other factor emerging from the PCA was Stay in Place. Contrasting with observation on intentions to move, agreement on Stay in Place *increased* with increasing time lived on the coast, $p<0.0001$ ($F [5, 736] = 5.93$), and increasing age, $p=0.04$ ($F [3, 738] = 2.81$). Household income and educational attainment did not have an effect on intention to stay in place. Those who had evacuated ($M=-0.064$, $SD=0.97$) were more likely to consider moving than those who stayed in place ($M=0.13$, $SD=1.05$); $t(642)=-2.13$, $p=0.03$. Gender and ethnicity did not show effects. Increasing distance from the Hurricane Matthew High Water Mark was weakly negatively related to expressed intentions to stay in place, $p=0.02$ ($F [1, 740] = 5.59$). Elevation above MHHW did not exhibit an effect. Again, this suggests that closeness to the ocean is more likely to result in intentions to stay in place.

It might be expected that the decision to move away from the coast or stay in place would be related to people's original reasons for living on the coast. For our respondents, those reasons fell into three groups—the attractions of the coastal lifestyle, attachment to family and friends, and practical issues such as the location of a job or the choice of a good place to raise children. Conversely, their reasons for considering moving away fell into three groups—loss of the coastal lifestyle they valued, loss of the family ties through children and friends moving away, and the pragmatics such as cost of living. Even so, there is strong agreement with any statements relating to staying in place and disagreement with those relating to moving away. Our interviews bore out and underscored people's reluctance to consider moving away (table 6.9).

It seems from the above analysis that respondents' motivations for moving or staying are complex, interrelated, and highly context dependent. It might also be expected that people's intentions to move away from the coast or stay in place would be affected by their expectations for climate-

Table 6.9. Interviewee intentions to stay or move away from the coast. Data by Meredith Welch-Devine.

	No.	
Will not consider moving	30	45.50%
Will consider moving if I suffer catastrophic damage	11	16.70%
Will consider moving for mild to moderate climate-related changes	14	21.20%
Will consider moving for other personal reasons	11	16.70%
TOTAL	66	

related change, in this case sea level rise or increased frequency of damaging storms. While intentions to move away are positively related to expectations of both sea level rise ($r[741] = 0.15$, $p<0.0001$) and severe storms ($r[741] = 0.17$, $p<0.0001$), those values are small. Intentions to stay in place show no significant relationship to either climate-related factor.

Our interviews of coastal residents yielded additional insights. In response to questions about their expectations for the future, many respondents replied to the effect of, "If it started happening every year I'd move" (e.g., C22/23, M20/21) or "If my home were completely washed away, I'd move" (e.g., C07, M24), and many of those respondents think it may happen at some time—just not necessarily within their lifetimes. Those who were directly impacted by Hurricane Matthew are clearly more motivated to move away:

> C11: I am not going to live on Tybee again. . . . Everything is changing, and a hurricane did come to Tybee, and it was a bad hurricane. Also, another thing that would happen was if it rained and it was a high tide, I would have to pull the sandbags in front of the . . . you know. I just can't live like that. And, seriously, for me? Once I experienced a flood, I don't want to do that again. I just am not going to put myself back there. And I feel really bad because I miss Tybee a lot. If I feel like it's in my heart, but . . . I can't.

> C05: I can see climate change in my backyard, with the amount of water that comes in on the spring king tide. It's no longer down the bank, it's up in my yard now, so my wife and I are moving. . . . My daughter lives in XXXX, and the primary reason is to be closer to them. But my wife and I have been through eight hurricanes now. . . . and this one was pretty bad for us. . . . And then the third thing is climate change, getting away from the coast where, as I understand it from the research I've read, hurricanes aren't getting more frequent, but they're getting bigger and heavier. Stronger storms. So that's why we're going up there.

Intentions to Move Away

The Theory of Planned Behavior anticipates that behavioral outcomes are shaped by the attitudes that respondents bring to decision-making about eventual migration. The first step in investigating these relationships is via correlations of Attitudes toward adaptation responses and migration-related Behavioral Outcomes—Intentions to Move or Stay in Place.

Respondents' intentions to move away were strongly positively related to respondents' attitudes of entitlement (r[741] = 0.25, $p<0.0001$), very strongly to their attitudes of personal responsibility (r[741] = 0.55, $p<0.0001$), and less strongly to their appreciation of government assistance (r[741] = 0.11). Their intentions to stay in place were moderately positively related to attitudes of entitlement (r[741] = 0.28, $p<0.0001$) and appreciation for government assistance (r[741] = 0.23, $p<0.0001$) but moderately negatively related to their attitudes toward taking personal responsibility (r[741] = -0.12, $p<0.0001$). While most of these correlation values are modest, Analysis of Variance reveals the strong relationship of intentions to move away with attitudes of personal responsibility, $r^2 = 0.30$, (F [1, 740] = 320.49, $p<0.0001$).

Mitigating Variables, Social Norms, and Perceived Behavioral Control

The behavioral intentions that were expressed were not solely shaped by respondents' values. The Theory of Planned Behavior proposes that respondents' behavioral intentions would reflect the opinions of those around them: family, friends, and trusted community figures who shape the way we behave. In addition, our perceptions of our own abilities to undertake actions shape the actions we consider making.

Our survey asked respondents to tell us what people like them and people they respect are thinking in regard to sea level changes and increased storm severity and frequency, i.e., social norms with respect to climate-related change (table 6.10). We similarly asked people about their capacity to respond, i.e., their perceived behavioral control (table 6.11).

In each case, we again performed Principal Components Analysis to identify a small number of factors to stand in for responses to these individual variables and to be used in subsequent analyses. Three factors emerged from the analysis of Social Norms, explaining 63.8 percent of variance. Reluctant to Move is composed of expressions where people disagree with the idea of moving. Stay in Place is composed of expressions that embrace staying in place. Climate Skeptic includes two expressions that question whether change is occurring. Cronbach's Alpha for the five items in Resistant to Move was 0.79, generally regarded as Good; 0.64 for

Table 6.10. Investigating social norms. Factor groupings: R = Reluctant to Move, S = Stay in Place, C = Climate Skeptic. Data by Brian Orland.

	Disagree / Strongly disagree	Agree / Strongly agree
Most people like me worry about having to leave their homes R	26.5%	49.3%
Most people like me believe sea level rise will force us out of our homes R	38.6%	30.3%
Most people whose opinions I value will move to a safer part of this community R	33.3%	29.5%
Most people like me are thinking of moving in the next five years R	42.7%	28.6%
Most people like me will choose to move to a new community R	41.2%	26.8%
Most people like me will do what is needed to stay in their homes S	8.7%	66.5%
Most people like me expect to be living in the same home twenty years from now S	30.9%	46.6%
Most people whose opinions I value expect to ride out any storms S	25.7%	45.6%
Most people whose opinions I value are not concerned about sea level rise C	30.1%	35.8%
Most people whose opinions I value expect the climate to remain as it is C	35.6%	30.2%

Climate Skeptic, a low score; and 0.57 for Stay in Place, which indicates the individual variables comprising the factor were not strongly related (Pairwise correlations 0.25 to 0.37), although the correlation probabilities were all significant at $p<0.001$.

Three factors also emerged from Perceived Behavioral Control, explaining 54.08 percent of variance. Cronbach's Alpha for the three items in Personally in Control was 0.58, generally regarded as a low score; 0.52 for Seek Advice, a poor score; and 0.54 for Victim of Circumstances, also poor. Pairwise correlations within factors are between 0.14 and 0.38, correlation probabilities were all significant at $p<0.001$

Accordingly, we examined the relationships between the factors comprising Social Norms and Perceived Behavioral Control and the actions that might be taken in response to climate-related change—Intentions to Move or Stay in Place. For each of these factors, tables 6.12 and 6.13 present the correlation values, r, and Analysis of Variance for each interaction with the factors of Social Norms and Perceived Behavioral Control.

Table 6.11. Investigating perceived behavioral control. Factor groupings: [C] = Personally in Control, [A] = Seek Advice, [V] = Victim of Circumstances. Data by Brian Orland.

	Disagree / Strongly disagree	Agree / Strongly agree
Any choice about moving is up to me [C]	12.5 percent	69.2 percent
I am confident that I'll be able to move if that becomes necessary [C]	14.5 percent	65.5 percent
I will not have a problem moving to a new community if that becomes necessary [C]	21.9 percent	57.1 percent
I'll research authoritative sources to decide if it is necessary to move [A]	11.4 percent	64.7 percent
I can wait until later to make any decision about moving [A]	11.2 percent	60.4 percent
I'll seek advice from people important to me before deciding to move [A]	16.7 percent	59.1 percent
It will NOT be easy for me to decide to move if the time comes [A]	29.5 percent	49.6 percent
I will be able to recover from any damage my home suffers [V]	24.7 percent	43.1 percent
I'm concerned that I'll be forced to move by unexpected events [V]	32.3 percent	38.0 percent
I will NOT be able to pay for protection to allow me to stay here [V]	31.0 percent	36.8 percent

Table 6.12. Behavioral outcomes: intentions to move away. **Bold indicates strong associations**. Data by Brian Orland.

		r	r^2	df	F	p
Social norm	**Reluctant to move**	**0.59**	**0.34**	**1,740**	**385.28**	**<0.0001**
	Expect to stay	-0.28	0.08	1,740	63.18	<0.0001
	Expect no change	0.06	0.003	1,740	2.48	0.12
Perceived behavioral control	Personally in control	0.19	0.04	1,740	27.75	<0.0001
	Will seek advice	-0.03	0.001	1,740	0.77	0.38
	Victim of circumstances	**0.40**	**0.16**	**1,740**	**137.26**	**<0.0001**

Table 6.13. Behavioral outcomes: intentions to stay in place. **Bold indicates strong associations**. Data by Brian Orland.

		r	r^2	df	F	p
Social norm	Resistant to move	0.02	0.00	1,740	0.39	0.39
	Expect to stay	**0.44**	**0.20**	**1,740**	**180.05**	**<0.0001**
	Expect no change	0.12	0.01	1,740	11.04	0.0009
Perceived behavioral control	Personally in control	-0.05	0.002	1,740	1.69	0.19
	Will seek advice	**0.41**	**0.17**	**1,740**	**147.90**	**<0.0001**
	Victim of circumstances	0.002	0.00	1,740	0.004	0.95

Figures 6.4 and 6.5 show these associations. In these four cases, the relationships are evident and strong. Residents' stated intentions to move away in response to anticipated change (figure 6.4) are strongly positively related to the same kind of decisions being made by the significant influences around them—family, friends, and respected community figures. They are also strongly positively related to respondents' feelings that they won't be able to pay for protections to allow them to stay and that they might be forced to move by unexpected events. Their intentions to stay in place (figure 6.5) are strongly positively related to the same kinds of intentions among those around them whose opinions they respect and value. They are also strongly positively related to Seek Advice, which is composed of agreements that they would do research or wait until later because it is not easy to make such decisions.

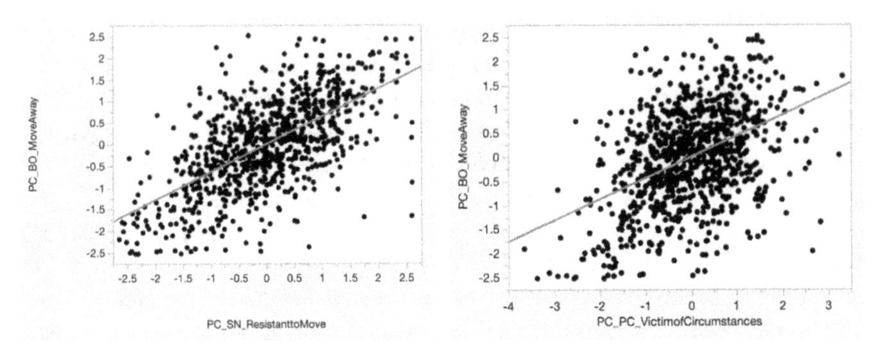

Figure 6.4. Intentions to move away versus (a) social norm, resistant to move, (b) perceived behavioral control, victim of circumstances. © Brian Orland.

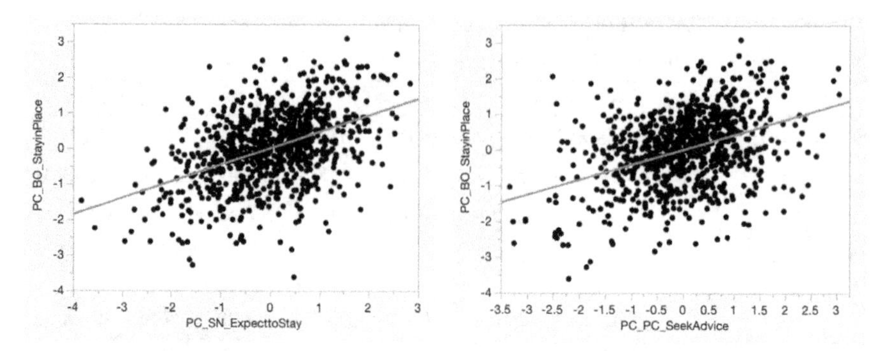

Figure 6.5. Intentions to stay in place versus (a) social norm, expect to stay, (b) perceived behavioral control, seek advice. © Brian Orland.

Discussion

The amenity and quality of life aspects of living on the coast contribute strongly to respondents' reasons for living on the coast and the changes that would make them consider moving. Natural beauty, the pace of life, recreational opportunities, and sense of connection all contribute strongly, and the potential loss of those as factors in deciding to move away all suggest that quality of life is a dominant consideration. While family and friend connections, growing up in the area, and being a good place to raise kids were positive contributors to wanting to stay on the coast, agreement on these was less emphatic. Economic issues such as investment opportunities and jobs were also less important. Demographic differences were related in expected but not substantial ways. Older, wealthier, and better-educated residents were more in agreement that lifestyle issues were important to their decisions to live on the coast than family, friend, and economic ones. African Americans and females found family and friends more important in their decisions. However, there were no demarcations of sufficient clarity or significance that might indicate an opportunity for policy intervention to incentivize migration to safer areas.

People's expectations for the future similarly appear little affected by demographic or locational factors (table 6.6). They are equally concerned about the potential effects of both sea level rise and increased storm damage on the region, and equally so regarding the personal impacts of storm damage, but are less concerned for the personal impacts of sea level rise. Hurricane Matthew's damage was significantly more widespread than flooding, and potential flood areas are a relatively small proportion of our entire six-county study area. We had also expected that location would have an effect on people's willingness to consider moving away, especially those closest to the water (distance to MHHW and elevation above MHHW) or who may have experienced flooding (distance to HWM). In fact, we were surprised that those more distant from the shore were more likely to consider migrating than those who lived closer, as shown by the relatively strong negative relationship between distance from the coast and the importance of lifestyle factors in choosing to live in the coastal region. The explanation lies in the roles that lifestyle amenities of the shore play in people choosing their shoreline or close-to-shoreline homes, and those considerations outweigh the practical issues of flood or storm damage hazard. It may be that wealthier people can afford to adapt to those hazards or that advanced age means that the next impact may not occur in their lifetime. Individual interview responses underscore those attitudes:

> P02: In fact, the storm hasn't changed our thinking about living here. We still look at it as a long shot. That's what life is about, really. You make your choices and hope they work out. Sometimes they do, and sometimes they don't. We played the odds by moving here, so after four years, we got hit. Maybe it'll be another hundred years. . . . I'm not a scientist, but based on what I read about climate change, chances are we'll probably have more disruptive weather, more violent weather. If it became a once-a-year or twice-a-year thing, we're both sixty-nine years old. We might think about moving somewhere, but I don't know where. . . . I think this island will be in serious jeopardy in the future. It's sad to say because it's such a beautiful place. But long term? Not during my lifetime, I don't think. I certainly think that this will not be a habitable island in the not too, too distant future.

Attitudes were strong regarding staying in respondents' current homes. They would take the necessary protective measures and not expect to relocate. This response may be shaped by their experience of weathering a hurricane (Matthew) that inflicted less damage than expected throughout the region. Very exposed locations were impacted badly, but those were few. There are strong feelings among our respondents that government should do more to protect homes and equally strong feelings that government should not help people move elsewhere. Government plays a challenging role in these coastal locations, where some do not accept that sea levels are rising (6 percent in this survey) or damaging storms becoming more frequent (5 percent), yet they are expected to plan for and invest in protecting communities.

While the actions that might be taken by respondents grouped strongly into intentions to move away or to stay, there were few demographic or distance factors that might have shaped those decisions. Older, wealthier, and better-educated residents were less inclined to move away, and African Americans were more likely to consider moving. Some of these observations run counter to the expectation that wealthier people would have the means to move away whereas African Americans are likely to have less means and the stronger ties to family expressed in our survey that might keep them in place.

The Theory of Planned Behavior framework was helpful in thinking about how the attitudes held by coastal residents would convert to intentions to act in response to climate-related change. What was surprising was the strength of the mitigating factors of Social Norms, the way that influential people around you shape your actions, and Perceived Behavioral Control, the extent to which people feel they have the means or ability to take action. The Social Norm "Reluctant to Move" was highly related to intention to move away, a factor composed of intentions to move but under duress—"if my home is threatened," and negatively "back to where I moved from," as was the Perceived Behavioral Control factor where respondents expressed their fear they will be forced to move by unexpected events and may not be able to afford protection. The Social Norm "Expect to Stay in Place" was highly related to intentions to stay in place. The Perceived Behavioral Control "Will seek advice" was also highly related—the constituent variables, "I'll research authoritative resources," "It will NOT be easy for me to decide . . . ," and "I can wait until later to make any decision," taken together resemble procrastination, and may thus lead to respondents remaining in place.

Conclusion

Our research has produced four major findings. First, people do not expect to migrate in the face of climate-related change; the overwhelming expectation is that people will do all they can to stay in place. Second, quality of place expressed as natural beauty and offering recreational opportunities plays a larger role in thinking about the future than demographic variables or economic or social ties. Third, those further from the ocean and the reach of storm flooding are more likely to consider future migration—the amenity of the coast is a strong "pull" factor. Finally, the oldest, wealthiest, and best educated residents are less inclined to move away from the coast than younger, less wealthy, and less educated cohorts.

The intent of this study was to explore less-studied aspects of the decisions shaping possible future migration away from the coast as well as to revisit those demographic and distance variables that are more familiar. In doing so, however, it raises further questions. One overarching observation relating to the title of the chapter is that people's individual values, attitudes, and resulting actions are hugely variable. We treated age as a four-level categorical variable (18–24, 25–44, 45–64, and 65+), household income as a seven-level variable (25,000 through to $150,000+), and educational attainment as a seven-level variable (from not finished high school to graduate degree). For each, responses toward several factors, such as reasons to live on the coast or to move away, attitudes to climate-related change, and intentions to move away, showed that the highest category of age, income, or education behaved differently. The oldest, wealthiest, and best educated frequently responded "out of line"—less influenced by coastal qualities of natural beauty and recreation, less intentions to move than younger, less affluent, and less educated cohorts. The significance of this lies in the fact that those characteristics tend to describe the people most likely to shape policy, direct resources, and argue persuasively for the outcomes they desire. We do not have detailed enough information to chase this idea further, but we should be concerned if the values and actions of decision-makers diverge from those of the public at large.

Although it should come as no surprise that quality of life and lifestyle issues are central to the attractiveness of the coast, we observed a perplexing trade-off, that resistance to moving away *increases* as distance to open water *decreases*. Homes and businesses may be exposed and vulnerable to sea level rise and storm damage, but until the problems arrive at the property there is no reason to leave the environment you value.

We heard these same kinds of seemingly contradictory, yet logical in their context, statements numerous times in our interviews. Sometimes the decision to stay in place is driven by attachment, sometimes by lack of means to move, sometimes by the knowledge it can easily be rebuilt, sometimes by inability to make the decision to move. Policy to date has often been based on simplification of maddeningly complex situations. Policies impose lines on the ground, define segregations of the population, and assume that attitudes once held are permanent and immutable. Our work demonstrates that simplified views are not adequate as a basis for policy development. Even so, while our survey responses hint at the various threads and pressures at play, the synoptic view of traditional quantitative analysis also hides the richness of local variability, including the local holdout who is nevertheless key to understanding and responding to the conundrum of how or whether to help communities that may prefer to be left alone:

> **M09:** I think it's [sea level rise] on the way. I think with everything getting warmer and warmer, the winters not being as cold, I think that's causing it. . . . But I'm not moving. I'm not moving anywhere, so definitely not that. . . . It's just home.

Acknowledgments

This material is based upon work supported by the National Science Foundation under Grant No. 1719532, The Wake of Hurricane Matthew: Vulnerability, Resilience, and Migration, and in part by an Institutional Grant (NA10OAR4170084) to the Georgia Sea Grant College Program from the National Sea Grant Office, National Oceanic and Atmospheric Administration, United States Department of Commerce. Institutional Review Board (IRB) approval for Study 00004472 by the University of Georgia IRB. We thank Jill Gambill, David Rickless, Danielle Valdes, and Arianne Wolfe for their assistance in data gathering, coding, and interpretation, as well as numerous coastal Georgia residents for participating in interview and survey phases of this project.

Brian Orland is professor emeritus at both the University of Illinois and Penn State University and former Rado professor of geodesign at the University of Georgia. His teaching and research focus on environmental perception, the understanding and representation of environmental impacts, and the design of information systems for community-based design and planning. Recent work includes the use of serious games, visualization and mobile devices for data collection, information dissemination, and citizen engagement in landscape design and planning. He also codirects a 220-university global collaboration called the "International Geodesign Collaboration."

Meredith Welch-Devine is assistant dean of the Graduate School, University of Georgia. Her primary research interests include climate change adaptation, collective management of common-pool resources, and policy and practice related to conservation and sustainability. She is currently exploring the impacts of extreme weather events on how people think about climate planning and adaptation and integrating political ecology and ethnoecology to more closely examine how people perceive and understand climate change, particularly how they use biodiversity as an indicator of change.

Micah Taylor is a PhD student at the University of Georgia specializing in emerging geospatial science, visualization technology, environmental psychology, and geodesign. His research addresses the role of visual media, including maps, 3D models, animations, virtual and augmented reality, and mobile applications in communicating complex aspects of the environment and environmental change. The goal of his work is to find more effective ways for citizens to be involved in regional design and planning through emerging technologies, using the principles of storytelling to enable more informed decisions for the environments in which they live.

References

Adger, W., Jon Barnett, Katrina Brown, Nadine Marshall, and Karen O'Brien. 2013. "Cultural Dimensions of Climate Change Impacts and Adaptation." *Nature Climate Change* 3: 112–117. doi: 10.1038/nclimate1666.

Ajzen, Icek. 1985. "From Intentions to Actions: A Theory of Planned Behavior." In *Action Control: From Cognition to Behavior*, edited by Julius Kuhl and Jürgen Beckmann, 11–39. Berlin, Heidelberg: Springer Berlin Heidelberg.

Balica, S. F., N. G. Wright, and F. van der Meulen. 2012. "A Flood Vulnerability Index for Coastal Cities and Its Use in Assessing Climate Change Impacts." *Natural Hazards* 64(1): 73–105.

Bilgili, Ö., M. Siegel. 2015. "To Return Permanently or to Return Temporarily? Explaining Migrants' Intentions." *Migration & Development*. DOI:10.1080/21632324.2015.1088241

Bilskie, Matthew, S. C. Hagen, Stephen Medeiros, A. T. Cox, Michael Salisbury, and David Coggin. 2016. "Data and Numerical Analysis of Astronomic Tides, Wind-Waves, and Hurricane Storm Surge along the Northern Gulf of Mexico." *Journal of Geophysical Research: Oceans* 121. doi: 10.1002/2015JC011400.

Bowser, G. C., and S. L. Cutter, 2015. "Stay or Go? Examining Decision Making and Behavior in Hurricane Evacuations." *Environment* 57(6): 28–41.

Bronen, R. 2015. "Climate-Induced Community Relocations: Using Integrated Social-Ecological Assessments to Foster Adaptation and Resilience." *Ecology and Society* 20(3): 36.

Buchanan, M. K., R. E. Kopp, M. Oppenheimer, and C. Tebaldi. 2016. "Allowances for Evolving Coastal Flood Risk under Uncertain Local Sea-Level Rise." *Climatic Change* 137(3): 347–62.

Bukvic, Anamaria, and Graham Owen. 2016. "Attitudes towards Relocation Following Hurricane Sandy: Should We Stay or Should We Go?" *Disasters* 41. doi: 10.1111/disa.12186.

Bukvic, Anamaria, Andrew Smith, and Zhang Angang. 2015. "Evaluating Drivers of Coastal Relocation in Hurricane Sandy Affected Communities." *International Journal of Disaster Risk Reduction* 13. doi: 10.1016/j.ijdrr.2015.06.008.

Bukvic, Anamaria, Hongxiao Zhu, Rita Lavoie, and Austin Becker. 2018. "The Role of Proximity to Waterfront in Residents' Relocation Decision-Making Post-Hurricane Sandy." *Ocean and Coastal Management* 154: 8–19. doi: 10.1016/j.ocecoaman.2018.01.002.

Burger, Joanna, and Michael Gochfeld. 2017. "Perceptions of Severe Storms, Climate Change, Ecological Structures and Resiliency Three Years Post–Hurricane Sandy in New Jersey." *Urban Ecosystems* 20(6): 1261–75. doi: 10.1007/s11252–017–0678-x.

Butler, William, R. E. Deyle, and C. Mutnansky. 2016. "Low-Regrets Incrementalism: Land Use Planning Adaptation to Accelerating Sea Level Rise in Floridas Coastal Communities." *Journal of Planning Education and Research* 36. doi: 10.1177/0739456X16647161.

Christensen, Rhonda, and G. Knezek. 2015. "The Climate Change Attitude Survey: Measuring Middle School Student Beliefs and Intentions to Enact Positive Environmental Change." *International Journal of Environmental and Science Education* 10: 773–88. doi: 10.12973/ijese.2015.276a.

Connell, J. 2016. "Last Days in the Carteret Islands? Climate Change, Livelihoods, and Migration on Coral Atolls." *Asia Pacific Viewpoint* 57(1): 3–15.

Crate, S. 2011. "Climate and Culture: Anthropology in the Era of Contemporary Climate Change." *Annual Review of Anthropology* 40: 175–94.

Cutter, S. L. 2016. "The Landscape of Disaster Resilience Indicators in the USA." *Natural Hazards* 80(2): 741–58.

Cunsolo Willox, Ashlee, Sherilee L. Harper, and Victoria L. Edge. 2012. "'Myword': Storytelling in a Digital Age; Digital Storytelling as an Emerging Narrative Method for Preserving and Promoting Indigenous Oral Wisdom." *Qualitative Research* 13(2): 127–47. doi: 10.1177/1468794112446105.

Dong, Yanan, Saiquan Hu, and Junming Zhu. 2018. "From Source Credibility to Risk Perception: How and When Climate Information Matters to Action." *Resources, Conservation and Recycling* 136: 410–417.

Duerden, Mat D., and Peter A. Witt. 2010. "The Impact of Direct and Indirect Experiences on the Development of Environmental Knowledge, Attitudes, and Behavior." *Journal of Environmental Psychology* 30(4): 379–392.

Dyckman C. S., C. St. John, and J. B. London. 2014. "Realizing Managed Retreat and Innovation in State-Level Coastal Management Planning." *Ocean and Coastal Management* 102: 212–23.

Farbotko, C., and H. Lazrus. 2012. "The First Climate Refugees? Contesting Global Narratives of Climate Change in Tuvalu." *Global Environmental Change* 22(2): 382–90.

Fishbein, Martin A., and Icek Ajzen. 1975. *Belief, Attitude, Intention and Behavior: An Introduction to Theory and Research.* Vol. 27. Reading, MA: Addison-Wesley.

Fletcher, C. S., A. N. Rambaldi, F. Lipkin, and R. R. J. McAllister. 2016. "Economic, Equitable, and Affordable Adaptations to Protect Coastal Settlements against Storm Surge Inundation." *Regional Environmental Change* 16(4): 1023–34.

Fussell, E., N. Sastry, and M. Vanlandingham. 2010. "Race, Socioeconomic Status, and Return Migration to New Orleans after Hurricane Katrina." *Population and Environment* 31(1–3): 20–42.

Grier, Peter. 2005. "The Great Katrina Migration." *Christian Science Monitor* 12: 14.

Groen, Jeffrey A., and Anne E. Polivka. 2010. "Going Home after Hurricane Katrina: Determinants of Return Migration and Changes in Affected Areas." *Demography* 47(4): 821–44. doi: 10.1007/BF03214587.

Groenewold, G., B. Bruijn, and R. Bilsborrow. 2012. "Psychosocial Factors of Migration: Adaptation and Application of the Health Belief Model." *International Migration* 50(6): 211.

Gutmann, Ethan, Roy M. Rasmussen, Changhai Liu, and Kyoko Ikeda. 2018. "Changes in Hurricanes from a 13-Yr Convection-Permitting Pseudo-Global Warming Simulation." *Journal of Climate* 31(9): 3643–57.

Hauer, Mathew E. 2017. "Migration Induced by Sea-Level Rise Could Reshape the US Population Landscape." *Nature Climate Change* 7(5): 321–325.

Hauer, Matthew, Jason Evans, and Deepak Mishra. 2016. "Millions Projected to Be at Risk from Sea-Level Rise in the Continental United States." *Nature Climate Change* 6(7): 691–95.

Hoffman, Susanna M. 2017. "Disasters and Their Impact: A Fundamental Feature of Environment." In *Handbook of Environmental Anthropology*, edited by H. Kopnina and E. Ouimet, 193–205. London: Routledge.

Hoogendoorn, Gijsbert, Jennifer Fitchett, and Bronwyn Grant. 2016. "Climate Change Threats to Two Low-Lying South African Coastal Towns: Risks and Perceptions." *South African Journal of Science* 112: 1–9. doi: 10.17159/sajs.2016/20150262.

King, D., D. Bird, K. Haynes, H. Boon, A. Cottrell, J. Millar, T. Okada, P. Box, D. Keogh, and M. Thomas. 2014. "Voluntary Relocation as an Adaptation Strategy to Extreme Weather Events." *International Journal of Disaster Risk Reduction* 8: 83–90.

Klabunde, A., and F. J. Willekens. 2016. "Decision-Making in Agent-Based Models of Migration: State of the Art and Challenges." *European Journal of Population* 32(1): 73–97.

Koenig, Todd A., Jennifer L. Bruce, Jim O'Connor, Benton D. McGee, Robert R. Holmes, Jr., Ryan Hollins, Brandon T. Forbes, Michael S. Kohn, Mathew F. Schellekens, Zachary W. Martin, and Marie C. Peppler. 2016. *Identifying and Preserving High-Water Mark Data: U.S. Geological Survey Techniques and Methods*. 3-A24 in Applications of Hydraulics. US Geological Survey. http://dx.doi.org/10.3133/tm3A24.

Kousky, C. 2014. "Managing Shoreline Retreat: A U.S. Perspective." *Climatic Change* 124(1): 9–20.

Landry, Craig E., Okmyung Bin, Paul Hindsley, John C. Whitehead, and Kenneth Wilson. 2007. "Going Home: Evacuation-Migration Decisions of Hurricane Katrina Survivors." *Southern Economic Journal* 74(2): 326–43. doi: 10.2307/20111970.

Lu, M. 1998. "Analyzing Migration Decisionmaking: Relationships between Residential Satisfaction, Mobility Intentions, and Moving Behavior." *Environment and Planning A: Economy and Space* 30(8): 1473–95. doi: 10.1068/a301473

Marino, E., and H. Lazrus. 2015. "Migration or Forced Displacement? The Complex Choices of Climate Change and Disaster Migrants in Shishmaref, Alaska and Nanumea, Tuvalu." *Human Organization* 74(4): 341–50.

McCaffrey, Sarah, Robyn Wilson, and Avishek Konar. 2018. "Should I Stay or Should I Go Now? Or Should I Wait and See? Influences on Wildfire Evacuation Decisions." *Risk Analysis* 38(7): 1390–1404.

Milfont, Taciano L., Laurel Evans, Chris G. Sibley, Jan Ries, and Andrew Cunningham. 2014. "Proximity to Coast Is Linked to Climate Change Belief." *PLOS One* 9(7): e103180. doi: 10.1371/journal.pone.0103180.

Morgan, Philip D., ed. 2010. *African American Life in the Georgia Lowcountry: The Atlantic World and the Gullah Geechee*. Athens, GA: University of Georgia Press.

Musser, Jonathon, Kara Watson, and Anthony Gotvald. 2017. "Characterization of Peak Streamflows and Flood Inundation at Selected Areas in North Carolina Following Hurricane Matthew, October 2016." Federal Emergency Management Agency: USGS.

Pham, Erika. O., Christopher T. Emrich, Zhenlong Li, Jamie Mitchem, and Susan Cutter. 2020. "Evacuation Departure Timing during Hurricane Matthew." *Weather, Climate, and Society* 12(2): 235–248.

Rahmstorf, Stefan. 2017. "Rising Hazard of Storm-Surge Flooding." *Proceedings of the National Academy of Sciences* 114(45): 11806–8. doi: 10.1073/pnas.1715895114.

Song, Jie, and Binbin Peng. 2017. "Should We Leave? Attitudes towards Relocation in Response to Sea Level Rise." *Water* 9: 941. doi: 10.3390/w9120941.

Stojanov, Robert, Barbora Duží, Ilan Kelman, Daniel Nemec, and David Procházka. 2016. "Local Perceptions of Climate Change Impacts and Migration Patterns in Malé, Maldives." *Geographical Journal* 183. doi: 10.1111/geoj.12177.

Theodori, A. E., and G. L. Theodori. 2015. "The Influences of Community Attachment, Sense of Community, and Educational Aspirations upon the Migration Intentions of Rural Youth in Texas." *Community Development* 46(4): 380–91. doi:10.1080/15575330.2015.1062035

Thomas, Merryn, Nick Pidgeon, Lorraine Whitmarsh, and Rhoda Ballinger. 2015. "Mental Models of Sea-Level Change: A Mixed Methods Analysis on the Severn Estuary, UK." *Global Environmental Change* 33. doi: 10.1016/j.gloenvcha.2015.04.009.

Trumbo, C. W., and G. J. O'Keefe. 2005. "Intention to Conserve Water: Environmental Values, Reasoned Action, and Information Effects across Time." *Society & Natural Resources* 18(6): 573–85.

Weller, S., R. Baer, J. Prochaska. 2016. "Should I Stay or Should I Go? Response to the Hurricane Ike Evacuation Order on the Texas Gulf Coast." *Natural Hazards Review*. doi: 10.1061/(ASCE)NH.1527–6996.0000217, 04016003.

Wilmot, Susan R. 2009. "Attitudes, Behavioral Intentions, and Migration: Resident Response to Amenity Growth-Related Change in the Rural Rocky Mountain West." PhD diss., Utah State University, Ogden, UT.

Wheeler, Benedict W., Mathew White, Will Stahl-Timmins, and Michael H. Depledge. 2012. "Does Living by the Coast Improve Health and Wellbeing?" *Health & Place* 18(5): 1198–201. doi: https://doi.org/10.1016/j.healthplace.2012.06.015.

Yoon, D. K. 2012. "Assessment of Social Vulnerability to Natural Disasters: A Comparative Study." *Natural Hazards* 63(2): 823–43.

Chapter 7

"The Times They Are A-Changin'" but "The Song Remains the Same"

Climate Change Narratives from the Coromandel Peninsula, Aotearoa New Zealand

Paul Schneider and Bruce Glavovic

Introduction

The imperative for coastal communities to implement proactive and sustained measures to adapt to climate change is now well established (Boyer, Meinzer, and Bilich 2017; IPCC 2019). A little over a decade ago, this imperative was nascent in many parts of the world, including Aotearoa New Zealand (ANZ) (Rouse et al. 2017), the locus of this research. Pockets of overt climate change denial persist, even in the face of obvious climate change–driven impacts. Local calls for adaptation action are at times met with passive indifference by governing authorities, and, in some jurisdictions, thinly veiled reluctance to act is cloaked in a veneer of tokenistic gestures. At other times, bold steps are taken to reduce coastal hazard risk. What drives local responses to the unfolding climate emergency facing low-lying coastal communities? How might local communities and their governing authorities be galvanized to take meaningful action to reduce exposure and vulnerability to climate change impacts? Addressing these questions necessitates in-depth understanding of local community dynamics, cultures, histories, and livelihoods and the interactions between coastal communities, civic leaders, and governing authorities. Such understanding cannot be developed on the basis of short-term

Notes for this chapter begin on page 195.

studies by outsiders. Yet, few longitudinal studies by researchers embedded in local communities have been undertaken to address such questions (Archer et al. 2017; Fawcett et al. 2017; Moreno and Shaw 2018), making it difficult to reveal the underlying nuances and drivers of adaptation action or inaction.

This ethnographic study of adaptation on the Coromandel Peninsula in ANZ traces the evolution of local climate change events, publications, and responses since 2009, focusing on local narratives about adaptation (cf. figure 7.7). We peer beneath the surface of rhetoric and superficial accounts that might otherwise be proffered in a one-off survey of a sample of the local community. Paul, the lead author, and his family have been part of a local Coromandel community for over a decade (since 2007), and Bruce has maintained a close association in his role as Paul's research supervisor and collaborator. We have thus been embedded in local realities for over a decade. Following the distinction between "thin" and "thick" descriptions popularized by Geertz (1973), we endeavor to present a "thick description" of local narratives, drawing on insights from and with community members and local stakeholders as we and they seek to make sense of climate change and face the challenge of escalating coastal hazard risk. Our description is complicated by the reality of multiple local narratives, divergent viewpoints, and contending voices within and between communities on the peninsula, within and between local government actors, and between the local authorities and the communities they seek to govern. In offering this thick description, we reveal the undeniable but, at times, "below-the-surface" influence of power and politics in shaping the trajectory of adaptation on the Coromandel Peninsula. The "story" has evolved in convoluted ways over the last decade, with adaptation responses waxing and waning. A persistent adaptation gap—the mismatch between rhetoric and institutionalized adaptation measures—has been deep and real. But recent initiatives driven by the Thames-Coromandel District Council (TCDC) signal the possibility of significant change: a shoreline management planning process involving local communities has been initiated, and it promises to address long-term coastal hazard risk. What precipitated this adaptation turn? And, given the vexed nature of climate change on the peninsula over the last decade, how deep is this adaptation move? As Bob Dylan mused, are we witnessing a time of change, or will the song remain the same, to invoke another music legend, Led Zeppelin?

We provide a brief description of our research approach before describing the Coromandel Peninsula setting and the institutional milieu within which adaptation has been and is being framed. On the surface, well-intentioned and robust legislation and policy provisions have been put in place to enable local communities to institutionalize anticipatory actions to reduce coastal hazard risk and chart climate-resilient development pathways. However, things are murkier below the surface. Critical scholars from diverse disciplines bemoan the ongoing prioritization of short-term private interests over concerns about citizen engagement, Māori (the indigenous people of ANZ) rights and *Mātauranga Māori* (ancestral Māori knowledge), public safety, and equitable and environmentally sustainable coastal development. Blame is often sheeted home to the ANZ experiment with neoliberalism—hamstringing adaptation efforts, privileging elite interests, marginalizing Māori, stultifying authentic local democracy, causing environmental degradation, and encouraging high-risk shoreline development. We draw insights from political ecology to move beyond a macrolevel critique of neoliberalism to reveal the localized influence of power and politics on environmental governance and adaptation responses in ANZ. The many stories shared with us over the last ten years attest to the complex, contested, and changing drivers of local responses to escalating coastal hazard risk. We conclude by imagining how Coromandel communities and their governing authorities might chart alternative pathways that would institutionalize more engaged, equitable, climate-resilient, and sustainable coastal development pathways.

This Ethnography

This research is grounded in a decade-long, and ongoing, ethnographic study of adaptation on the Coromandel Peninsula, ANZ (Schneider and Glavovic 2019). A community-based participatory research approach (Cvitanovic et al. 2019) characterizes our effort to "give voice" to local narratives about climate change and coastal hazard risk along with the barriers and enablers for developing effective strategies to reduce risk. We sought to conduct research *with* rather than *on* research participants; to probe the often invisible and unspoken elements of everyday life from the vantage point and lived experience of participants; and also to delve deeply into local nuances and context as opposed to being satisfied with a "thin description" of unfolding realities. We used an ethnographic approach to probe the multifaceted realities prevailing in the diverse communities of the peninsula. Our goal has been to craft a "thick description," grounded in a deepening understanding of the cultures, histories, and livelihoods of the peninsula, of how local people, from their own perspective, view climate change, behave, and interact, leveraging our insights as researchers and with Paul actively participating in the life and seasonal rhythms of the Coromandel.

Trust was built over time with many key stakeholders as we sought to reveal the "local reality and the climate change adaptation dilemma" (Schneider 2014) and "contrasting climate change perceptions" (Schneider and Glavovic 2019; Schneider, Glavovic, and Farrelly 2017). Sixty-two in-depth interviews were carried out to understand how risk governance, resilience planning, and coastal adaptation were envisaged and being undertaken. Local participants included Māori (including *kaumātua*—elderly Māori of standing in an extended family group, i.e., *whānau*, and community members); elected councilors/politicians; local and regional council management; planning and engineering staff, as well as professionals contracted to the local council; independent specialists; and a wide range of community members, including shoreline residents, farmers, activists, and people involved in the spectrum of Coromandel livelihoods. Insights were also gained from interviews with key figures involved in climate change issues at the national level. Interviews based on open-ended questions were conducted at locations selected by research participants. Interviews were recorded, transcribed, and thematically analyzed for commonalities and differences. Braun and Clarke's (2014) six-phase approach to thematic analysis was used, comprising (1) data familiarization, (2) initial sorting of issues for code generation, (3) the identification of themes, (4) the review of themes, (5) the definition of themes, and (6) the identification of research findings. Participants were asked to reflect on key themes, including barriers and opportunities, coastal issues, risk, extreme events, future prospects, coastal management issues, and governance roles and responsibilities. Together, these interviews and the stories shared help to build a holistic understanding of the complexity of social and biophysical coastal interactions, rooted in participants' intimate connections to the peninsula, with many having lived here for decades. A council planner interviewed fittingly described the merits of this approach:

> We've got records of people living in communities for generations. There are diary notes, photographs, and much more. The familiarity of a place. . . . You can draw on stories from that place. . . . Anybody who has been living in a particular locality for decades has stories to tell, and there's good sound common sense. . . . And of course there's always the photographs and stuff that get brought out from under the bed in the shoebox. (Council planner 2012, pers. comm.)

Photographs, or photo elicitation, were also used as "tools" during the interview process, often prompting what Pink (2008: 2) describes as "inevitably collaborative" storytelling. We also systematically reviewed news media for stories about climate change and adaptation on the peninsula. And we reviewed and analyzed scholarly publications and the "gray literature," including government and nongovernment reports, plans, and policy documents. Hence, we can share over a decade of multifaceted and continuously evolving stories about adaptation from the Coromandel Peninsula.

The Coromandel Setting and Institutional Context

The Geography of the Coromandel Peninsula

The Coromandel Peninsula is located off the North Island's east coast. It is one of ANZ's favorite holiday destinations (Davison 2011). The Coromandel offers four hundred kilometers of "iconic and diverse coastline" (TCDC 2018: 1): white sandy beaches, clear water, forest-clad hills, and a feeling of remoteness and wilderness, despite being only a couple hours' drive from the main centers of Auckland, Hamilton, and Tauranga. Communities comprise permanent residents, those on holiday, and Māori settlements. Each of the fifty-plus settlements has its own distinctive features, history, and lifestyle characteristics (TCDC 2016). Despite their geographic proximity, the communities have widely divergent values and views, including countervailing perceptions about climate change (Schneider et al. 2017). These differing values, views, and perceptions reflect stark differences in culture, historical experience, worldviews, interests, and socioeconomic and political standing.

Over the summer period (December to March), the peninsula is transformed by an influx of holidaymakers. Over the last fifty years, larger communities, like Whitianga, whose population swells by 600 percent in summer, morphed from quiet coastal villages into resort towns. Whitianga's resident population of 4,368 in 2013 (Statistics New Zealand 2013) is projected to reach a permanent population of up to 6,000 by 2040, located mainly on low-lying coastal land (Monin 2012), which is experiencing coastal squeeze and facing escalating coastal hazard risk. Seventy percent of the Coromandel's beach areas and dunes are developed within one hundred meters of the sea (ARC 2004), in areas prone to coastal hazards (TCDC 2015). The legacy of historical development decisions, dramatic increases in property values, ongoing pressure to develop the shoreline, and continuing approval of property development in "natural" coastal areas (greenfield development), some of which are high-risk locations, make this region a "hotspot" for climate change impacts and constrain adaptation prospects (Schneider et al. 2017).

Expected climate change impacts on the Coromandel include more frequent and intense heavy rainfall events and a rise in sea level (EW and TCDC 2003; MFE 2013, 2014) that push the sea inland, thus affecting

Figure 7.1. Coromandel beachfront property development at Hahei. © Paul Schneider.

low-lying areas and estuaries. Under all climate change scenarios, sea level will continue to rise during the twenty-first century and beyond, and the rate of sea level rise will very likely be faster than in the past few decades (MFE 2017). Without well-planned and managed adaptation responses, many existing coastal defenses, such as sea walls, will be breached (Gluckmann 2013; MFE 2008; NIWA 2008).

The most obvious manifestation of changing coastal conditions on the Coromandel is "major erosion along several areas of our coastline" (TCDC 2014: 2). Local beachfront property owners tend to use rocks and ad hoc structures, such as wooden walls/planks and concrete, in desperate attempts to protect their properties against erosion. However, legislative and policy provisions and guidance by central government (Department of Conservation 2010; MFE 2017) discourages the construction of coastal defenses, as these interfere with dynamic coastal processes. They can also have a detrimental effect on public amenity values, and they tend to exacerbate erosion further down the coast. With private property increasingly exposed to adverse impacts, there have long been calls by many locals for a "right" to combat the erosion by constructing protective works to prevent deterioration of their low-lying coastal properties. In contravention to evolving legislative and policy stipulations, these calls have been echoed over the past three mayoral terms. In 2014, the then district mayor (TCDC 2014: para. 5) supported beachfront property owners calling for protection of coastal properties, stating: "Protection work need[s] to be done sooner rather than later because every time we wait, we're losing more of our coastline." As a result, beaches such as Buffalo Beach, a "two-mile stretch of gleaming white sand" is now backed by "rock protection works" continually costing local ratepayers "about NZ$120,000 per annum" in maintenance (EW 2006: E2). To add insult to injury, the "rock protection works have led to a "loss of high tide beach" (EW 2006: B10).

Until very recently, a deep and persistent adaptation gap prevailed on the Coromandel Peninsula—with palpable climate change denial commonplace and apparent at the highest level of the TCDC, the predominant response to coastal hazard risk tended toward reliance on protective measures (Schneider and Glavovic 2019; Schneider et al. 2017). But "change is in the air." Following the adoption of a Coastal Management Strategy and Coastal Hazards Policy in 2018, the TCDC embarked on an ambitious shoreline management planning process that signals a volte-face in adaptation prospects:

> All of our coastal communities will be relied upon to tell us their coastal stories, pass on their knowledge of coastal environments, engage in discussions and work through solutions. We will work with communities at the grassroots level to inform, be informed by and collaborate in identifying objectives, issues and solutions. In recognising the coastal environment as taonga [treasures], we will work directly with mana whenua [those with power or authority over tribal lands] to ensure that SMPs [Shoreline Management Plans] reflect their objectives. (TCDC 2020a: 2)

What brought about this turnaround, and how deep does it go? To answer this question, one first needs to understand the institutional milieu that has shaped adaptation in ANZ.

The Institutional Setting for Adaptation

ANZ is a small, developed island nation in the southwest Pacific Ocean, with a population of about five million people concentrated along the shoreline. The country was settled by Māori in the late thirteenth century. Indigenous ways of life were dramatically affected by European contact from the late 1700s. The Treaty of Waitangi, the country's founding document, was signed in 1840 by about five hundred Māori chiefs and representatives of the British Crown. Fundamentally different understandings about the Treaty were written into the English and Te Reo Māori versions of it. Following the 1860s wars, which represented a gross imbalance of manpower and weaponry, lands were confiscated, access to traditional resources was cut off, and a series of persistent injustices invoked that continue to influence the well-being and prospects of Māori. The Waitangi Tribunal, established in 1975, was given authority to discern the meaning of the Treaty. An Office of Treaty Settlements was established to resolve claims of Treaty breaches, make restitution, and restore Crown relationships with *iwi* (Māori extended kinship group or tribe). Principles of the Treaty now underpin many laws, including statutes that shape coastal hazard risk and adaptation to climate change, and, among other things, require local government to consult local Māori as *tangata whenua* (people with ancestral roots in and authority over a particular locality).

From the 1980s, the country embarked on a bold neoliberal-inspired reform of the political economy and public sector management, moving from an interventionist state to deregulation and market self-regulation, privatization of the public sphere, and government and public spending cutbacks. Fragmented legislation and local government bodies were consolidated. The reform was deeper than parallel reforms in many other Western liberal democracies. Public decision-making responsibilities were devolved to local government, with paradoxical strengthening of provisions for public participation. Shortcomings were exposed by the global financial crisis, the devastating Canterbury earthquakes, and a series of social, economic, technological, and environmental stresses and shocks in the late 2000s and early 2010s. These were compounded by central government pressure on local authorities to exercise fiscal discipline in the face of these global trends and local stressors, while councils and ratepayers were simultaneously expected to assume ever greater financial responsibility for major infrastructure works, including deferred maintenance, and to make provision for environmental protection, and tourism and urban growth, without enabling government support. Recentralization of public decision-making responsibility and a strident agenda of debt reduction, containment of public services, and fiscal austerity were championed by the three-term center-right government under the New Zealand National Party from 2008. The relationship between local councils and the communities they govern reached a crossroads in the early 2010s (Asquith 2012). A robust legislative framework and strong local government managerial capabilities and financial autonomy were in place. Increasing attention was focused on managing natural hazard risk. But local action on climate change was stifled, local government's role in fostering community well-being was curtailed, many environmental aspirations were unrealized, and some lamented the caliber of elected members, community disengagement, and the "deepening democratic deficit"

(Asquith 2012; Cheyne 2015). From 2017, the election of a Labour-led coalition government with the Green Party and New Zealand First (which held the balance of power in ANZ's mixed-member proportional representation electoral system) saw a reinvigoration of climate action and restoration of well-being provisions through amendments to the Local Government Act.

Since the early 1990s, coastal hazard risk management has been governed principally through the Resource Management Act 1991 (RMA). The RMA comprises a hierarchy of regulatory provisions that guide how coastal hazards should be addressed. Since 2016, the amended RMA has included management of natural hazard risk as a matter of national significance—a regulatory provision invoked once the destructive potential of natural hazards was tragically underscored by the Canterbury earthquake sequence in 2010–11. A series of floods and other extreme events, the impacts of climate change, and the risks posed by rising sea level compelled more focused attention on natural hazard risk over the last decade, with a more intense spotlight on coastal adaptation in recent years (Rouse et al. 2017).

The RMA requires local government to "have regard to the effects of climate change" (RMA §7[i]). Several other statutes relevant to coastal hazard risk management and climate change adaptation include the Local Government Act (LGA) (with provisions for community well-being, Long Term Plans, and community engagement), the Local Government Official Information and Meetings Act 1987 (includes Land Information Memoranda for matters, like hazards, affecting private property), the Building Act 2004 and Building Code (addressing among other things the safety of buildings and flood standards), and the Civil Defence Emergency Management (CDEM) Act 2002 (including provisions for national and regional preparedness, disaster response, recovery, and risk reduction). Implementation of legislation is supported by national-level policies, including provisions explicitly focused on adaptation principally through the New Zealand Coastal Policy Statement (NZCPS). The NZCPS was first promulgated in 1991 under the RMA, was reissued in 2010, and is currently supported by national guidance for local government on coastal hazards and climate change (MFE 2017). The orientation of the NZCPS is precautionary, and it requires councils to have a planning horizon of at least one hundred years. The requirement to avoid increasing risks due to natural hazards and climate change (objective 5) is elaborated in policies 24–27, which include locating new development (including infrastructure) away from areas prone to coastal hazards, consideration of managed retreat for existing development exposed to natural hazard risk, and restoring natural defenses against coastal hazards.

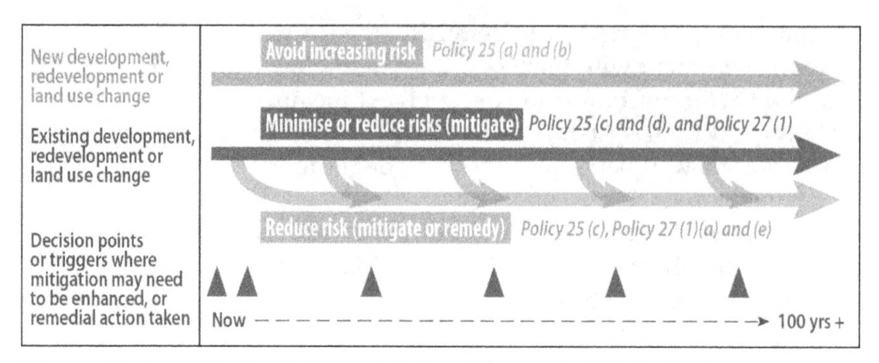

Figure 7.2. New Zealand Coastal Policy Statement 2010 decision context for coastal areas exposed to coastal hazards and climate change. Note: uses terminology from RMA 1991 §5(2) (c). Source: MFE 2017.

Implementation of land-use planning provisions, emergency management, and the provision of basic services are devolved to regional and local (district and city) councils. For the Coromandel Peninsula, the former is the Waikato Regional Council (WRC); the latter is the Thames-Coromandel District Council (TCDC). Among other environmental and emergency management responsibilities, the WRC manages coastal erosion and flooding as well as water quality and quantity. The TCDC manages, among other things, land use planning, building control, emergency management, storm water, wastewater and water supply, infrastructure, and local roads. The avoidance, reduction, and mitigation of natural hazard risk is regulated through policies, plans, and rules at both regional and local council levels, through RMA Regional Policy Statements (RPS) (together with Regional Coastal Plans) and District Plans, as well as provisions in the statutes mentioned previously. The NZCPS and the Waikato RPS require that the District Plan identify coastal hazards and restrict subdivision, use, and development within areas subject to coastal hazards over a one-hundred-year timeframe, including increased hazard risk due to climate change (figure 3). Nonstatutory planning instruments, including Asset Management Plans and Hazard Mitigation Plans, are additional "tools" that can be used by the WRC and TCDC to facilitate adaptation. And Iwi Management Plans can be used by Māori authorities to address environmental and climate change concerns.

In summary, one might conclude that ANZ has robust statutory provisions in place to reduce coastal hazard risk, adapt to climate change, and build community resilience. More generally, Māori and citizens can participate meaningfully in governance endeavors that foster community well-being, public safety, and sustainability. Why then is there a growing body of scholarship critical about the state of environmental and natural hazards governance in ANZ? Why are protests about climate change inaction taking place on the Coromandel Peninsula and elsewhere in the country, and why is a court case pending against the TCDC on the grounds that the council decided not to sign the Local Government Leaders' Climate Change Declaration? It is necessary to probe recent scholarship and local action to address these questions.

Critical Reflections on Environmental and Natural Hazards Governance in Aotearoa New Zealand

In recent decades, the institutional setting for environmental and natural hazards governance at the local level in ANZ has been reshaped by complex interactions between seemingly unrelated and often "distant forces"—from globalization and neoliberal-inspired restructuring to opportunities for citizen engagement in local decision-making, political recognition of Māori customary ownership and natural resource management, as well as disasters and climate change (Cheyne 2015; Dinica 2018; Haggerty 2007; Haggerty, Campbell, and Morris 2009; Harmsworth, Awatere, and Robb 2016; Livesey and McCallum 2019; Memon and Kirk 2012; Schneider et al. 2017).

Concern about the closing down of opportunities for authentic local democratic engagement in ANZ has persisted over the last two decades (Bond and Thompson-Fawcett 2007; Cheyne 2015; Dinica 2018; Grey and Sedgwick 2015; Gunder and Mouat 2002). Notwithstanding a range of formal opportunities for participation in planning processes, in practice, meaningful participation is often restricted, dissent annulled, and dominant neoliberal norms bolstered, whether in the context of water governance (Kirk, Brower, and Duncan 2017), coastal and marine projects (Le Heron et al. 2019; Šunde et al. 2018), or oil and gas initiatives (Diprose, Thomas, and Bond 2016) and decisions about coal mining (Bond and Fougère 2018; Fougère and Bond 2016). Opening up democratic space necessitates openness to countervailing perspectives and even dissent. For Gunder and Mouat (2002), the RMA planning system provides little opportunity for meaningful choice or freedom to resist—compounded by the decision-makers and arbiters of appeal being the "oppressors." In so doing, the planning system institutionalizes "symbolic violence and victimization." In practice, exclusion from the planning system is commonplace, despite lauded provisions for citizen engagement (Cheyne 2015). In part at least, the system is driven by institutional performativity and efficiency, with unintended adverse public consequences (Gunder 2003;

Gunder, and Mouat 2002). Technical, scientific, and legal expertise is privileged in expert-driven analytical frameworks that lead to top-down decision-making and stakeholder tensions, and the inherently political nature of such judgments is masked (Tadaki, Allen, and Sinner 2015). While refinements to the "letter of the law" might be necessary, they are by no means a solution, because laws are a product of historically embedded societal norms and dominant discourses (Bond and Fougère 2018). Neoliberal restructuring, including the shift to a "contract state," has profoundly affected the community and voluntary sector by, among other things, enabling community and voluntary sector inclusion in devolved local affairs but paradoxically limiting democratic debate and engagement by engendering a climate of fear, compulsion, and exclusion (Grey and Sedgwick 2015).

New public management practices are seen by some to be at the root of the inability of local government to secure authority and autonomy over freshwater governance, paradoxically deepening the very problems these practices were meant to solve (Kirk et al. 2017). Innovations in collaborative governance have done little to shift the overarching political economy goal to maximize primary production export-led growth (Kirk et al. 2017). Even major collaborative environmental governance "experiments" in the freshwater arena have led scholars to conclude that such efforts are "less than democratic, less than fair, and less than good for the environment" (Brower 2016; Roberts et al. 1995). Based on an investigation of environmental decision-making about freshwater in Canterbury, ANZ, Thomas and Bond (2016) argue that the practice of democracy is barely distinguishable from authoritarianism in the prevailing neoliberal regime. They identify the potential for counterhegemonic actions. Mediating the power differentials that exist between role players, especially between Māori, local and central government, local communities, and others holding political and economic power, is a real struggle but is foundational for realizing good public outcomes across the terrestrial and marine realms of ANZ (Barrett et al. 2019; Le Heron et al. 2019; Šunde et al. 2018).

Where planning authorities, resource users, and *iwi/hapū* (i.e., tribe/sub-tribe) build meaningful relationships, good community outcomes can be delivered using statutory provisions as well as tools such as Iwi Management Plans and co-management arrangements (Harmsworth et al. 2016; Makey and Awatere 2018; Thompson-Fawcett, Ruru, and Tipa, 2017). Where such relationships are weak, the planning process can be a drain on *iwi* and *hapū* and the resultant plans of little value (Thompson-Fawcett et al. 2017). Such interactions are, however, invariably complicated by the history of colonization. Even when Treaty settlements return land to Māori tribal authorities, planning authority is retained by local and central government, and planners have to come to terms with decolonizing their practices in what Livesey and McCallum (2019) describe as a "settler colonial planning system."

The influence of power and politics—and social vulnerabilities—that may otherwise be invisible in day-to-day community life were starkly exposed in the devastating earthquake experience in Greater Christchurch in 2010–11 and in many other seismic and hydro-meteorological extreme events over the last decade. Living through the Christchurch earthquakes led Hayward (2013) to conclude that the notion of resilience needs to be expanded to include compassion—expressed as shared vulnerability—and political resistance, and that these are the bedrock of community recovery. A narrow framing of resilience limits postdisaster recovery to a return to the status quo, but an expanded view of resilience opens up the possibility of transformation and community empowerment that challenges the dominant neoliberal discourse (Cretney and Bond 2014; Cretney, Thomas, and Bond, 2016; Uekusa 2018). Public participation in recovery work is essential, but the ANZ experience reveals the limits of prevailing practices, which can verge on tokenism and paradoxically narrow the scope for authentic democratic engagement (Cretney 2018). The Christchurch disaster experience shows that governance practices grounded in neoliberalism have a profound depoliticizing impact—closing down the space for democratic engagement—but it need not extinguish hope for grassroots recovery and the potential for resistance and repoliticization (Cretney 2019). The value of culturally grounded responses was underscored by the Māori response to the Christchurch earthquakes (Kenney and Phibbs 2014, 2015).

Adaptation to climate change is an inevitable part of local democracy, community development, and well-being ambitions, and local government engagement with communities is integral to realizing these ambitions: it is not a one-off project consultation exercise (Simon, Diprose, and Thomas 2020). Community members and environmental activists in ANZ have mobilized to challenge "business as usual" practices, including state interventions, which drive inequitable and unsustainable development, and dangerous levels of global warming (Diprose et al. 2017; Diprose et al. 2016). Bond and colleagues (2015) argue that challenging the prevailing postpolitical hegemony requires vibrant contestatory politics and that instances of such contestations tend to have paradoxical outcomes—barely nudging and perhaps even reinforcing the dominant hegemony while offering a glimmer of hope about the potential for community protest and dissent to prompt change. Climate change may even create opportunities for strengthening communities and local democracy by leveraging emerging networks and emancipatory discourses, notwithstanding the real risks and challenges climate change poses (Diprose et al. 2017).

This brief survey of scholarship critical of prevailing environmental and natural hazards governance practices in ANZ contrasts sharply with the picture painted in the earlier synopsis of formal institutional provisions that promise to deliver local democracy, Māori rights, public safety, equity, and sustainable development in the face of climate change. A radical critique of neoliberalism helps to situate this dichotomy, but a deeper dive below the surface of such a macrolevel structural critique is necessary to unravel it. Delving into the microlevel adaptation narratives of and actions/inactions on the Coromandel Peninsula over the last decade enables such an exploration.

Political ecology is a useful framing device to craft this story because it sharpens the focus on power and politics, along with their impact on ecological values and environmental outcomes (Wolf 1972). Research with a political ecology orientation can help to "reveal winners and losers, hidden costs, and the differential power that produces social and environmental outcomes" (Robbins 2012: 20). Power and politics have a fundamental bearing on adaptation efforts (Tschakert et al. 2016). Policy provisions may become ineffective in enabling communities to adapt, not least because they are impeded by inequitable local power relations (Biesbroek et al. 2014; Huitema et al. 2016). Local leadership strongly influences the way adaptation unfolds over time (Termeer, Dewulf, and Biesbroek 2017; Termeer et al. 2011). The trajectory of adaptation is thus shaped by "entangled socio-political contestations, biophysical change, livelihood desires, struggles for authority to govern change, and desires for social and political recognition by both those promoting programs and recipients of them" (Nightingale 2017: 12). How have entangled narratives and adaptation moves evolved over the last few decades on the Coromandel Peninsula?

Reflections on Over a Decade of Adaptation Experience on the Coromandel Peninsula

There is a wide diversity of views about climate change and adaptation within and between communities, levels of government, and other stakeholders associated with the peninsula. What is more, the adaptation story is continuously evolving. Its trajectory has been strongly influenced by the voice of the community and Māori in shaping local action, local leadership, power struggles, extreme meteorological events and their impacts, the legislative and policy landscape, and the growing body of scientific evidence and media reports—all embedded in a recent history of globalization and the neoliberal political economy of ANZ. In this section, we offer a chronology of stories, events, evidence, and practices to reveal the complex assemblage of factors shaping adaptation on the peninsula. Perhaps "The Times They Are A-Changin'"; perhaps "The Song Remains the Same."

Climate Change Leadership, Science, and the Law: 2000–2010

Notwithstanding scientific consensus about dangerous levels of global warming driven by greenhouse gas emissions that has been documented in IPCC Assessment and Special Reports approved by governments around the world from 1990,[1] it was not until 2002 that ANZ enacted the Climate Change Response Act, accompanied by the Resource Management (Energy and Climate Change) Amendment Act in 2004. The latter meant that the RMA was amended by the addition of provisions expressly addressing climate change. In short, climate change was well and truly on the table—a problem that would require action from all levels of government. Six years after the RMA amendment S.7(i)—having particular regard to the effects of climate change—the NZCPS stipulated a one-hundred-year-plus planning time horizon together with a range of other anticipatory provisions to enable local adaptation. These provisions were introduced at the time of the 2010–11 Canterbury earthquake sequence—a time when the entire country was galvanized to take natural hazard risk seriously. Local and regional government began to prepare for a future characterized by escalating coastal hazard risk in a changing climate (cf. figure 7.7).

In 2004, the Thames-Coromandel District elected its first female mayor, who was also at the time New Zealand's youngest mayor. Philippa Barriball (2004–10) chaired the subcommittee for local government on climate change. She took pride in helping coastal communities understand the importance of the issue:

> When educating the public about the anticipated climate change impacts, I don't try to sell climate change to people. I'm trying to sell what is important to people and how this might be at risk. I tap into the emotive side of people and ask questions like, "Which is your favourite beach? How would it affect you if we were to build a rock wall around it? (Barriball 2010, pers. comm.)

In hindsight, Mayor Barriball was ahead of her time with regard to climate change. There was little publicly accessible ANZ-centered climate science available, and there were virtually no newspaper articles explicitly about the Coromandel Peninsula, coastal risk, and climate change until 2013, when "The Hidden Cost of Seaside Living" was reported in a local newspaper (figure 7.3).

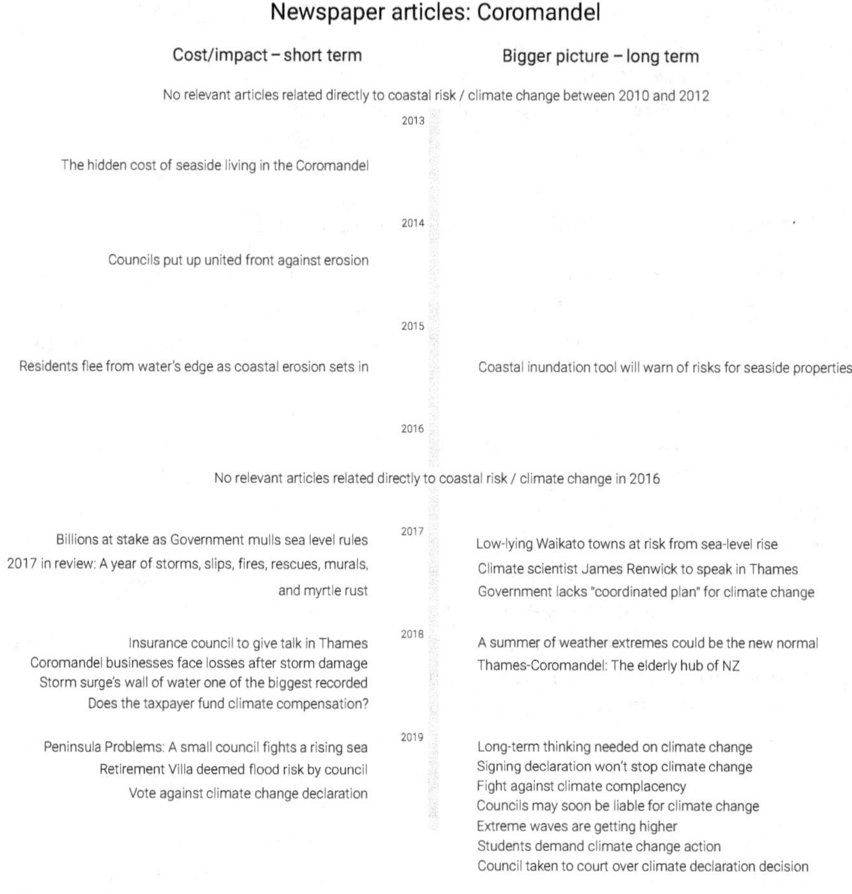

Figure 7.3. Newspaper articles covering coastal risk/climate change on the Coromandel since 2010. © Paul Schneider.

Following the election of Mayor Glenn Leach in 2010 (he held the seat until 2016), local leadership on climate change essentially came to a halt. Mayor Leach said:

> I look at it and say, if there is a problem, we can only take leadership from what comes out of central government. (Leach 2012, pers. comm.)

Climate Change Denial and Protection Works versus Storms, Science, Law, and the Media: 2010–18

Mayor Leach seems to have been uninfluenced by the mounting calls for action, including central government's regulatory provisions (such as the

2010 NZCPS) and guidance for local authorities on adapting to climate change (MFE 2017); the prime minister's chief science advisor's report on New Zealand's changing climate and oceans; the IPCC AR5 Working Group II assessment (published in 2013–14); and a 2015 Parliamentary Commissioner for the Environment (2015) report on preparing New Zealand for rising seas.

Notwithstanding the foregoing, climate change denial was prevalent on the Coromandel even in the early 2010s. According to a beachfront property owner, "We can worry about it once it takes shape" (beachfront property owner 2012, pers. comm.). Another said, "Don't know. Do I care? . . . Mmmh, don't know. All the global warming and sea rising, I don't think it's a major issue really. Although if anybody is affected . . . [pauses] . . . we are" (beachfront property owner 2012, pers. comm.). Others, including TCDC staff, emphasized the need for urgent action: "The question that has got to be asked: Is it acceptable for people to continue being exposed to these sorts of hazards?" (senior council planner 2010, pers. comm.). A local coastal scientist spoke about the influence of power and politics on local coastal decision-making:

> In theory we could say, "Get your bloody seawalls off our land," but politically that would never happen in a hundred years. Any one of those guys has more access to political power than half of the rest of the community put together. I took that power on once in an environment court case with a very good environment court judge and a reasonable commission. They dealt to us very harshly. You don't buy a beachfront property if you are poor, so people have a lot of economic and political power. These people are simply the movers and shakers in our society. Beachfront properties, that's how it is . . . erosion reaches their boundary and it gets stopped and environmental regulations go out the window, public interests go out the window. (Local coastal scientist pers. comm. 2010)

A Waikato Regional Council hazards and emergency management officer (pers. comm. 2012) explained that while "council is subservient to the policies . . . [the] whole area is still a bit grey . . . so often you [still] see developments . . . [where] the developer will go through the environment court."

Councilors on the Coromandel demanded action to "protect" the coast, which, to them, was not happening fast enough. Local councilor Murray McLean gained popularity when he called for "decisive action" and "not just another meeting" before he took matters into his own hands (Preece 2012).

Major storms in 2015 and 2016 caused further damage along sections of the peninsula, and, together with the publication of a Royal Society report

Figure 7.4. Beachfront property development at Te-Whanganui-o-Hei/ Mercury Bay. © Paul Schneider.

on climate change implications for New Zealand (Royal Society of New Zealand 2016), added further weight to the local reality of climate change and the adaptation imperative. The need to take escalating coastal hazard risk more seriously was further underscored by the inclusion of "significant risks from natural hazards" as "a matter of national importance" in amendments to the RMA (§6[h]) in 2016.

Newly elected Mayor Sandra Goudie, who took over from Mayor Leach in 2016, faced mounting local concern about and central government attention on climate change from the beginning of her term. Several cyclones had caused severe damage to coastal infrastructure and communities. Cyclone Cook hit the region in April 2017, underscoring the relevance of the Ministry for the Environment's "Coastal Hazards and Climate Change" guidance published in the same year. Projected climate change impacts and escalating coastal hazard risk were widely considered a priority for coastal communities and their governing authorities. The weight of scientific evidence, regulatory provisions, and local experience made climate change denial increasingly untenable. A major storm in January 2018 stands out as a focusing event (Birkland, 2019) that could mark a turning point in the long-standing adaptation impasse on the peninsula.

PROPERTY HAURAKI HERALD, APRIL 13, 2012 23

EROSION WORK: Emergency work was undertaken at Brophy's Beach in Whitianga last week to try to halt erosion.

Erosion needs quick fix

By DOROTHY PREECE

Easterly weather and high tides pounding the foreshore at Brophy's Beach reserve in Whitianga caused council to instigate overnight emergency protection before Easter.

On March 27 Mercury Bay Community Board member June Bennett reported stormwater infrastructure was exposed and causing a hazard at Brophy's Beach. The board requested Thames Coromandel District Council take action "under emergency" and recommended "soft option" geotech sandbags, not rocks.

On Tuesday, April 3 nature took over and council workers were called out overnight, dumping rocks to try to hold the erosion.

Area manager Lesley McCormick said the emergency action needed a retrospective resource consent. She was now scoping the sandbag project as a long-term option.

"Once we have the project scope agreed by the board, I can then look at how it can be funded," Ms McCormick said.

Councillor Murray McLean said in recent years the Brophy's Beach foreshore had been protected by dragging sand back up the beach.

"Now all the sand is gone, there is nothing to drag back and we need another solution quickly," he said.

> "Once we have the project scope agreed by the board, I can then look at how it can be funded." – Lesley McCormick

NEWS HAURAKI HERALD, AUGUST 3, 2012 5

BIG ISSUE: A digger doing remedial work at Buffalo Beach, Whitianga on Wednesday.

Action demanded on coast erosion

By DOROTHY PREECE

Councillors Tony Fox and Murray McLean have voiced strong complaints about what they say is a lack of action on coastal erosion issues in Mercury Bay.

At the Mercury Bay Community Board meeting on Tuesday, the councillors digressed from a discussion on the draft Community Board Plans. They said coastal erosion was the biggest issue facing the Mercury Bay ward and it was being ignored.

"We have seen three metres of grass disappear from our waterfront and I want to know when someone is going to take action, instead of these endless reports," Mr Fox said.

"We don't need any more reports, we want conclusions and decisive action," Mr McLean said.

Area manager Sam Marshall said he would arrange for the board to meet with consultants and Waikato Regional Council staff to try to expedite the remedial work.

"It would have to be a different approach, not just another meeting," Mr McLean said.

Board chairwoman Alison Henry said coastal erosion should be tagged as a priority in the proposed Community Board Plans for 2013-14.

Figure 7.5. Local newspaper articles reporting on emergency action taken to defer coastal erosion. © Paul Schneider

5 January Storm Surge and a Plan for Action versus Protests and a Court Case: 2018–20

The 5 January 2018 storm surge hit the Coromandel's west coast rather unexpectedly. A low-pressure system had been anticipated, but the extent of the associated storm impacts went beyond predictions. Water levels reached 2.8 meters above normal spring tide, only 0.2 meters short of the highest recorded level in 1938. The WRC described the storm surge as a one-in-two-hundred-year event. This extreme event was almost entirely due to coastal influences. Thames, the region's biggest town with a population of 6,693 at the last census in 2013, regularly experiences river flooding as a result of high rainfall events. Storm surge flooding, however, had not been experienced in the living memory of most residents. The Thames Coast highway, the only road along the Thames Coast, was extensively and severely damaged. In many places, the road was eroded up to the centerline. Many coastal properties were damaged. One home was rendered uninhabitable, and nine homes were moderately damaged on the western coast (Thames Coast) of the peninsula. There were thirteen uninhabitable and seventy-two moderately damaged houses along the Kaiaua/Miranda coast.

The New Zealand Transport Agency (NZTA), responsible for highways, decided to repair the Thames Coast highway as quickly as possible to enable traffic flow on the peninsula's only north-south route on the west coast. An NZTA spokesperson said, "We've had to come in and reconstruct the rock wall to protect the road, to protect it from further damage coming into the winter season . . . so we've imported just over 100,000 tons of armor rock. . . . If we'd decide to do nothing, we lose the highway" (NZTA spokesperson 2018, pers. comm.). How can this short-term decision be reconciled with the NZCPS requirement to make decisions with a one-hundred-plus-year timeframe in mind, given rising sea levels? Regardless, on the ground, the reality of coastal hazard risk was unmistakable. And the prospect of escalating coastal hazard risk in the face of climate change was too obvious for most people to ignore, especially after a further two cyclones in 2018 and the publication of the Climate Change Adaptation Technical Working Group recommendations on adaptation to climate change (CCATWG 2018). One local councilor described the storm event as a "game changer" (local councilor 2018, pers. comm.).

In June 2018, after extensive community consultation, the TCDC issued its 2018–28 Long Term Plan (LTP)—a key provision under the LGA that details community well-being outcomes that the council aims to achieve—and associated priority investments, services and projects, and associated costs. Annual Plans provide a more detailed breakdown of work proposed over each successive year. The 2018–28 LTP for the Coromandel details projected expenditure on measures to reduce coastal hazard risk and adapt to climate change and follows the most recent government regulatory provisions and guidance. To give effect to these provisions, and synchronous with approval of the LTP, the TCDC approved a Coastal Management Strategy (to guide use and protection of the coastal environment, within a partnership framework, based on coastal management principles, objectives, policies, and implementation measures) and a Coastal Hazards Policy (which describes how to sustainably manage the effects of coastal hazards on the district's coastal foreshore). Furthermore, a three-year process to develop Shoreline Management Plans (SMPs) was initiated in 2019 to define flooding and coastal erosion risks to people, and the social, cultural, economic, and natural environment of the district over the next century and beyond, "through active involvement of all key stakeholders" (TCDC 2019: para. 14). For implementing the Coastal Management Strategy, NZ$2.6 million was budgeted in the LTP, NZ$1.9 million of which was awarded to Royal HaskoningDHV, headquartered in the Netherlands, to develop the SMPs. These SMPs are "part of the focus on ensuring our communities are engaged, prepared, protected and safe in the long-term" (TCDC 2019: para. 17).

These actions signal a volte-face in the TCDC's approach to coastal management and coastal hazard risk: "The Times They Are a-Changin'," or could it be that "The Song Remains the Same"?

In a Radio New Zealand interview in February 2019 with Mayor Goudie (Gudsell 2019), the interviewer asked the mayor: "Don't you believe you have an obligation to tell your ratepayers whether you believe climate change is happening?" Her response was: "No, I don't." On 2 March 2019, the TCDC voted 6–3 against signing the Local Government New Zealand Leaders' Climate Change Declaration—an aspirational document signed by fifty-nine mayors and regional council chairs declaring that councils have an important part to play in tackling climate change (nineteen are yet to sign).

School students voiced their anger over the council decision. One of the students, Helena Mayer, pointed out that the actions that the TCDC "are

Figure 7.6. Fifteen-year-old high school student Helena Mayer and Coromandel residents Nancy and Eric Zwaan protesting in front of the Thames-Coromandel District Council office in Thames (day 12 of 38). © Beaton, 2019.

currently taking in response to climate change are not leading the way, nor do they reflect the urgency of the disasters that have been predicted" (Tantau 2019: para. 12). A local climate change activist, Sheena Beaton, interviewed in 2020, said that she and the group Hauraki Coromandel Climate Action (HCCA)

> spoke at every council meeting leading up to the decision [to not sign the Local Government Leaders' Climate Change Declaration]. And then the decision was made, but in a pretty shonky sort of fashion. It was clear that the penny hadn't dropped, and there were red flags right from the start. For example, Sandra [the mayor] had written her own report. . . . It was kind of underhanded and, you know, Sandra had clearly swayed the decision to go [the way it did]. (Sheena Beaton 2020, pers. comm.)

In 2019, the Local Government Funding and Financing Productivity Commission draft report was published, and the Climate Change Response (Zero Carbon) Amendment Bill was passed in parliament. The pressure was and continues to be on the mayor and the TCDC to translate 2018 promises and investment into effective local action that creates "solutions . . . to help our communities adapt to coastal hazards and risks" (TCDC 2020b: para. 1). The rhetoric is encouraging, with Mayor Goudie observing that participatory planning or "coastal panels" are to "be the engine for our Shoreline Management Plan project, which is all about building resilient coastal communities" (TCDC 2020b: para. 10).

On 3 July 2019, the HCCA applied to the High Court for a judicial review of the decision by the TCDC not to sign the Local Government New Zealand Climate Change Declaration.

Through the 2019 winter (3 July–23 August), climate change protesters held a vigil in front of the TCDC offices, holding a banner inscribed: "We are in a climate and ecological emergency." The protest was then moved "to the main street as part of the #FridaysForFuture[2] movement." When asked if, during the over thirty-eight days of standing in front of the local council building, anyone from the TCDC attempted to engage in dialogue, climate activist Sheena Beaton replied that "staff were advised not to talk" to her and that the mayor "wouldn't talk" to her (pers. comm. 2020).

In March 2020, the TCDC's response and applications to the court were rejected, and the matter is now proceeding to a substantive hearing. An HCCAG member (2020, pers. comm.) explained that "[the TCDC] tried to ping us for costs and stuff, but that was thrown out . . . so basically the case is proceeding in front of the High Court."

To complicate the story, some interviewees paint a picture of a local council in strife as a result of local politics and power struggles, com-

A risk-shaping pathway: key events for Aotearoa New Zealand 2010–2019

Natural hazard events | Key climate change publications | Policy actions

2004 — RMA S.7(i) - particular regard to the effects of climate change
2010
Canterbury earthquake
2011 — NZ Coastal Policy Statement 100+ year time horizon needs to be considered
Christchurch earthquake
2012
2013
Chief Science Adviser Report on NZ's changing climate and oceans
IPCC AR5 Working Group II assessment
2014
2015 — PCE Report
NZ storm
NZ storm
2016 — Royal Society Report on climate change implications for NZ
RMA S.6(h) - significant risks from natural hazards become matters of national importance
2017
Ex cyclone Debbie
Change of Government
Edgecumbe flood
MFE Coastal Hazards and Climate Change Guidance
2018
Ex cyclone Fehi and Gita
CCATWG Recommendations on adaptation to climate change
2019
Local Government Funding and Financing Productivity Commission Draft Report
Climate Change Response (ZERO Carbon) Amendment Bill

Figure 7.7. Key events, publications, turning points, and policy actions affecting climate change action and adaptation on the Coromandel. © Paul Schneider.

pounded by staff turnover and organizational turmoil: "With a bit of history from previous councilors and officers, and quite a bit of shake up and restructure . . . a lot of 'us' and 'them' sort of thing" (former local council staff member 2018, pers. comm.), with "high staff turnover and politics going on in there" (coastal specialist 2020, pers. comm.). A coastal specialist interviewed to gauge the extent to which this picture was accurate quickly pointed out that the situation is "problematic, to say the least." The high staff turnover resulted in the project coordinator of the vital SMP process leaving "this gaping hole" with "no one with any coastal expertise within council." This informant expounded further:

> For whatever reason, and I don't know the internal politics of it, they haven't appointed anyone. There are probably not that many coastal engineers, in inverted commas, around. I think they're quite keen to have an engineer. That's because the budget is attached to the infrastructure team within council, and the person leading the infrastructure team in council is an engineer, and they're all engineers. And it's that culture of engineers. Therefore only an engineer can do it. But there are other ways of thinking about that. (Coastal specialist 2020, pers. comm)

To complicate the story further, preparation of the SMPs has been outsourced to an international consultancy, which makes it difficult to fully account for the intricacies of local community dynamics and realities—notwithstanding subcontracted local specialists:

> Council has this outsourced model, completely outsourced model, where they had this massive budget and they've got these international consultants to do this coastal hazard modeling and stuff, which is where the bulk of the budget goes. And they sort of seem to be going down the track of every other coastal management coastal adaptation program ever made. (Coastal specialist 2020, pers. comm.)

According to this key informant, the consultants responsible for developing the SMPs "don't understand what the Treaty [of Waitangi] is; they don't understand the RMA or the Local Government Act . . . any of the context within which they're working and they're not communicating [well]." Building partnerships and collaborating with local *iwi* will be key to the success of the SMPs. But, as suggested by this key informant, "there is the *iwi* partnership as well, but they don't really want to progress that." This is cause for concern given that Māori have faced historical injustices and grievances that continue to cast a shadow over social-ecological debates, including responding to climate change.

The local council–Māori relationship appears entrenched in hardened positions, and reconciliation may be difficult to achieve. On the one hand, the extent to which the local council is seeking genuine partnership with local *hapū* and *iwi* has been questioned, while on the other hand, there is long-standing and deep mistrust, as signaled by this insight:

> We don't trust any of the government departments. We don't need them. . . . Council is just a pack of arseholes there to gather revenue . . . you can fight them and they'll go away, but not for long; they come back from a different angle. (Māori research participant 2014, pers. comm.)

The overall challenge of communication, coordination and engagement is clearly vexed. The previously quoted coastal specialist (2020, pers.

comm.) views collaboration between the local and regional councils as essential to the SMP process but observed that

> they [TCDC] don't want to partner with the regional council. I don't know where that's coming from. I've been told it's a Mayoral directive. So [they are] gonna end up with these plans. . . . And, you know, . . . [they are] not going to have the buy-in from the people who are able to implement the bloody plans. And so [they are] going to end up in the same space, I think, as where [they] have always been. But with sort of a shinier cover.

The "shinier cover," however, is unlikely to be sufficient to address the concerns raised by some community members, specialists, and activists about the need for urgent and authentic adaptation action, as explained in 2018 by local lawyer Denis Tegg (2018, pers. comm.):

> What really shocked me when I got started to research this . . . you know, I sort of thought that maybe Thames was at risk . . . but when those coastal inundation maps came out from the Regional Council. . . . Once I got my head around how to use that tool, I started to get really concerned. Subsequently I stumbled on this risk census from NIWA. It has nationwide figures, and you can drill down into individual towns and extract the figures. When I did that, it just blew me out of the water. I mean, and there was even a graph in there showing you know the 15 most at-risk towns and cities in New Zealand. And Thames is number 8. I mean come on! I mean, why is it that? Why aren't we shouting that from the rooftops? You know, that rather than sort of pretending there's not a problem we should be, you know, thinking: "God we've got a problem here," like Dunedin and Hawkes Bay are recognizing it, and are putting their hand up for funding, and all the rest of it, and we're just sort of pretending it's not a problem.

In a conversation with a local coastal specialist in 2020 (pers. comm.), the following came to light:

> TCDC certainly made a good start with their SMPs in terms of addressing coastal risk. The criticism is that they are behind in terms of the timing. But then again better late than never. . . . There are some issues. A key one is the extent to which they are—and planning to—engage with iwi. They are expecting iwi to come in at a later stage. That's not the way to gain confidence if you set up the structure and then later ask iwi "how are you going to fit in'? The extent to which they are engaging with iwi just isn't robust at the moment. . . . Another one is the overall governance that they've approved, which sees the local council and the regional council separated out and where they don't have joint decision-making. We've seen from other examples, such as in the Hawkes Bay, that joint governance oversees the project and is really important. But TCDC have said that this is "their" project and that they don't want to collaborate with anybody else. There's quite a bit of to-ing and fro-ing. They drafted this governance

> structure where they had two versions. One that involved regional council and one that didn't. But local councilors didn't see this, they were only shown one version, which is the one where regional council is separate.

A local activist pondered how the TCDC might take responsibility to "avoid the unthinkable" (implying regular destructive coastal storms and dangerous global warming), which "surely would be in everybody's interest" (local climate activist 2020, pers. comm.). She said that the unwillingness of the TCDC to take meaningful adaptation and mitigation action

> just astounds me. I know she [TCDC mayor] is appealing to the farming community and all that, but I think there's gotta be a way to also get through to them. The farmers rely on the environment just as much as everybody else, and they're feeling it now with the drought . . . there's just so many connections.

With the onset of winter in mid-2020, over a decade on from the initiation of this ethnographic research, where do things stand? A one-off survey of perceptions by an outsider is unlikely to have revealed the historically embedded and complex interplay of worldviews, contradictions, contestations, and complex interactions that cut across (and within) geographies, sectors, communities and time, and ultimately the waxing and waning of this adaptation story. Depending on who one might speak to: "The Times They Are A-Changin'" or "The Song Remains the Same."

Conclusions

We set out to address two overarching questions: What drives local responses to the unfolding climate emergency facing low-lying coastal communities? And how might local communities and their governing authorities be galvanized to take meaningful action to reduce exposure and vulnerability to climate change impacts? Our decade-long ethnography exposes a complex, contested, and still-unfolding adaptation cacophony on the Coromandel. An adaptation deficit or gap prevailed until two years ago—notwithstanding decades of mounting scientific evidence and robust regulatory provisions at the national level intended to enable local coastal hazard risk reduction, adaptation, and resilience building. A severe storm in January 2018 appears to have been instrumental in mobilizing local council to initiate participatory shoreline management plans with an outlook of one hundred years. A cursory scan of recent actions might lead one to conclude, much like Bob Dylan mused during an earlier time of social and political change: "The Times They Are A-Changin." On closer inspection, these recent adaptation moves are part of a more convoluted narrative. Even as the local council initiated its SMPs, the TCDC councilors, led by the mayor, voted against signing a declaration to signal that the council would take leadership to address climate change. This prompted some community members, including local activists, schoolchildren, and local specialists, to express their dismay, using protest and even court action against the council to expose what they felt was a perpetuation of climate change denial and inaction. For them, instead of this being a time of change, "The Song Remains the Same," to invoke Led Zeppelin.

What drives such countervailing and contradictory action and inaction? Our thick description of adaptation on the Coromandel Peninsula exposes the pervasive influence of power and politics in shaping local choices that are historically embedded in an assemblage of disparate local and distant forces that intersect in complex ways, including the lingering effects of colonialism that impinge on Māori rights, ancestral knowledge, and practices, and the impact of neoliberal-inspired restructuring and continued privileging of private economic interests. Absurdly, protecting those interests, literally, adversely impacts community well-being and risk exposure—and deepens the democratic deficit at the local level, even if ameliorated for a moment by visionary local leadership.

How might the communities and governing authorities of the Coromandel be galvanized to chart pathways that institutionalize more engaged, equitable, climate-resilient, and sustainable coastal development pathways? Our still-evolving ethnography suggests that there is no panacea, no blueprint of "pathway choices" from which adaptation courses can be chosen and "engineered." Turbulence, surprise, and shock are inevitable, and these will be compounded by continued technological, social, and environmental change. Adaptation trajectories are emergent with a complex array of intersecting factors having influence. What is clear is that continuing the trajectory of the last few decades—of climate change denial and reliance on protective measures and emergency responses—will entrench exposure and vulnerability. Reinvigorating local democracy is foundational for charting a more promising pathway. Paradoxically, this involves a struggle against the prevailing hegemony, even in ANZ. It necessitates vigorous contestatory politics at the local level, bolstered by an enabling regulatory setting, and more engaged and visionary community and council leadership. Perhaps questions to consider in preparing the SMPs have less to do with risk modeling and analysis and more to do with the shared vulnerabilities of all Coromandel inhabitants—including nonhuman inhabitants and all those past, present, and future: How would *tangata whenua* institutionalize risk reduction and resilience building measures to last a century or more? How might the most marginalized be given voice in the process of addressing climate change and adaptation?

When the inhabitants of the Coromandel look back one hundred years from now, what will they say about the choices made today, and tomorrow, and the next day?

Acknowledgments

With thanks, the authors acknowledge support and funding from the Aotearoa–New Zealand Earthquake Commission (EQC), Massey University, and the CALENDARS project. The views expressed here are solely those of the authors and neither reflect nor are necessarily endorsed by these organizations. The authors thank all research participants for sharing their stories and providing insight into this adaptation narrative. We thank Axel Kreuter (Medical University of Innsbruck) for his design work on figures 7.3 and 7.7. Further, we thank Trisia Farrelly for providing insights into the scholarly and real-life world of ethnography and political ecology.

Paul Schneider is a natural hazard risk and resilience research fellow at Massey University's Resource and Environmental Planning Programme. Paul's research focuses on climate change adaptation, risk, resilience, environmental issues, politics, and power. Paul's PhD on climate change adaptation "Beyond Technical Fixes and Business as Usual" combined environmental planning with political ecology. His current research focuses on the way institutions represent seasons and how to build adaptive capacity to seasonal change. The social and ecological setting of Aotearoa New Zealand's Coromandel Peninsula remains Paul's key research focus.

Bruce Glavovic is a professor at Massey University. He has degrees in economics and agricultural economics, environmental science, and urban and environmental planning. He has worked for over thirty years in academia, environmental consulting, and government, mainly in New Zealand, South Africa, and the United States. His research centers on the role of governance in building resilient and sustainable communities, with a focus on coastal communities and the roles of land-use planning, collaboration, and conflict resolution and bridging the science-policy-practice interfaces. Recent coedited books include: *Climate Change and the Coast: Building Resilient Communities* (2015), and *Adapting to Climate Change: Lessons from Natural Hazards Planning* (2014). Bruce is the 2016–19 co-chair of the Scientific Steering Committee (SSC) of Future Earth Coasts. Bruce was coordinating lead author of the chapter on sea level rise and its implications for low-lying islands, coasts, and communities in the IPCC's *Special*

Report on the Ocean and Cryosphere in a Changing Climate (2017–19), and is a lead author in the IPCC's *Sixth Assessment Report*.

Notes

"The Times They Are A-Changin'" is the title track of the third studio album by Bob Dylan, released on 13 January 1964. The song captures the spirit of the prevailing social and political upheaval. "The Song Remains the Same" is a song by Led Zeppelin released on the 1973 album *Houses of the Holy*, and also lends its name to the band's concert film released on 28 September 1976.

1. The Kyoto Protocol was adopted in 1997 when the parties to the Framework Convention agreed to the imposition of specific targets for thirty-six developed countries, including New Zealand.
2. #FridaysForFuture is a movement that began in August 2018 after fifteen-year-old Greta Thunberg held a vigil in front of the Swedish parliament for three weeks to protest against the lack of climate action (https://fridaysforfuture.org).

References

ARC (Auckland Regional Council). 2004. "Hauraki Gulf State of the Environment Report—Coastal Hazards of the Gulf." Retrieved 9 May 2006 www.arc.govt.nz.

Archer, L., J. D. Ford, T. Pearce, S. Kowal, W. A. Gough, and M. Allurut. 2017. "Longitudinal Assessment of Climate Vulnerability: A Case Study from the Canadian Arctic." *Sustainability Science* 12(1): 15–29.

Asquith, A. 2012. "The Role, Scope and Scale of Local Government in New Zealand: Its Prospective Future." *Australian Journal of Public Administration* 71(1): 76–84.

Barrett, P., P. Kurian, N. Simmonds, and R. Cretney. 2019. "Community Participation in the Development of the Ōngātoro/Maketū Estuary Project: The Socio-ecological Dimensions of Restoring an Interconnected Ecosystem." *Aquatic Conservation: Marine and Freshwater Ecosystems* 29(9): 1547–60.

Biesbroek, G. R., C. J. A. M. Termeer, J. E. M. Klostermann, and P. Kabat. 2014. "Analytical Lenses on Barriers in the Governance of Climate Change Adaptation." *Mitigation and Adaptation Strategies for Global Change* 19(7): 1011–32.

Birkland, T. A. 2019. "Agenda Setting and the Policy Process: Focusing Events." Oxford Research Encyclopedia of Politics. Retrieved 28 September 2020 from https://doi.org/10.1093/acrefore/9780190228637.013.165.

Bond, S., G. Diprose, and A. McGregor. 2015. "2Precious2Mine: Post-politics, Colonial Imaginary, or Hopeful Political Moment?" *Antipode* 47(5): 1161–83.

Bond, S., and L. Fougère. 2018. "Prising Open the Postpolitical Spaces of Planning Regimes: A Reply to ER Alexander." *Planning Theory* 17(4): 647–52.

Bond, S., and M. Thompson-Fawcett. 2007. "Public Participation and New Urbanism: A Conflicting Agenda?" *Planning Theory & Practice* 8(4): 449–72.

Boyer, M. A., M. Meinzer, and A. Bilich. 2017. "The Climate Adaptation Imperative: Local Choices Targeting Global Problems?" *Local Environment* 22(1): 67–85.

Braun, V., and V. Clarke. 2014. "Thematic Analysis." Encyclopedia of Quality of Life and Well-Being Research. Retrieved 20 February 2021 from https://link.springer.com/referenceworkentry/10.1007 percent2F978–94–007–0753–5_3470.

Brower, A. L. 2016. "Is Collaboration Good for the Environment? Or, What's Wrong with the Land and Water Forum?" *New Zealand Journal of Ecology* 40(3): 390–97.

CCATWG (Climate Change Adaptation Technical Working Group). 2018. "Adapting to Climate Change in New Zealand." Ministry for the Environment. Retrieved 20 February 2021 from https://www.mfe.govt.nz/sites/default/files/media/Climate percent20Change/ccatwg-report-web.pdf.

Cheyne, C. 2015. "Changing Urban Governance in New Zealand: Public Participation and Democratic Legitimacy in Local Authority Planning and Decision-Making." *Urban Policy and Research*: 1989–2014.

Cretney, R. M. 2018. "Beyond Public Meetings: Diverse Forms of Community Led Recovery Following Disaster." *International Journal of Disaster Risk Reduction* 28: 122–30.

———. 2019. "An Opportunity to Hope and Dream: Disaster Politics and the Emergence of Possibility through Community-Led Recovery." *Antipode* 51(2): 497–516.

Cretney, R. M., and S. Bond. 2014. "'Bouncing Back' to Capitalism? Grass-Roots Autonomous Activism in Shaping Discourses of Resilience and Transformation Following Disaster." *Resilience* 2(1): 18–31.

Cretney, R. M., A. C. Thomas, and S. Bond. 2016. "Maintaining Grassroots Activism: Transition Towns in Aotearoa New Zealand." *New Zealand Geographer* 72(2) 81–91.

Cvitanovic, C., M. Howden, R. Colvin, A. Norström, A. M. Meadow, and P. Addison. 2019. "Maximising the Benefits of Participatory Climate Adaptation Research by Understanding and Managing the Associated Challenges and Risks." *Environmental Science & Policy* 94: 20–31.

Davison, I. 2011. "Holiday Poll Rates North the Best." *New Zealand Herald*. Retrieved 28 December 2011 from www.nzherald.co.nz/coromandel/news/article.cfm?l_id=123&objectid=10775557.

Department of Conservation. 2010. "New Zealand Coastal Policy Statement." Wellington, NZ: Department of Conservation.

Dinica, V. 2018. "Public Engagement in Governance for Sustainability: A Two-Tier Assessment Approach and Illustrations from New Zealand." *Public Management Review* 20(1): 23–54.

Diprose, G., S. Bond, A. C. Thomas, J. Barth, and H. Urquhart. 2017. "The Violence of (In)action: Communities, Climate and Business-as-Usual." *Community Development Journal* 52(3): 488–505.

Diprose, G., A. C. Thomas, and S. Bond. 2016. "'It's Who We Are': Eco-nationalism and Place in Contesting Deep-Sea Oil in Aotearoa New Zealand." *Kōtuitui: New Zealand Journal of Social Sciences Online* 11(2): 159–73.

EW (Environment Waikato). 2006. "Buffalo Beach Coastal Erosion Management Strategy." Retrieved 14 February 2010 from www.waikatoregion.govt.nz/PageFiles/4816/tr06–20.pdf.

EW (Environment Waikato) and TCDC (Thames-Coromandel District Council). 2003. "Thames Coast Flood Risk Assessment." Retrieved 15 January 2008 from www.ew.govt.nz/PageFiles/1142/TePuru.pdf.

Fawcett, D., T. Pearce, J. D. Ford, and L. Archer. 2017. "Operationalizing Longitudinal Approaches to Climate Change Vulnerability Assessment." *Global Environmental Change* 45: 79–88.

Fougère, L., and S. Bond. 2016. "Legitimising Activism in Democracy: A Place for Antagonism in Environmental Governance." *Planning Theory* 17(2): 143–69.

Geertz, C. 1973. *The Interpretation of Cultures*. New York: Basic Books.

Gluckmann, P. 2013. "New Zealand's Changing Climate and Oceans: The Impact of Human Activity and Implications for the Future." Office of the Prime Minister's Science Advisory Committee. Retrieved 4 July 2000 from www.pmcsa.org.nz/wp-content/uploads/New-Zealands-Changing-Climate-and-Oceans-report.pdf.

Grey, S., and C. Sedgwick. 2015. "Constraining the Community Voice: The Impact of the Neoliberal Contract State on Democracy." *New Zealand Sociology* 30(1): 88–110.

Gudsell, K. 2019. "Climate Change Declaration 'Politically Charged.'" Morning Report, Radio New Zealand. 20 February 2019. Retrieved 22 February 2019 from www.rnz.co.nz/news/national/382892/climate-change-declaration-politically-charged-thames-coromandel-mayor-sandra-goudie.

Gunder, M. 2003. "Passionate Planning for the Others' Desire: An Agonistic Response to the Dark Side of Planning." *Progress in Planning* 60(3): 235–319.

Gunder, M., and C. Mouat. 2002. "Symbolic Violence and Victimization in Planning Processes: A Reconnoitre of the New Zealand Resource Management Act." *Planning Theory* 1(2): 124–45.

Haggerty, J. 2007. "'I'm Not a Greenie but . . .': Environmentality, Eco-populism and Governance in New Zealand Experiences from the Southland Whitebait Fishery." *Journal of Rural Studies* 23(2): 222–37.

Haggerty, J., H. Campbell, and C. Morris. 2009. "Keeping the Stress Off the Sheep? Agricultural Intensification, Neoliberalism, and 'Good' Farming in New Zealand." *Geoforum* 40(5): 767–77.

Harmsworth, G., S. Awatere, and M. Robb. 2016. "Indigenous Māori Values and Perspectives to Inform Freshwater Management in Aotearoa-New Zealand." *Ecology and Society* 21(4).

Hayward, B. M. 2013. "Rethinking Resilience; Reflections on the Earthquakes in Christchurch, New Zealand, 2010 and 2011." *Ecology and Society* 18(4).

Huitema, D., W. N. Adger, F. Berkhout, E. Massey, D. Mazmanian, S. Munaretto, and C. J. A. M. Termeer. 2016. "The Governance of Adaptation: Choices, Reasons, and Effects; Introduction to the Special Feature." *Ecology and Society* 21(3).

IPCC (Intergovernmental Panel on Climate Change). 2019. *IPCC Special Report on the Ocean and Cryosphere in a Changing Climate*. Edited by H.-O. Pörtner, D. C. Roberts, V. Masson-Delmotte, P. Zhai, M. Tignor, E. Poloczanska, K. Mintenbeck, A. Alegría, M. Nicolai, A. Okem, J. Petzold, B. Rama, and N. M. Weyer.

Kenney, C., and S. Phibbs. 2014. "Shakes, Rattles and Roll Outs: The Untold Story of Māori Engagement with Community Recovery, Social Resilience and Urban Sustainability in Christchurch, New Zealand." *Procedia Economics and Finance* 18: 754–62.

———. "A Māori Love Story: Community-Led Disaster Management in Response to the Ōtautahi (Christchurch) Earthquakes as a Framework for Action." *International Journal of Disaster Risk Reduction* 14: 46–55.

Kirk, N., A. L. Brower, and R. Duncan. 2017. "New Public Management and Collaboration in Canterbury, New Zealand's Freshwater Management." *Land Use Policy* 65: 53–61.

Le Heron, E., J. Logie, W. Allen, R. Le Heron, P. Blackett, K. Davies, A. Greenaway, B. Glavovic, and D. Hikuroa. 2019. "Diversity, Contestation, Participation in Aotearoa New Zealand's Multi-use/user Marine Spaces." *Marine Policy* 106: 103536.

Livesey, B., and T. A. McCallum. 2019. "'Returning Resources Alone Is Not Enough': Imagining Urban Planning after Treaty Settlements in Aotearoa New Zealand." *Settler Colonial Studies* 9(2): 266–83.
Makey, L., and S. Awatere. 2018. "He Mahere Pāhekoheko Mō Kaipara Moana–Integrated Ecosystem-Based Management for Kaipara Harbour, Aotearoa New Zealand." *Society & Natural Resources* 31(12): 1400–1418.
Memon, P., and N. Kirk. 2012. "Role of Indigenous Māori People in Collaborative Water Governance in Aotearoa/New Zealand." *Journal of Environmental Planning and Management* 55(7): 941–59.
MFE (Ministry for the Environment). 2008. "Climate Change Effects and Impacts Assessment." Retrieved 12 January 2010 from www.mfe.govt.nz/publications/climate-change/climate-change-effects-and-impacts-assessment-guidance-manual-local-6.
———. 2013. "Climate Change Information New Zealand: Doing Our Fair Share." Retrieved 6 August 2015 from www.climatechange.govt.nz/reducing-our-emissions/targets.html.
———. 2014. "Adapting to Sea Level Rise." Retrieved 6 August 2018 from www.mfe.govt.nz/climate-change/adapting-climate-change/adapting-sea-level-rise.
———. 2017. "Coastal Hazards and Climate Change: Guidance for Local Government." Retrieved 10 August 2018 from www.mfe.govt.nz/publications/climate-change/coastal-hazards-and-climate-change-guidance-local-government.
Monin, P. 2012. "Hauraki-Coromandel Places—Mercury Bay." Te Ara: The Encyclopedia of New Zealand. Retrieved 8 March 2014 from www.teara.govt.nz/en/hauraki-coromandel-places/page-6
Moreno, J., and D. Shaw. 2018. "Women's Empowerment Following Disaster: A Longitudinal Study of Social Change." *Natural Hazards* 92(1): 205–24.
Nightingale, A. J. 2017. "Power and Politics in Climate Change Adaptation Efforts: Struggles over Authority and Recognition in the Context of Political Instability." *Geoforum* 84: 11–20.
NIWA (National Institute of Water and Atmospheric Research). 2008. "Regional Impacts." Retrieved 9 August 2010 from www.niwa.co.nz/our-science/climate/information-and-resources/clivar/impacts.
Parliamentary Commissioner for the Environment. 2015. "Preparing New Zealand for Rising Seas." Retrieved 20 September 2019 from www.pce.parliament.nz/media/1380/preparing-nz-for-rising-seas-web-small.pdf.
Pink, S. 2008. "Mobilising Visual Ethnography: Making Routes, Making Place and Making Images." *Forum Qualitative Sozialforschung / Forum: Qualitative Social Research* 9(3).
Preece, D. 2012. "Action Demanded on Coast Erosion." *Hauraki Herald*. Retrieved 3 August 2012 from www.stuff.co.nz/waikato-times/news/hauraki-herald.
Robbins, P. 2012. *Political Ecology: A Critical Introduction*. 2nd ed. New York: Blackwell.
Roberts, M. W., W. Norman, N. Minhinnick, D. Wihongi, and C. Kirkwood. 1995. "Kaitiakitanga: Maori Perspectives on Conservation." *Pacific Conservation Biology* 2: 7–20.
Rouse, H., R. Bell, C. Lundquist, P. Blackett, D. Hicks, and D. King. 2017. "Coastal Adaptation to Climate Change in Aotearoa-New Zealand." *New Zealand Journal of Marine and Freshwater Research* 51(2): 183–222.
Royal Society of New Zealand. 2016. "Climate Change Implications for New Zealand." Retrieved 8 March 2017 from www.royalsociety.org.nz/what-we-do/our-expert-advice/all-expert-advice-papers/climate-change-implications-for-new-zealand.

Schneider, P. 2014. "Local Reality and the Climate Change Adaptation Dilemma: Beyond Technical Fixes and Business as Usual." PhD diss., Massey University, Manawatu, New Zealand.

Schneider, P., and B. Glavovic. 2019. "Climate Change Adaptation in New Zealand." In *Oxford Research Encyclopedia of Natural Hazard Science*. Oxford: Oxford University Press.

Schneider, P., B. Glavovic, and T. Farrelly. 2017. "So Close Yet So Far Apart: Contrasting Climate Change Perceptions in Two 'Neighboring' Coastal Communities on Aotearoa New Zealand's Coromandel Peninsula." *Environments* 4(3): 65.

Simon, K., G. Diprose, and A. C. Thomas. 2020. "Community-Led Initiatives for Climate Adaptation and Mitigation." *Kōtuitui: New Zealand Journal of Social Sciences Online* 15(1): 93–105.

Statistics New Zealand. 2013. "2013 Census Quickstats about a Place: Whitianga." Retrieved 12 July 2019 from http://archive.stats.govt.nz/Census/2013-census/profile-and-summary-reports/quickstats-about-a place.aspx?request_value=13633&parent_id=13632&tabname=#13633.

Šunde, C., J. Sinner, M. Tadaki, J. Stephenson, B. Glavovic, S. Awatere, A. Giorgetti, N. Lewis, A. Young, and K. Chan. 2018. "Valuation as Destruction? The Social Effects of Valuation Processes in Contested Marine Spaces." *Marine Policy* 97: 170–78.

Tadaki, M., W. Allen, and J. Sinner. 2015. "Revealing Ecological Processes or Imposing Social Rationalities? The Politics of Bounding and Measuring Ecosystem Services." *Ecological Economics* 118: 168–76.

Tantau, K. 2019. "Thames Student Asks Council to Consider Climate Action Plan Ahead of Protest Rally." *Stuff*. Retrieved 17 March 2020 from www.stuff.co.nz/waikato-times/news/hauraki-herald/115395714/thames-student-asks-council-to-consider-climate-action-plan-ahead-of-protest-rally.

TCDC (Thames-Coromandel District Council). 2014. "The Costs of Protecting Our Coastlines." Retrieved 3 April 2015 from www.tcdc.govt.nz/Your-Council/News-and-Media/News-and-Public-Notices/News-Archived-Articles/July-2014/The-costs-of-protecting-our-coastlines/.

———. 2015. "Long Term Plan 2015–2025." Retrieved 1 December 2015 from www.tcdc.govt.nz/Download/?file=/Global/FINAL percent20Long percent20Term percent20Plan percent20as percent20at percent2013th percent20July percent2015.pdf.

———. 2016. "Communities and Wards." Retrieved 10 January 2017 from www.tcdc.govt.nz/Your-Council/Communities-and-Wards/.

———. 2018. "Preparing Our Communities for Coastal Hazards." TCDC Latest News and Public Notices. Retrieved 12 July 2018 from www.tcdc.govt.nz/Your-Council/News-and-Media/News-and-Public-Notices/News-Archived-Articles/May-2018/Preparing-our-communities-for-coastal-hazards/.

———. 2019. "Protecting Our Coast." Retrieved 10 January 2020 from www.tcdc.govt.nz/Your-Council/Documents-incl-Bylaws-Policies-and-Strategies/Coastal-Management-Strategy/.

———. 2020a. "Shoreline Management Plans." Thames: TCDC.

———. 2020b. "Shoreline Management Plans—Call for Volunteers." Retrieved 10 May 2020 from www.tcdc.govt.nz/Your-Council/Council-Projects/Current-Projects/Shoreline-Management-Plans/.

Termeer, C. J. A. M., A. Dewulf, and G. R. Biesbroek. 2017. "Transformational Change: Governance Interventions for Climate Change Adaptation from a Continuous Change Perspective." *Journal of Environmental Planning and Management* 60(4): 558–76.

Termeer, C. J. A. M., A. Dewulf, H. Van Rijswick, A. Van Buuren, D. Huitema, S. Meijerink, T. Rayner, and M. Wiering. 2011. "The Regional Governance of Climate Adaptation: a Framework for Developing Legitimate, Effective, and Resilient Governance Arrangements." *Climate Law* 2(2): 159–79.

Thomas, A. C., and S. Bond. 2016. "Reregulating for Freshwater Enclosure: A State of Exception in Canterbury, Aotearoa New Zealand." *Antipode* 48(3): 770–89.

Thompson-Fawcett, M., J. Ruru, and G. Tipa. 2017. "Indigenous Resource Management Plans: Transporting Non-Indigenous People into the Indigenous World." *Planning Practice & Research* 32(3): 259–73.

Tschakert, P., P. J. Das, N. Shrestha Pradhan, M. Machado, A. Lamadrid, M. Buragohain, and M. A. Hazarika. 2016. "Micropolitics in Collective Learning Spaces for Adaptive Decision Making." *Global Environmental Change* 40: 182–94.

Uekusa, S. 2018. "Rethinking Resilience: Bourdieu's Contribution to Disaster Research." *Resilience* 6(3): 181–95.

Wolf, E. 1972. "Ownership and Political Ecology." *Anthropological Quarterly* 45(3): 201–5.

Chapter 8

Climate Change and East Africa's Past

Three Cautionary Tales

A. Peter Castro

According to the most recent assessment by the Intergovernmental Panel on Climate Change (IPCC), Africa is one of the most vulnerable continents to the environmental and social processes unleashed by global warming. Human-driven climate change already affects African ecosystems, "and future impacts are expected to be substantial" (Niang et al. 2014: 1202). The impacts of global warming on the African continent are by no means uniform. The IPCC predicts reduced rainfall for parts of northern and southern Africa, but other areas, such as the Ethiopian highlands, may experience increased precipitation. Vulnerability arises from a high dependence on rainfed farming and herding, which represents 98 percent of agricultural production in sub-Saharan Africa (Niang et al. 2014: 1212). Even with rapid urbanization in recent years, economies and livelihoods still mainly rely on natural resources. Although they are among the people least responsible for anthropogenic climate change, Africans are disproportionately at risk from it (Castro, Taylor, and Brokensha 2012). Their governments and societies are responding to the challenge, though they possess what the IPCC diplomatically calls a low capacity for action. Nevertheless, much is known about the way forward for climate change adaptation. The IPCC wisely observes that "no single adaptation strategy exists to meet the needs of all communities and contexts in Africa" (Niang et al. 2014: 1226). All strategies will need to integrate equity and social justice concerns to achieve widespread and effective action. Some success stories already exist in this regard (Niang et al. 2014).

Notes for this chapter begin on page 220.

The IPCC's Africa report also warns about "maladaptation risks," policies and actions that may undermine social and environmental conditions. These risks can occur at various levels, whether in the form of national policies promoting economic growth at the expense of agro-environmental stability or communities cutting trees on steep slopes. In my experience in East Africa, international- and national-level actions do the greatest harm, but locally based deeds usually receive the most attention. In Darfur, Sudan, media attention often emphasized ethnic animosity and local inability to cope with drought, for example, in overlooking how colonial, postcolonial, and international actions marginalized and polarized the region (Castro 2018). Understanding the articulation of national- and global-level processes with rural communities is indispensable for adapting to climate change in socially and environmentally sustainable ways. East Africa contains an unappreciated, even sometimes unknown, history of state interventions related to claims about climate change.

This chapter presents three cautionary tales from East Africa's past related to climate change, state interventions, and rural communities. These tales of conflict and displacement take place in Kenya, Somalia, and Ethiopia, respectively, drawn from my academic and applied research in those countries. I see the three cases as relevant for gaining perspective on present-day climate change maladaptation risks. The first tale is inspired by recent evictions occurring in Kenya and other places for nominally green goals such as carbon sequestrating (REDD Monitor 2019). Environmental ends are seen as justifying unjust means, as customary forest dwellers and users are deemed squatters and forced to uproot. As will be shown, these actions are similar to ones taken in the early days of colonial Kenya, when British officials seized large tracts of occupied forests using an environmental crisis narrative as justification. The second case is prompted by concern about media and political discourse regarding climate refugees. Instead of illuminating the predicament of displaced peoples, the term climate refugee gets used to convey dystopian narratives, or ones in which refugees get reduced to abstractions, whether as victims or threats (Høeg and Tulloch 2019). I saw refugees and pastoralists portrayed in similar ways in Somalia during the early 1980s, to their detriment. The third tale deals with the repeated use of resettlement in Ethiopia, which has been driven by fears about relentless environmental decline in the highlands. Although the practice has waned in recent years, awareness is needed of the challenges and difficulties posed by resettlement. This tale comes out of my involvement with the BASIS Collaborative Research Support Program from 1999 to 2007. In all three cases I show how people attempted to be more than pawns in power games, resisting oppression or pursuing opportunities. Much of the pain and trouble encountered in the three tales could have been prevented by treating people with greater respect and by including them in public decision-making. This message is a vital one for climate change adaptations.

Climate Change and Crisis Narratives in Kenyan Forests

My first case begins with the British entry into Kenya during the Scramble for Africa in the late 1800s. European explorers and representatives from the Imperial British East Africa Company (IBEAC) followed trails into its interior first forged by Swahili and Arab traders from the coast. Their caravans had long relied on the Kikuyu highlands to replenish food supplies. Early Europeans were often astounded by the bountifulness of Kikuyu agriculture. For example, Ernest Gedge (1892: 526), who accompanied an 1891 expedition to the southern slopes of Mount Kenya, wrote: "Crossing the stream and entering the Kikuyu country, the caravan found itself in a densely populated district; the villages lying on the slopes of the hills, which were a mass of luxuriant crops, beautiful trees, and sparkling streams." The highlands benefited from orographic rainfall and fertile soils. The frequent difficulty caravans had in obtaining supplies after leaving the highlands underscored the area's value. Yet the Kikuyu were sometimes unwilling to provide food to the outsiders, so IBEAC officials, whose royal charter bestowed on them administrative authority, used "high-handed methods" such as violence to obtain it (Sorrenson 1968). In 1895, Great Britain declared Kenya a protectorate, and soon the railroad reached into its interior on the southern edge of Kikuyu country. As the British became more familiar with the highlands, they coveted the area. Sir Charles Eliot (1905), protectorate commissioner from 1901 to 1904, was a proponent of making Kenya a self-supporting colony based on exports by European enterprises. It would become "a white man's country." To fulfill this goal, the British extended their rule northward into lands held by the Kikuyu and other Africans along the ridges of the Nyandarua Range and Mount Kenya.

W. Scoresby Routledge, an anthropologist, was among the early European trekkers into central Kenya in 1902. He later recalled that "the province was . . . practically unknown and its people unsubdued" (Routledge and Routledge 1910: ix). After initial research among the Kikuyu, Routledge left, later returning with his wife Katherine for additional ethnographic fieldwork. In 1910, they published the first extensive study of the Kikuyu. In line with previous accounts, the Routledges (1910: 6) emphasized the Kikuyu's agricultural prowess, though they identified a downside to it: deforestation: "In the heart of Kikuyu . . . scarcely a tree remains. As far as the eye can reach, in all directions, spread one huge garden." They claimed that British rule, with its supposed peace and stability, served to accelerate clearing. Kikuyu elders were quoted about the environmental impact of this situation: "'In old days,' we were told, 'there were many big trees and few people and much rain. Now the big trees are all dead and like earth, so there is little rain'" (Routledge and Routledge 1910: 7). The situation was alarming, the Routledges felt, as deforestation threatened the watershed, timber supply, and soil quality for future white settlement. They chided officials for not intervening, especially since land clearing occurred near a government outpost, but noted that foresters recently seemed to be taking action.

From the time of W. Scoresby Routledge's first entry into the Kikuyu highlands to his book's publication, much had changed in the region (Castro 1995). Small but well-armed colonial forces had repeatedly hammered the Kikuyu and Embu peoples, killing many and confiscating large numbers of livestock. The British conquest was fought almost ridge to ridge in some places, and harsh punitive expeditions ensured no future armed resistance would occur (Sorrenson 1968). Officials increasingly received applications for large tracts of farmland and forests. Everything seemed up for grabs, yet the European presence was still light on the ground. The memoir of Richard St. Barbe Baker (1931), who rose to become Kenya colony's assistant chief conservator of forests, indicated that foresters took seriously the claim that Kikuyu land clearing diminished rainfall. Many officials also believed that the Kikuyu practiced a very destructive form of shifting cultivation, destroying forests and soils. These views were rooted in ignorance and prejudice. The Kikuyu were sophisticated resource users whose techniques such as intercropping, agroforestry, incomplete clearing, and limited tilling allowed some plots to remain under almost continuous cultivation. They also utilized numerous communal and farm-level strategies to retain woodland and trees as part of their economic and social lives (Castro 1995). Their resource management practices were not easily detected by officials who assumed them to be "childlike, simple and impetuous" (Baker 1931: 23).

I have found the claim that cutting trees results in drier conditions to be widespread, though not always widely accepted, in East Africa. One often hears it repeated as part of conservation crusades. Mesfin (1991: 38) recorded that Ethiopia's officially atheistic Marxist regime subjected rural people to "a rather heavy campaign" to make them believe in such a link instead of viewing drought as God's will.[1] In the midst of one of the worst droughts to hit East Africa, Lester Brown and Edward Wolf (1985) tried to highlight the link between deforestation and drought as part of a broader model of population-induced climate change. They argued that Africa's apparent drying trend in the mid-1980s, which resulted in famines in Ethiopia and Sudan (though not in Kenya, where the government responded quickly), came from human pressures on the land. This approach differs from contemporary concerns about anthropogenic greenhouse gas emissions and the role of forests as sinks or sources of their emissions. Brown and Wolf's idea never gained much traction; in many ways it suggested that drought-stricken poor people were responsible for their own lack of rain. But their calls for more commitment of international aid to Africa offered a gentler neo-Malthusianism than Garrett Hardin's (1985) contention at the time that Ethiopia's "overpopulated" poor be allowed to starve to death to restore carrying capacity balance.

Nowadays, human-induced climate change is understood as largely driven by industrial and high consumption lifestyles associated with the world's prosperous populations, not its poor (Castro et al. 2012). What is today called the AFOLU sector—agriculture, forestry, and other land use—contributes less than one-fourth of anthropogenic greenhouse gas emissions, much of it tied to large-scale land clearing rather than the action of smallholders (Smith et al. 2014). Researchers recognize that land use affects weather by altering biophysical processes such as evapotranspiration and albedo, but outcomes are complex due to the multiple variables involved. Lawrence Kiage's (2013) review of human impacts on African land degradation questions the received knowledge on the topic, concluding that many of the processes affecting agro-ecological conditions are still not well understood.

The Kikuyu elders quoted by the Routledges may have been influenced by a series of unusual calamities that struck parts of central Kenya in the late 1800s. These events included drought in 1897/98, locust infestations, smallpox outbreaks, and localized famine in 1898/99 (Muriuki 1974). The latter, affecting the southern Kikuyu highlands, was attributed to trading too much food to caravans and colonial outposts, leaving families unprepared to deal with poor harvests. Many places depopulated by famine and disease were soon occupied by Europeans. Ironically, the land clearing witnessed by the Routledges was largely driven by the increased trade in food. Not surprisingly, British officials decided that large tracts of forest could not be left in the hands of these "ignorant, reckless tribesmen." Due to lobbying by foresters, colonial authorities proved reluctant to privatize large tracts of central Kenyan forests as well, fearing land speculation (Castro 1995). Ultimately, the crisis narrative about the forests resulted in their being placed under state control, utilizing the system of management pioneered in colonial India.

The creation of state forest reserves around Mount Kenya starting in 1910 displaced thousands of Africans (Castro 1995). The British misinterpreted Kikuyu settlements in the forest as encroachment. The reserve boundaries were also essentially lines of convenience, placing large numbers of homes, farms, and grazing grounds within them. Evicted people received no compensation, only warnings not to return. No protests occurred—the punitive expeditions were fresh memories. The injustice was not forgotten, and by the early 1930s unrest over land appropriation compelled the British to convene a commission to adjudicate African claims against settlers and the government. With regards to Mount Kenya, Chief Njega of Ndia, a stalwart of African rights but also a public servant of the colonial state, testified about evictions he carried out two decades earlier: "The huts were uncountable, very, very many" (quoted in Castro 1995: 56). Some British officials sided with the African claims about Mount Kenya and other areas, yet little compensation was forthcoming. The lost forests and other lands remained a source of tension for the rest of the colonial era.

Once the Kenya Land Commission ended in the mid-1930s, colonial authorities sought to intensify land use in the African areas, since the boundaries of the White Highlands were now regarded as sacrosanct. Concerns about climate change again influenced events. A series of East African droughts in the late 1920s and early 1930s generated fears that Kenya was becoming drier. America's soil conservation campaign against the Dust Bowl proved inspirational (Anderson 1984). Officials cajoled and coerced African farmers to terrace, plant fodder grasses, enclose their herds, and other measures that became highly unpopular. Both the simmering land grievances and the anti–soil conservation sentiment helped fuel the anticolonial Mau Mau uprising during the 1950s (Castro and Ettenger 1994). The high cost of defeating it convinced Great Britain to grant Kenya its independence in 1963.

Concerns about climate change and forest evictions have emerged again in Kenya. Many of today's issues reflect its colonial heritage. For decades Kenya retained the old system of state forestry in which local communities possessed no formal role in forest management. Significant reforms took place in 2005, with the creation of a new forest service and legislation allowing for local co-management of state forests. By this time, however, several serious issues afflicted the forestry sector, including corruption, mismanagement, lack of control over forest areas (which had become farms or were converted to other uses), and related issues (Kamau et al. 2018). Recognition of the troubled state of Kenya's forests coincided with a surge in global appreciation of the environmental services provided by trees, especially watershed protection and carbon sequestration. Kenyan officials and international donors sought to address forest sustainability through a number of conservation projects.

One of the targeted sites is Embobut Forest in Kenya's western highlands. This forest is regarded by the Sengwer people as their homeland.

Before colonization, the Sengwer moved seasonally between the wooded Cherangany Hills, where Embobut is located, and the Trans Nzoia plains. White settlers disrupted the pattern during the colonial era when they seized the plains. Officials also sought to impose regulatory measures over the Cherangany forests but permitted the Sengwer to continue their occupancy (KNCHR 2018). The government's toleration of the forest dwellers ended during the 1980s, with evictions initiated. The Sengwer opposed them, seeking to retain their homes and farms. An inconclusive struggle lasted for two decades when evictions escalated and turned increasingly violent. This shift was triggered by a World Bank–financed conservation project in 2007, followed by the European Union's (EU) supported Water Tower Protection and Climate Change Mitigation and Adaptation Project beginning in 2016. Both projects encountered criticism globally due to the evictions, and the EU suspended its project following the killing of a Sengwer man (KNCHR 2018).

Attempts to address Sengwer claims have been mired in numerous issues, including whether they are indigenous inhabitants and questions about compensation (Lynch 2016). The recent report by the Kenya National Commission on Human Rights calls for the respecting of the Sengwer's human rights and the ending of evicting, with indigenous communities treated as integral partners in forest conservation (KNCHR 2018). Yet a recent Kenyan government forest task force acknowledged the Sengwer as indigenous status but claimed their current agricultural practices are at odds with sustainable forestry management. Their report called for relocating the traditional forest dwellers adjacent to the reserves (Kamau et al. 2018). The Sengwer's travails are by no means unique. Instead, forest-dwelling populations worldwide face such challenges, as their holdings are increasingly valued for carbon sinks and other conservation purposes (see REDD Monitor 2019). Although agencies nowadays often possess procedures for local participation, compliance can be spotty, as the initiative for doing so usually rests with the parties promoting the intervention. They often get to decide who gets deemed a "primary stakeholder." Too many times, however, agencies learned the hard way that anyone who can interfere with what you are trying to accomplish is indeed a primary stakeholder.

"Environmental Refugees" and Pastoralists in Somalia: Implications for Representations of Climate Refugees

Somalia, situated along the northeastern Horn of Africa, serves as the setting for the second tale. From August to December 1983, I served as a consultant for CARE-Somalia on its Hiran Refugee-Reforestation Project, a $4 million endeavor financed by the United States Agency for International Development (USAID). The project reflected the outcome of Somalia's disastrous Ogaden War and a realignment of Cold War allegiances in East Africa. The Ogaden region, with its mainly Somali pastoral population, was conquered by Imperial Ethiopia in the late 1800s. Separatism long simmered, and political instability following the overthrow of Emperor Haile Selassie by the Marxist Derg government in 1974 offered opportunity for action. The Western Somali Liberation Front, aided by Somalia's Marxist ruler Mohamed Siad Barre, launched an uprising in 1977. Once a staunch American ally, Ethiopia now turned to the Soviet Union, Cuba, and other communist countries, who supported it against their former Somali comrades. After some initial success, Somali forces suffered defeat by April 1978 (Cooper 2015). Ethiopia's retaking of the Ogaden sparked a huge refugee flow, creating a humanitarian crisis. In the days before compassion fatigue engulfed the public (see Moeller 1999), *National Geographic* readers were startled to see a thin, bareheaded child on the July 1981 cover, with lead articles titled "Somalia's Hour of Need" and "The Dispossessed."

The Barre regime initially devoted considerable attention to the unfolding refugee crisis, but its capacity for action was soon overwhelmed (Lewis 2002). Somalia ranked among the world's poorest nations. A generally hot, dry, and drought-prone land, its economy centered on a nomadic pastoralism generally well adapted to low rainfall and high climatic variability (Thurow et al. 1989). Only limited climatic data existed, and it suggested that a decline in rainfall in recent decades, affecting pastures (Nelson 1981; Krokfors 1983). Moreover, the country had endured a severe shock in the mid-1970s: a prolonged and harsh drought, regarded among the worst ever. Famine emerged in the hardest-hit areas. About twenty thousand people and five million herd animals died (Simons 1995). Another two hundred thousand pastoralists became destitute, and officials tried to resettle many of them on state farms and in fishing communities. The massive arrival of Ogaden refugees further strained government resources. Reliable statistics are unavailable, but perhaps more than a million people entered Somalia, a nation of perhaps four million inhabitants in 1980 (Jordan 1981; Lewis 2002). International relief soon arrived, exceeding $100 million, spawning a refugee industry that had lasting consequences, many of them negative, for the country.

CARE was a nongovernmental organization that received financing from the United States and Europe to carry out development rather than relief activities. American aid, which halted when Barre's socialist regime came to power in 1969, resumed after he broke with the Soviet Union. In the Hiran project, USAID planners wanted to address deforestation and denuded conditions around refugee camps. Provisioning building poles, fencing, and fuelwood for refugees took a heavy toll on vegetation, which was also impacted by livestock from the camps. Fear existed that the refugees were causing desertification. As noted in the Kenyan case, many experts believed that land use practices induced localized climate change. For example, the Independent Commission on International Humanitarian Issues (1985: 82) observed: "What is certain is that environmental deterioration, once set in motion, can become self-reinforcing. . . . The loss of vegetation cover adversely affects the amount of rainfall, and as the former depends on rain its own decline is also then speeded up." Land clearing was usually associated with agriculture, but rapid urbanization resulting from refugees flows now posed an ecological threat. CARE-Somalia received funding to establish tree plantations adjacent to refugee camps near Beledweyne. Organized as a food-for-work scheme, refugees hired by CARE were supposed to receive the equivalent two dollars in sugar per day as payment. My scope of work included examining local willingness to work on the project, since camp residents already received food aid by virtue of their refugee status. Some planners felt that the refugees would be uninterested. They also worried that refugees from pastoralist backgrounds might be unwilling to plant trees or engage in other cultivation-related tasks such as weeding. Labor recruitment and retention were seen as major obstacles to project success.

Journalist Mary Harper (2012: 12) has observed that outsiders often find Somalia "a hard place to understand." The refugee camps of the early 1980s offered a prime example. The Barre regime, which included Ogaden refugees integrated into its administration, deliberately fostered confusion (Maren 1997). Officials tried to maintain high levels of foreign aid by limiting basic information about refugees, including their numbers, composition, and whereabouts. They were successful for a long time. Attempts to conduct camp censuses in 1981 by international agencies were thwarted. The United Nations High Commission for Refugees and other entities eventually accepted a planning figure of seven hundred thousand refugees, with specific numbers set for each camp, which never changed (Lewis 2002). Michael Maren (1997), involved in early relief efforts, believed the actual numbers were as low as four hundred thousand at that time. My fieldwork in three Hiran camps indicated that the planning figures were unrealistically high. Other key aspects about the refugees, such as their ethnic or clan identities, or even why they left the Ogaden, were unknown (Holcomb 1981). Answers were not easily obtained, in part because officials restricted access to the camps. I met aid officers in Mogadishu who were denied permission to travel outside of the capital. Even those outsiders who regularly worked in the refugee camps, such as food distribution monitors, were not granted unrestricted movement. International officials sometimes displayed open disregard for camp residents. On one occasion I saw a caravan of fourteen vehicles drive into a camp, slow down, and then turn around without stopping, to the astonishment and anger of its inhabitants.

My field research, carried out with Hussein Ma'ow, faced strong scrutiny, constant interference, and, occasionally, intimidation. Questions about kin affiliation, for example, were forbidden by officials, as they might "encourage tribal thought." We completed our surveys through persistence and diplomacy. The refugees themselves often pushed back against attempts to silence them, though the oppressive apparatus of the Barre regime extended even into the most forlorn corners of the camps. Our studies revealed that the issue was not whether refugees would work, as they were very willing, but who got the jobs. The patronage networks that permeated the country extended into the camps, including determining who received excess food aid. The project planted many trees and provided income but also generated resentment, as favored groups received more favors. The project's efforts to promote environmental sustainability actually helped fray the social fabric. In presenting my first detailed ethnographic report to an American aid official in Mogadishu, I received a sobering response. Although appreciative of the work and sympathetic to the findings, she said that probably little could be done to alter the situation in the camps. The main concern of US policy was having access to the former Soviet port and military facilities at Berbera. In fact, her observation explained much about the indifference often shown by international agencies (Maren 1997). This is not to say that the Hiran project was unalterable. Recommendations such as paying refugees in cash rather than sugar were accepted. But the condition and fate of refugees as a whole were not a priority for aid policymakers.

During my stay in Somalia, I had several informal conversations with fellow expatriates who worked in varying aspects of the international aid industry about the nature of the refugee crisis. These were individuals drawn from Asia, Europe, and North America, with different professional backgrounds. As Harper (2012) would say, we found much around us difficult to grasp. The politics of food distribution, marked by inflated camp figures and other irregularities, made for a complicated analysis. Unlike Somalia in later years, almost all the food sent into the country reached inside the camps, but from there diversion to the military and others took place. For some, food aid seemed necessary for both relief and long-term national survival, Reliable figures seemed unavailable at the time, though one study later claimed that Somalia only produced 60 percent of the food per capita in the period 1980 to 1982 than it had a decade earlier (ICIHI 1985: 67). Some colleagues regarded Somalia's chronic food insecurity, the large and prolonged refugee crisis, and the overflow of food it provided as evidence of pastoralism's decline as a viable livelihood. Anecdotal evidence pointed to Somali nationals receiving refugee rations. Instead of living off the land, the Somalis were seen as choosing a cozy state of dependence, reliant on international relief. My surveys for CARE disputed this perspective as a caricature, missing the dynamism of household strategies.

Many people regarded the pastoralists, with their nomadic, traditional ways, as incapable of meeting the needs of a modern economy. The pastoralists, with their nomadic, traditional ways, appeared incapable of meeting the needs of a modern economy. The troubles with pastoralism supposedly went beyond food aid creating a disincentive to work. Those influenced by Garrett Hardin's (1968) notion of the tragedy of the commons viewed the herders as inherent threats to sustainable rangeland management. Rangeland deterioration and desertification were assumed to be widespread and expanding. Although Hiran region experienced widespread flooding in late 1983, the country was generally perceived as undergoing a drying trend, adding to pastoralism's woes (Nelson 1981; Krokfors 1983). Thus, the Ogaden conflict served as both a signal of distress and a trigger for movement prompted by a drying land and a dying way of life. The Barre regime regarded pastoralists as economic and environmental liabilities, promoting sedentarization and carrying out other interventions with limited success (Shepherd 1988; Lewis 2002).

My anthropological training inclined me to see nomadic herders as a misunderstood and underappreciated group. The few encounters I had with pastoralists supported the view of them as knowledgeable and sophisticated resource managers. I once attended a meeting in Hiran region to witness the introduction of an internationally funded livestock and agricultural development project to a group of elders. The project, which was ready to start, intended to use aerial spraying to eradicate tsetse flies, which transmit trypanosomiasis to both humans and cattle. The fly-infested areas, which included large tracts near the Shebelle River, were only lightly used by the local population and nomads. As the project staff explained to the elders, spraying would permanently open these tracts to both herding and farming under the project's guidance. The elders were being hired by the project to facilitate its acceptance and to spread the word about its virtues. Most of the elders seemed pleased, rising one by one to thank the project staff for its initiative. One man stood, and after offering his gratitude for the project, then surprised everyone by saying that he was troubled by it. He agreed that most of the time the tsetse fly prevented use of the land by people and livestock. In times of severe drought, however, the tsetse fly receded, allowing herds to move in and take advantage of the fresh grazing grounds. If these places were open permanently, he asked, where would the people and animals go when drought came? The room erupted in argument among the elders. Meanwhile, the stunned project staff had no response. They had not anticipated this possibility. Unfortunately, the elder's statement was not unique, as people elsewhere also complained, usually to no avail, about similar projects that set up permanent water points in customary wet season grazing areas, upsetting local rotational grazing systems by keeping livestock around in dry seasons (Miskell, personal communication, 8 August 2019).

On another occasion, Hussein Ma'ow and I talked with a local man about environmental issues. He described the mid-1970s drought as a horrible time. Even the Shebelle River stopped flowing for a while. The Shebelle is one of only two permanent streams in the country, and its drying up seemed incredible given recent flooding. People relied on the moisture from its floods to plant crops, a practice known as recessional agriculture. I queried the man about weather lore and was surprised to hear about the nighttime sky and stars. Despite Hussein's explanations, I could not grasp the connection between weather prediction and astrological phenomenon. Only years later, when John Miskell directed me to Muusa H. I. Galaal's (1970) brilliant "Stars, Seasons and Weather in Somali Pastoral Traditions," I obtained a clearer sense of the man's ideas. Galaal highlighted the multifaceted importance of weather knowledge, including the role of the stars and moon in Somali pastoralist interpretations and predictions of meteorological events. Herders and cultivators sought to integrate this information into decisions about their movements to avoid as much disaster and ruin as possible.

Barre's increasingly corrupt and brutal dictatorship fell in 1991 after a bloody civil war. The breakaway Republic of Somaliland emerged in the north, a peaceful, democratic nation, lacking recognition (Lewis 2008). An internationally supported federal government operates in Mogadishu, but it competes with Islamist extremists for control. Despite the lack of centralized authority, some economic sectors thrived, including, at times, herding (Little 2003; Harper 2012). But conflict, insecurity, and limited infrastructure, especially in the context of climate change, have undermined lives and livelihoods. The country's capacity for climate change adaptation also suffered. Somalia appears to be experiencing increased "frequent episodes of both excessive and deficient rainfall" (Ogallo, Ouma, and Omondi 2017: 47). In other words, both droughts and floods occur more frequently. Humanitarian and media accounts in recent years documented the toll of this pattern. More than 250,000 people died during the 2010–12 drought (UN News 2013). In 2019, Somalia had 2.6 million displaced people, many of them facing acute hunger due to a series of poor rains and a general lack of productive assets (UNOCHA 2019). Yet flooding along the Shebelle River in late 2019 affected more than a half million people, with Beledweyne among the hardest-hit places (Mumin and Burke 2019).

In recent years, media reports highlighting Somalia's "climate refugees" have appeared. Some of these have been exemplary, offering cogent analyses of the complex predicament of displaced Somalis (for example, see Goldbaum 2018). Others have been less convincing in their portrayals. As the earth experiences increasingly intemperate weather, rising sea levels, and similar manifestations of climate change, more and more people will be compelled to seek refuge. The legal status of being a refugee has yet to be defined, though a recent ruling by the United Nations Human Rights Committee suggests that recognition may be coming (UNHCR 2020). Media report will be vital in shaping public perceptions about what will likely become a highly contentious issue. Several analysts warn about accounts that foster dystopian and apocalyptic narratives (Bettini 2013) of climate refugees, or otherwise frame them in ways furthering their marginalization (Høeg and Tulloch 2019). My Somalia experience made me cautious about such representations. In these times of bullying nationalism, when opportunistic leaders and fearful citizens look for foreign scapegoats, misperception and mistreatment of refugees and displaced people will go hand in hand. Social scientists, journalists, and activists need to contribute more forcefully and effectively to such debates.

A "Climate So Unpredictable That It Cannot Now or in the Future Support the Present Population": Drought, Famine, and Resettlement in Ethiopia

The above quoted passage comes from Sandra Steingraber's (1988: 18) study of the Derg regime's 1984 plan to resettle 1.5 million people from drought-stricken northern highlands to Ethiopia's southwest. At the time, its leadership under Mengistu Haile Mariam was celebrating ten years of Marxist revolution. They had deposed a long-serving emperor discredited by a slow response to famine in the highlands. Now Chairman Mengistu faced a similar situation in some of the same provinces. His government's sluggish response to starving villagers was matched by the global community's lethargy regarding Ethiopia's predicament (Gill 1986). Once the Derg recognized the need for drastic action, they viewed resettlement as a means of offering famine relief while also advancing revolutionary transformation. Their plan was, according to Alula Pankhurst (1992: 50), "the most complex and ambitious operation in the history of the Ethiopian state." The government had been shamed by the need to rely on massive international relief. Resettlement was portrayed by officials as an act of national self-determination, a means to diminish dependence on foreign aid. By the end of 1984, with the famine now a global news story, authorities designated it as their highest priority (Pankhurst 1992).

A key assumption underlying resettlement was that the highlands faced a permanent environmental and social crisis (Steingraber 1988). The land was perceived as becoming so unproductive, its climate so capricious, that farming was increasingly untenable. The magnitude of the 1984–85 famine appeared to offer confirmation: eight million rural Ethiopians on the verge of starvation, with perhaps more than six hundred thousand dying (Gill 2010). Officials worried that this was only the start of things to come. The Derg asked the UN Food and Agriculture Organization (FAO), a technical assistance agency, to carry out a large-scale study of land degradation in the highlands. Its final report issued in 1986 emphasized that a truly grave situation existed: "Too much land has already been irreversibly lost from productive use" (FAO 1986: xvii). Other areas were in an advanced stage of land degradation. Population growth ensured further pressure on resources. According to the FAO (1986: 232), "If present trends continue unchecked, today's children may be likely to see over a third of the Highlands become incapable of supporting cropping in their lifetime." The likely and appalling outcome would be "increasingly frequent and severe famine" (FAO 1986: 232). From this crisis narrative, resettlement was regarded as both necessary and urgent. Officials saw it as an opportunity to fulfill a range of major goals: modernizing agriculture, promoting collectivization, and eliminating unemployment (Pankhurst 1992). Critics noted that resettlement also further enhanced the state's power over the peasantry, as uprooted villagers depended on its agencies for their livelihoods (Clay, Steingraber, and Niggli 1988).

Organized resettlement, whether in development or disaster settings, is exceptionally challenging, requiring careful planning, substantial investment for preparation and services, active settler involvement in decision-making, and overall appreciation of the substantial multidimensional risks (Scudder 2018). Most schemes have been disappointments, sometimes significantly so. Prior attempts at state-sponsored resettlement in Ethiopia had failed (Pankhurst 1992). Nevertheless, the Derg launched its plan, conducting a massive propaganda campaign to entice volunteers. Officials used coercion as well. More than half a million people, most of them from Wollo Province, had been moved by January 1986, when the government suspended the program. Kurt Jansson (1987), head of the UN's relief operation in Ethiopia, saw many problems with it, including rushed planning, inadequate logistical support and infrastructure, and use of force in the selection and treatment of settlers. Others reported hunger, disease, the splitting apart of families, refusal to allow settlers to return to their home areas, and human rights abuses (Clay et al. 1988). Tensions arose in the resettlement areas from host populations, who were seldom consulted about the moves and sometimes envious of the resources devoted to them. Pankhurst's (1992) ethnography documented resettlement's complexities, conveying the interplay between villager agency and state attempts to dominate their affairs. The program resumed in late 1987, on a smaller scale, but soon ended. The Derg's attention had turned to a more ambitious national policy of villagization, compelling more than twelve million people to move from dispersed rural homes to centralized, collectivized settlements (Gill 2010). These efforts at social engineering, which ended with the Derg's ouster in 1991, received widespread criticism. James Scott (1998: 250), for example, stated: "The new settlements nearly always failed their inhabitants as human communities and as units of food production." The coming to power of the Ethiopian People's Revolutionary Democratic Front (EPRDF) did not necessarily mean the end of resettlement as a solution to the highlands' environmental and development issues.

Between 1999 and 2007, I carried out research in Ethiopia as part of the BASIS Collaborative Research Support Program. The project, led by Peter Little and the late Workneh Negatu, brought together a multidisciplinary team of American and Ethiopian scholars. We focused on food security issues, including recovery from drought, in the eastern highlands of Amhara State. Some of the research sites in South Wollo and Oromiya Zones experienced the famines of the early 1970s and 1984–85. Many painful memories endured in those communities. Resettlement also occurred. Despite increases in population and limited application of the land use measures recommended by FAO's 1986 study, the apocalyptic collapse of agriculture had not occurred. Nevertheless, these communities were among the poorest in the nation. As in the past, rural Wollo attracted little investment for infrastructure due to the lack of cash crops, minerals, or other valuable resources. Its economy relied on rainfed agriculture. Most rural households had few productive assets and extremely low incomes: farm plots were very small, as were herds, and off-farm opportunities were scarce. A survey conducted by the BASIS team in 2001 revealed that more than four-fifths of the South Wollo and Oromiya households earned less than fifty dollars per capita, far below the national average (Little et al. 2006). A major food crisis in 1999–2000, triggered by prolonged drought, repeated poor harvests, and a sluggish response by international agencies, revealed Ethiopia's considerable vulnerability to weather shocks (Hammond and Maxwell 2002). Relief arrived in time to prevent starvation, with large-scale food aid distribution lasting into 2001. Households lost a lot of livestock through distress sales and death. A main finding of the study was that people heavily relied on their own self-help networks of family and friends to recover from shocks such as the 1999–2000 drought (Little et al. 2006).

Livelihoods depended on the weather, and the region had always known a substantial degree of climatic variability. As a woman from one of the higher, colder communities put it: "Sometimes there is too much rain, or frost, or too much sun." Intemperate times were followed by periods of comfort and abundance. But long-standing weather patterns that served as the basis for crop choices and the labor calendars from the highest mountains to the lowest valleys were changing, Worry existed as the onset and duration of the wet seasons became less predictable. A man from the warm lowlands in Oromiya Zone, for example, said that local elders regularly foretold the coming of the rains by the winds, but this practice no longer proved reliable. Once the wet seasons arrived, the rains were erratic, with significant dry gaps occurring that affected crops. Several people used phrases such as "lacking confidence in" or "being suspicious about" the continuation of the wet season. The temperatures, including in the usually cool higher areas, were hotter. Meteorological data supported these local perceptions regarding shifting weather patterns (Rosell and Holmer 2015). As I have documented elsewhere (Castro 2012), households responded by altering what and when they planted, with varying and generally limited success. They also increasingly engaged in selling firewood, trade, local wage work, and migration, with job seekers moving as far as Djibouti and Saudi Arabia.

The EPRDF government under Meles Zenawi responded to the threat of food insecurity in several ways, including improving the famine early warning system, implementing a safety net program for food aid, issuing land titles to rural households, encouraging use of green revolution technology, expanding rural infrastructure, and promoting national economic growth. Facing another drought in 2003, officials also revived another initiative: resettlement. Officials intended to move two million volunteers from the highlands to the western lowlands. The national government announced in 2009 that more than a million people had been transferred, with over 90 percent of them attaining self-sufficiency. Journalist Peter Gill (2010) was told by local authorities in a resettlement site that 20 percent of the volunteers had returned home since the program began. He also reported that many settlers still required food aid since they had been allocated poor quality land. Shumete Gizaw (2013) found that many of the resettled households felt that their lives were better than before. At the same time, their lives were dogged by many of the same problems that previously marred the program, including inadequate planning, logistical arrangements, and delivery of services. The new wave of resettlement also encountered a new problem largely created by the EPRDF's own policies: ethnic tensions. The EPRDF, seeking to eliminate what it felt to be Amhara cultural hegemony disguised as Ethiopian civic nationalism, reorganized the country into ethnically defined states. This action heightened identity politics. Resettlement now involved not the transfer of Ethiopians from one place to another but highland Oromos and Amharas into the homeland of other ethnic groups, generating hostility among the host population.

Resettlement's unpopularity among many highland households was evident during the BASIS research project. Several people were initially reluctant to be interviewed, as they feared that we were connected to the resettlement program. Their accounts about experiencing or resisting resettlement in the mid-1980s left no doubts about why the program was often disliked, if not detested. During an interview in 2002, a woman who lived in a village in Oromiya Zone, for example, recalled fleeing with her family to a food camp during the 1984 famine. Her husband and two of their children died there. She and her two remaining children were recruited for resettlement in Illubabor Province. Despite receiving food aid as part of the resettlement process, she disliked her new setting. Officials would not issue identity cards to the settlers, thus preventing them from leaving. This woman was determined to "escape," however, and food aid offered the means. She saved food, including cooking oil, which she sold to some of the Illubabor people. After four months, the woman had collected sufficient money for her and the children to use for transport and survival. They slipped past the measures designed to keep them in the resettlement site. Arriving back in her home community, the woman discovered that someone else now occupied the land. Fortunately, it was her late husband's brother, who returned the farmland. As is typical in Ethiopia, women do not plow land in their community, so she sharecropped out her land while also earning by collecting and selling firewood. The woman remarried but again became widowed, without having established close ties to her in-laws. Her eldest son left for Djibouti, where other villagers had gone, hoping to send remittances. She never heard from him again. The woman was among the poorest households in her community.

The strong local attachment to land was revealed by a woman from Gerado, a rural area adjacent to the city of Dessie in South Wollo Zone. She was also interviewed in 2002. When the famine of 1984 began, the woman was nursing an infant. Her husband was away, near Jimma in southwestern Ethiopia, pursuing trade. As she and the livestock grew thinner, their food supplies dwindling, the woman sent him a request for money, but nothing was forthcoming except a message "to struggle in any way that you can." The woman said wryly, "He was not much help." Local officials told her and others to report for resettlement. The woman had no desire to leave. Instead, she relied on a common drought-response strategy: selling wood. Rural Ethiopians often bring poles and firewood to market or offer bundles along roadsides. In an act of desperation, she dismantled her house, selling its poles and timber to the Gerado food relief camp. The woman described carrying her infant on her back as she sold wood to buy wheat. To preserve her land claim, she erected a small shelter covered by a thatch roof. The woman said with pride that she "would not be pushed from the area."

Other interviewees recalled deaths of people and livestock due to the drought. Many people had left Gerado to seek relief or resettlement. The peasant association organized by the government allocated their land to others. Not long before the drought had started, the Derg had established a producer's cooperative to farm some of the area's best land, including irrigated fields. The cooperative was not fondly recalled locally, as it took a sizable share of their harvests on behalf of the government. Yet, households, including the woman's, gained access to irrigated plots through membership. Despite the rhetoric of revolution, the peasant association and producer cooperatives were run entirely by men. The woman reported struggling to maintain her presence, and her land. She insisted on taking part in the cooperative, but some men pushed back. The woman remembered them saying, "She is a woman, she should not be allowed to participate." They refused to accept her. Others "begged" on her behalf, including her brother. "It was not easy." Ironically, women were an integral if essentially invisible part of the cooperative's farming activities, fulfilling tasks such as weeding, tending to oxen, and transporting harvested products. Even as she maintained the plot, the woman encountered difficulties getting her land plowed. She missed one wet season due to the unavailability of oxen before finally securing help. With the resumption of the rains, it took three years to rebuild her house. She stayed married despite her husband's limited material support.

In recent years, the Ethiopian government's focus on attaining rapid economic growth resulted in the shifting of resettlement priorities to megaprojects, such as dam building in the Lower Omo Valley, or agribusiness investments in Western Gambella. Urban expansion, especially in Addis Ababa, also resulted in displacement and resettlement as well. These cases are beyond the scope of this study, but it bears noting that controversy has occurred in these various settings. Resettlement continued to be a highly challenging social process, one often mired in top-down planning and undermined by insufficient commitment of resources. In recent years, the national leadership has undergone significant change. Hope exists for more inclusive public decision-making. In addition, while still committed to rapid economic growth, the national government's climate change adaption plan aims for attaining carbon neutrality by 2030. If Ethiopia can achieve both human uplift and environmental sustainability in the coming decade, it will emerge as a leader among nations. This would be a fitting situation for a nation whose people have, to paraphrase Mesfin (1991), "suffered under God's environment."

Conclusion

Human-induced climate change offers one of the greatest social and environmental challenges of our times. In global dialogue about what to do, the voices of rural Africans have been seldom heard. They are often portrayed as hapless victims of, or localized contributors to, changing climates, too poor or uneducated to offer insights into mitigation and adaptation strategies. The three cautionary tales presented here from Kenya, Somalia, and Ethiopia illustrate the dangers of poorly conceived and nonparticipatory responses to climate change and other environmental processes. Not only have these resulted in ineffective actions, they have often served to harm people, violating their human rights and undermining their livelihoods. Hopefully we can learn from these errors rather than repeat them, promoting an effective vision of inclusive climate change adaptation as suggested by the IPCC in its report on Africa (Niang et al. 2014).

Acknowledgments

I thank John Miskell, Andy Korn, and the editors for their comments. I am also grateful to the institutions that supported the original research that served as the basis for this chapter: in Kenya, the National Science Foundation, the Intercultural Studies Foundation, the University of California, and the Appleby-Mosher Fund at Syracuse University; in Somalia, CARE and the National Endowment for the Humanities; and in Ethiopia, the BASIS Collaborative Support Program's Horn of Africa Program. I wish to acknowledge colleagues and field assistants who contributed to the original fieldwork covered here: Cyrus Kibingo, George Muriithi, and John Githuri Gichobi in Kenya, Hussein Ma'ow in Somalia, and Mengistu Dessalegn Debela, Kassahun Kebede, and Dilu Shaleka in Ethiopia. I alone am responsible for the content, including errors, in this chapter.

A. Peter Castro is a professor of anthropology and a Robert D. McClure Professor of Teaching Excellence in the Maxwell School of Citizenship and Public Affairs at Syracuse University. He is an environmental and economic anthropologist, focusing on community resource management, climate change, livelihoods, and natural resource conflict management. His publications include coediting *Climate Change and Threatened Communities: Vulnerability, Capacity, and Action* (2012), which featured his case studies of eastern Amhara, Ethiopia, and central Darfur in Sudan. He wrote *Facing Kirinyaga: A Social History of Forest Commons in Southern Mount Kenya* (1995), and also coedited *Negotiation and Mediation Techniques for Natural Resource Management: Case Studies and Lessons Learned* (2007) and *Natural Resource Conflict Management Case Studies: An Analysis of Power, Participation and Protected Areas* (2003).

Note

1. He noted that the government campaign was largely unsuccessful. In the early 2000s, many rural Ethiopians in the highlands still interpreted drought as God's will (Castro 2012).

References

Anderson, D. 1984. "Depression, Dust Bowl, Demography, and Drought: The Colonial State and Soil Conservation in East Africa during the 1930s." *African Affairs* 83(332): 321–43.

Baker, R. St. B. 1931. *Men of Trees*. New York: The Dial Press.

Bettini, G. 2013. "Climate Barbarians at the Gate? A Critique of Apocalyptic Narratives on 'Climate Refugees.'" *Geoforum* 45: 63–72.

Brown, L., and E. Wolf. 1985. "Reversing Africa's Decline." Worldwatch Paper 65.

Castro, A. P. 1995. *Facing Kirinyaga*. London: Intermediate Technology Publications.

———. 2012. "Social Vulnerability, Climate Variability, and Uncertainty in Rural Ethiopia: A Study of South Wollo and Oromiya Zones of Eastern Amhara Region." In *Climate Change and Threatened Communities*, edited by A. P. Castro, D. Taylor, and D. W. Brokensha, 29–40. Rugby: Practical Action Publishing.

———. 2018. "Promoting Natural Resource Conflict Management in an Illiberal Setting: Experiences from Central Darfur, Sudan." *World Development* 109(9): 163–71.

Castro, A. P., and K. Ettenger. 1994. "Counterinsurgency and Socioeconomic Change: the Mau Mau War in Kirinyaga, Kenya." *Research in Economic Anthropology*, 15: 63–101.

Castro, A. P., D. Taylor, and D. W. Brokensha, eds. 2012. *Climate Change and Threatened Communities*. Rugby: Practical Action Publishers.

Clay, J. W., S. Steingraber, and P. Niggli, eds. 1988. *The Spoils of Famine*. Cambridge: Cultural Survival.

Cooper, T. 2015. *Wings over Ogaden*. Solihull: Helion.
Eliot, C. 1905. *The East Africa Protectorate*. London: Arnold.
FAO (Food and Agriculture Organization). 1986. *Ethiopia Highland Reclamation Study Final Report*. Vol. 1. Rome: Food and Agriculture Organization of the United Nations.
Galaal, M. H. I. 1970. *Stars, Seasons and Weather in Somali Pastoral Tradition*. Mogadishu: Ministry of Education.
Gedge, E. 1892. "A Recent Exploration of the River Tana to Mount Kenya." *Geographical Society Proceedings* 14: 513–33.
Gill, P. 1986. *A Year in the Death of Africa*. London: Paladin.
———. 2010. *Famine and Foreigners*. Oxford: Oxford University Press.
Goldbaum, C. 2018. "Somalia's Climate Refugees: The New Humanitarian." 21 February. Retrieved 4 August 2019 from https://www.thenewhumanitarian.org/feature/2018/02/21/somalia-s-climate-change-refugees.
Hammond, L., and D. Maxwell. 2002. "The Ethiopian Crisis of 1999–2000: Lessons Learned, Questions Unanswered." *Disasters* 26: 262–79.
Hardin, G. 1968. "The Tragedy of the Commons." *Science* 162(3859): 1243–48.
———. 1985. "Overpopulation Begets Hunger: Food Gifts Don't Help Poor." *Hackensack Record*, 8 November, A-29.
Harper, M. 2012. *Getting Somalia Wrong?* London: Zed Books.
Høeg, E. and C. D. Tulloch. 2019. "Sinking Strangers: Media Representations of Climate Refugees on the BBC and Al Jazeera." *Journal of Communication Inquiry* 43(3) 225–48.
Holcomb, B. 1981. "Somali Refugees or Refugees in Somalia? The Oromo Flight from Ethiopia." *Cultural Survival Quarterly Magazine* 5(2).
ICIHI (Independent Commission on International Humanitarian Issues). 1985. *Famine*. New York: Vintage Books.
Kiage, L. M. 2013. "Perspectives on the Assumed Causes of Land Degradation in the Rangelands of Sub-Saharan Africa." *Progress in Physical Geography* 37(5): 664–84.
KNCHR (Kenya National Commission on Human Rights). 2018. *An Interim Report of the High-Level Independent Fact-Finding Mission to Embobut Forest in Elgeyo Marakwet County*. Nairobi: Kenya National Commission on Human Rights.
Jansson, K. 1987. "Section 1: The Emergency Relief Operation—An Inside View." In *The Ethiopian Famine*, edited by K. Jansson, M. Harris, and A. Penrose, 1–77. London: Zed Books.
Jordan, R. P. 1981. "Somalia's Hour of Need." *National Geographic* 159(6): 748–55, 765–75.
Kamau, M. W. et al. 2018. *A Report on Forest Resources Management and Logging Activities in Kenya*. Nairobi: Republic of Kenya, Ministry of Environment and Forestry.
Krokfors, C. 1983. "Environmental Considerations and Planning in Somalia." In *Proceedings of the Second International Congress of Somali Studies*, edited by T. Labahn, 3:293–312. Hamburg: Helmut Buske.
Lewis, I. M. 2002. *A Modern History of Somalia*. 4th ed. Oxford: James Currey.
Lewis, I. 2008. *Understanding Somalia and Somaliland*. New York: Columbia University Press.
Little, P. D. 2003. *Somalia*. Oxford: James Currey.
Little, P. D., M. P. Stone, Tewodaj Mogues, A. P. Castro, and Workneh Negatu. 2006. "'Moving in Place': Drought and Poverty Dynamics in South Wollo, Ethiopia." *Journal of Development Studies* 42: 200–25.
Lynch, G. 2016. "What's in a Name? The Politics of Naming Ethnic Groups in Kenya's Cherangany Hills." *Journal of Eastern African Studies* 10(1): 208–27.
Maren, M. 1997. *The Road to Hell*. New York: The Free Press.
Mesfin Wolde Mariam. 1991. *Suffering Under God's Environment*. Berne: African Mountains Association and Geographica Bernensia.

Moeller, S. D. 1999. *Compassion Fatigue*. New York: Routledge.
Mumin, A. A., and J. Burke. 2019. "'We Have Nothing': Somalia Floods Raise Spectre of Famine." *The Guardian*, 19 November. Retrieved 14 April 2020 from https://www.theguardian.com/world/2019/nov/19/we-have-nothing-somalia-floods-raise-spectre-of-famine.
Muriuki, G. 1974. *A History of the Kikuyu, 1500–1900*. Nairobi: Oxford University Press.
Nelson, H. D., ed. 1981. *Somalia*. 3rd ed. Washington, DC: United States Government.
Niang, I., et al. 2014. "Africa." In *Climate Change 2014, Part B: Regional Aspects, Contribution of Working Group II to the Fifth Assessment Report of the Intergovernmental Panel on Climate Change*, edited by V. R. Barros et al., 1199–265. Cambridge: Cambridge University Press.
Ogallo, L. A., G. Ouma, and P. Omondi. 2017. "Changes in Rainfall and Surface Temperature over Lower Jubba, Somalia." *Journal of Climate Change and Sustainability* 1(2): 38–50.
Pankhurst, A. 1992. *Resettlement and Famine in Ethiopia*. Manchester: Manchester University Press.
REDD Monitor. 2019. Retrieved 4 August 2019 from https://redd-monitor.org/.
Rosell, S., and B. Holmer. 2015. "Erratic Rainfall and Its Consequences for the Cultivation of Teff in Two Adjacent Areas in South Wollo, Ethiopia." *Norsk Geografisk Tidsskrift-Norwegian Journal of Geography* 69(1): 38–46.
Routledge, W., and K. Routledge. 1910. *With a Prehistoric People*. London: Arnold.
Scott, J. 1998. *Seeing like a State*. New Haven, CT: Yale University Press.
Scudder, T. 2018. *Large Dams*. Singapore: Springer.
Shepherd, G. 1988. *The Reality of the Commons: Answering Hardin from Somalia*. ODI Social Forestry Network Paper 6d.
Simons, A. 1995. *Networks of Dissolution*. Boulder, CO: Westview.
Smith, P., et al. 2014. "Agriculture, Forestry and Other Land Use (AFOLU)." In *Climate Change 2014, Mitigation of Climate Change: Contribution of Working Group III to the Fifth Assessment Report of the Intergovernmental Panel on Climate Change*, edited by O. Edenhofer et al., 811–922. Cambridge: Cambridge University Press.
Sorrenson, M. P. K. 1968. *Origins of European Settlement in Kenya*. Nairobi: Oxford University Press.
Steingraber, S. 1988. "Resettlement in 1985–1986: Ecological Excuses and Environmental Consequences." In *The Spoils of Famine*, edited by J. Clay et al., 16–65. Cambridge: Cultural Survival.
Shumete Gizaw. 2013. "Resettlement Revisited: The Post-resettlement Assessment in Biftu Jalala Resettlement Site." *EJBE* 3(1): 22–57.
Thurow, T. L., D. J. Herlocker, and A. A. Elmi. 1989. "Development Projects and Somali Pastoralism." *Rangelands* 11(1): 35–39.
UN News. 2013. "Somalia Famine Killed Nearly 260,000 People, Half of Them Children." Retrieved 4 August 2019 from https://news.un.org/en/story/2013/05/438682-somalia-famine-killed-nearly-260000-people-half-them-children-reports-un.
UNHCR (United Nations Human Rights Committee)/USA. 2020. "UN Human Rights Committee Decision on Climate Change Is a Wake-Up Call, According to UNHCR." 24 January. Retrieved 15 April 2020 from https://www.unhcr.org/en-us/news/briefing/2020/1/5e2ab8ae4/un-human-rights-committee-decision-climate-change-wake-up-call-according.html.
UNOCHA (United Nations Office for the Coordination of Humanitarian Affairs). 2019. "Somalia Humanitarian Dashboard, June 2019." Retrieved 4 August 2019 from https://reliefweb.int/report/somalia/somalia-humanitarian-fund-2019-dashboard-23-july-2019.

Chapter 9

"Our Existence Is Literally Melting Away"

Narrating and Fighting Climate Change in a Glacier Ski Resort in Austria

Herta Nöbauer

Introduction: The Anthropocene of the Glacier Ski Resort

This chapter engages with a particular "anthroposcene"[1] in the cryosphere environment of the Austrian Alps that is characterized by environmental and climate change due to human activities (Sillitoe 2021). In the study I propose that glacier ski resorts illustrate an illuminating example of an anthroposcene because of the deep environmental changes they are experiencing due to tourism and climate change. Taking Austria's highest glacier ski resort, which rises to nearly thirty-five hundred meters, as an ethnographic case study, I will explore the relation between skilled workers and retreating glaciers and snow cover. The latter two have been described as some of the most significant signs of global climate change (IPCC 2013). The very long-term nature of the time frame within which they develop, along with their extraordinary capacity for balancing the atmosphere and providing vast water reserves for Europe, make alpine glaciers highly important signals and actors as well as outstanding, sensitive places. However, ongoing human activities and changes in climate conditions are transforming them into "endangered species" (Carey 2007) and troubled places, which, in turn, affect the life and work of humans in manifold ways.

Notes for this chapter begin on page 240.

Human-environmental dynamics in alpine cryosphere environments have only received scholarly attention in the last decade or so (Strauss 2009; Dunbar et al. 2012; Orlove et al. 2014; Elixhauser 2015; Huggel et al. 2015; Beniston et al. 2018). The interest has been prompted primarily by the profound changes and risks occurring in the alpine cryosphere due to the changing climate. However, until now there has been almost no anthropological enquiry focusing on glacier ski areas. This is perhaps surprising, because, as I argue, they provide a good entry point for learning more about the multifaceted processes, discrepancies, and paradoxes shaping human-cryosphere relations in the European Alps (Nöbauer 2021).

Glacier ski areas provide numerous sites that are specifically defined for use by a variety of people. The chapter highlights their significance as "occupationscapes" (Hudson et al. 2011) to local people. These scapes are "defined as landscapes formed and performed through histories of occupational behaviour" (Hudson et al. 2011: 21). While this concept articulates the structural and political dimensions of landscape formation through labor, it also shares certain similarities with the phenomenological approach to landscape as proposed by Ingold in his theory of "task-

Figure 9.1. The "occupationscape" of the Pitztal glacier ski resort, summer 2015. © Herta Nöbauer.

scape." According to Ingold, "tasks are the constitutive acts of dwelling," and "the entire ensemble of activities, in their mutual interlocking" (1993: 158), designate the "taskscape." Or, to put it in Hudson and colleagues' (2011: 29) words, in the taskscape "the habitual practices of humans form familiar patterns which can become landscapes or places."

Although all eight glacier ski resorts in Austria have been affected by retreating glaciers since the 1990s, albeit to different extents, they are considered and marketed as the sole remaining "future snow reservations" by scientists and even more so by tourist managers. Against this stance I argue that such future-oriented models hardly correspond to the experiences of skilled workers in Austria's highest glacier ski resort. Based on my anthropological fieldwork in this alpine cryosphere environment, I illustrate how vanishing glaciers and melting permafrost are profoundly affecting the landscape and the daily work and identities of these male workers and managers. By tracing the various narratives that emerge from their engagement with the troubled cryosphere and exploring the extensive range of practices applied to counter it, I examine the ways in which staff are experiencing and interpreting the profound changes in snow and icescapes. In doing so, I show that these dramatic environmental changes directly affect the workers' concerns about regional stability, their job security, and senses of identity. At the same time, I further argue, the practices applied to solving the problems caused by "unreliable" snow and ice constitute "technological fixes" (Rosner 2004). These in turn invite us to learn more about hegemonic cultural ideas about adaptation to climate change, and cultural change more generally.

However, before I elaborate on the alpine cryosphere "occupationscape" in more detail I first describe the site of my ethnographic study and the emergence of modern snow tourism there. How snow has gained preeminent economic value in the Pitztal Valley and how it has become a cultural resource for identity construction in Austria more generally are explained in the same section. The subsequent section engages with the prevailing local narratives about human-environmental relations. Through these, the self-understandings and attitudes of local people, particularly glacier workers, in relation to their environment are revealed. Presenting these narratives is especially important because, as I show in the final section, they provide a polyphonic set of explanations of the daily practices of snow and ice management that the glacier workers employ in order to provide "reliable snow" in times of economic and climate pressures. Moreover, I demonstrate that the narratives and techniques reveal a political ecology of mountain landscapes and of skiing. In the conclusion, I briefly round off my ethnographic analysis.

Toward an Anthropology of Snow in the High Alpine Environment

Austria's highest glacier ski area is located in a high alpine valley named the Pitztal, in the province of Tyrol, in western Austria. The area, which is home to numerous small and large glaciers, is part of the huge and impressive Ötztaler Alps, which shape the border between Austria and Italy. I have conducted research on the anthropology of snow and issues of vertical globalization[2] in this forty-kilometer long valley since 2012. Although I have visited (and still visit) the Pitztal glacier resort during each of my fieldwork stays, making a broad range of contacts, formal fieldwork in the glacier resort itself was carried out during the 2014 winter season and in the summer and early autumn season in 2015. The warmer season in particular is the most intensive time for the workers who maintain the overall infrastructure of the resort and prepare tourist snowscapes by engineering the mostly gray and brown-red landscape. Although a seemingly paradoxical choice when researching snow and ice issues, the timing was actually perfect for learning more about the occupational engagement in the alpine cryosphere.

Work in the Pitztal: From Poor Agrarian Livelihood to Prosperous Global Tourism

As in many mountainous regions in Austria and elsewhere, in the Pitztal Valley snow provides the major rationale for the regional economy and identity. Snow constitutes the most important occupationscape both for the local permanent residents and the significant numbers of seasonal migrants who currently come mainly from Eastern European countries. Of the more than four hundred ski resorts in Austria, three are located in this valley, and one of these is the glacier ski area. The resorts were established between the late 1960s and the early 1980s, the glacier resort being the latest one. Although some mountaineering in the Pitztal had already started in the nineteenth century (Pechtl 2005), modern winter tourism was initiated by provincial politicians together with local inhabitants in the 1960s, a decade later than in neighboring regions. They shared the socioeconomic and political aim of reviving the Pitztal, which at the time was extremely poor, and of securing its economy. This ambitious goal was expressed in the following commitment by local people in 1966: "We will put our existence, our future, and all of our energy into tourism" (Hochzeiger Bergbahnen 2009: 6).[3] The four political communes of the Pitztal, which currently have a total of around seventy-five hundred permanent residents, have undergone profound socioeconomic change since then.

They have been transformed from an extremely poor high alpine region that in the past was primarily based on agriculture into one whose wealth today derives from the service-based tourist economy. Tourism is supplemented by some small-scale trade and farming, mainly as a sideline, and alpine pasture farming, which, again, is mostly integrated into tourism. Tourism has brought stability to residence levels and even an increase in the population in the past few decades. However, while there is some in-migration in the mixed-economy northern region of the valley, more recently out-migration has increased again in the valley's most high-alpine southern part, which depends exclusively on tourism. This poses a certain threat to the Pitztal, as it does to other alpine regions in Austria and elsewhere. Therefore, the commitment to winter tourism continues to have great economic, social, cultural, and affective power in the Pitztal.

Memories of their poor livelihoods are still very vivid for many people in the Pitztal. They speak of a past full of deprivation and extreme physical hardship; brutally hard and cold winters; strong out-migration, including children's seasonal labor migration in order to prevent them from starving; lack of electricity, tarmac roads, and motor vehicles until the late 1950s; and many other material signs associated with modernity, progress, and comfort (Pechtl 2015). Those memories constitute a powerful narrative of anxiety about "falling back into the past." In particular, the fear is currently exploited by the tourist industry and ski resort companies to explain and legitimize ongoing plans to expand the glacier ski resort into hitherto "untouched" and protected glacier landscapes in order to establish Europe's largest glacier ski resort.

Counting more than one hundred employees, the glacier ski resort company is one of the biggest employers in the valley. The great majority of employees are local people, mainly men. Unlike the first ski resort that was established in the Hochzeiger area, which is entirely in the hands of local shareholders, the glacier ski resort is owned by a private, nonlocal Austrian company with a variety of investors. However, the land on which the ski resort operates is owned by the adjacent political commune of St. Leonhard, which, back in the 1980s, assigned the right to use the land for an unlimited period to the company, providing that it offered as many jobs as possible to local people. However, the divergence between land use and ownership causes various ambiguities and conflicts today. The glacier resort's ski season lasts eight months, usually from early September to early May—a period double that of ski resorts at lower elevations. The early season is dedicated to the training for competitive alpine and cross-country skiers from various countries, including Austrian champions. Due to its long season, the glacier ski resort is the most prosperous of all three resorts in the Pitztal, generating the greatest economic value. Its economic strength and the social power that derives from it seem to be the main reasons of why locals who might otherwise be critical of interventions in the landscape maintain a public silence.

"The White Gold": Skiing into Modernity and Globalization

Tourism has indeed brought modernity, prosperity, and globalization to the Pitztal.

Modern winter tourism has changed the value of snow, transforming it into a commodity now known as "white gold" throughout the European Alps (Denning 2015). Snow as a commodity and all associated infrastructure and imaginations now circulate within global economic and cultural flows—in addition to the global mobility of the many thousands of tourists from all over the world who consume snowscapes. All in all, turning natural snow into "white gold" was part of a broader process of configuring and pushing modernity and capitalist economy. It brought together specific ideas about nature (combining a romanticized aesthetic and the consumption of mountains), progress, labor, leisure, travel and mobility, infrastructures, subjectivity, body practices, and the (self-)governance of modern subjects (Tschofen 1999; Berthoud 2001; Bätzing 2003; Müllner 2017). More particularly, the assemblage implied novel technology for modifying snow (such as snow groomers and snowmaking machines), new infrastructure for accessing the mountains and accommodation, a new financial (credit) policy for acquiring that very same expensive infrastructure, and, last but not least, new styles of alpine skiing on prepared, flattened pistes, themselves previously unusual (Gross and Winiwarter 2015). The new ways of enjoying and consuming mountains and snow allowed an ever-growing number of people to literally "ski into modernity"[4] (Denning 2015) and, I wish to add, increasingly into contemporary globalization.

Performing Identity through Snow

Concurrently, after World War II, skiing was elevated to an important skill of (self-)discipline through which Austria's national identity could be represented and performed (Horak and Spitaler 2003; Tschofen 2004; Müllner 2013). International sporting events such as skiing championships (glorifying Austria's ski heroes and presenting magnificent snow landscapes on postcards and TV), the tourist industry, and national education politics (which institutionalized obligatory ski weeks in Austrian schools) contributed extensively to this identity construction. While the significance of skiing as a national attribute is contested today, its relevance for performing regional identity is still accepted. As I witnessed during my fieldwork, teaching proper skiing techniques even to small children is an integral part of education. Embodying the perfect skiing technique is perceived as a social prerequisite for becoming "a true Pitztaler and Tyrolean." Adults conceive of this embodied knowledge as a significant marker of regional difference in Austria. Accordingly, many of them distinguish between themselves as "the Westerners," who live in the mountains, and the others, the "Easterners," who live in the far-off city of Vienna. This stark oversimplification echoes the historical competition in constructing Austria as a nation after the decline of the Habsburg Empire in the early twentieth century. Notably, the attempts at nationalizing the alpine landscape[5] and, vice versa, naturalizing the nation were at that time directed against the hegemony of understanding the nation as a modern project of "urbanization" (Johler 2002: 102).

However, modern snow tourism—whether it is called mass tourism or the democratization of tourism—has in the past few decades come to be troubled and contested for several reasons. The two main ones are, perhaps not surprisingly, related to environmental issues: on the one hand, the extensive ways of modifying and engineering the alpine cryosphere for the purposes of tourism have attracted growing criticism from regional, national, and trans- and supranational environmentalist associations and institutions. They have the shared aim of protecting this sensitive location by means of "soft" or environmentally friendly tourism. On the other hand, the changing climate is increasingly affecting the alpine cryosphere and with it the glacier ski resort in particular. How the local staff of the Pitztal glacier ski company are narrating, experiencing, and coping with the changing climate in their daily work will be the topic of the next section.[6]

"Going Along with Nature": Local Narratives about Human-Environmental Relations

Before presenting the most important techniques for coping with both the paramount economic importance of and climate change in snow- and icescapes, it is important to describe the three predominant local narratives about humans' relationship with their environment. Doing so, I outline a framework for understanding local and glacier workers' attitudes toward the challenges brought by the changing economy and climate. As I will show, those narratives contain a diachronic and synchronic structure directed at the past and present alpine "occupationscape." Moreover, I argue, they reveal a political ecology of mountain landscapes and of skiing (Stoddart 2012).

The first narrative I call the "traditional agricultural adaptation narrative." By employing this diachronic narrative, local people connect with an alpine agricultural livelihood in which the relationship to the land as a natural resource appears quintessential. The narrative implies a comprehensive knowledge of the alpine environment and its landscapes and of how to cope with and care for the land. Adults in different social positions and situations repeatedly draw on this narrative when emphasizing, "We have always gone along with nature (and not against it)." They thus assert that they know how to deal with the hardship and dangers but also the benefits of and changes in the environment with regard to making a meaningful living out of it. However, tourism and EU agrarian politics have modified this sense of a deep relationship to the land and landscape. People hold both tourism and EU politics responsible for their shift from being agrarian land caretakers to nonagrarian landscape caretakers who are now occupied with protecting a landscape that is run in a nonecological way from "rewilding."[7] While the recent changes brought enormous relief from physical hardship, some older men and women expressed regret at having lost their former identity as autonomous mountain-farmers in their talks with me.

A few glacier workers also repeatedly voiced anger and cynicism about certain urban tourists exhibiting a lack of understanding of alpine livelihoods. "Quite a few urban visitors think the mountains are a romantic place, of course one with wireless internet access. They expect us to protect our environment according to the way they think it should be. But they have no idea what it really means to live here!" (field note, 15 September 2015). Such feelings reveal the ambivalence of being situated between the agricultural and touristic approaches to the environment and therefore lead us to the next narrative.

I call the second, synchronic narrative the "touristic adaptation narrative," in which local people perceive their relationship to the environment through the lens of the tourist economy. Such an understanding of the environment equates to what Escobar (1999), in his anti-essentialist political ecology of "regimes of nature," has termed "capitalist nature." It targets the modification, commodification, and, even more, the domination of nature and its resources for touristic and capitalist ends. Noteworthy here is the distinct time structure that organizes the tourist landscape and the perception of it. In contrast to agriculture, tourism is determined by modern ideas of leisure and holidays and thus implies different understandings of seasons. It further involves distinct expectations of having snow at specific times and in specific places. Many local glacier company staff stressed the imperative of controlling and managing the environment and the uncertainty regarding the weather by employing a large package of snow and ice management techniques (see below) in order to provide the prime prerequisite of "snow reliability" to the tourists. Heinrich, who is in charge of piste preparation and security, illustrated this approach: "Regardless of whether this is climate change or not, if glaciers and snow are retreating like this, we have to go along with that fact and make all possible efforts to provide snow using various methods in order to guarantee tourism" (interview, 6 August 2015). However, they admitted that one can never control nature entirely.

The third narrative has evolved with respect to the ever-increasing importance of technology and infrastructure in high alpine environments, which has, in Escobar's (1999) terms, generated a "technonature." I call this the "technological adaptation narrative." As a "regime of nature," it combines with "capitalist nature" and thereby aims at using, accessing, controlling, modifying, and, where required, dominating nature. The narrative, in turn, encompasses contexts in which tourist and community domains and interests coincide (such as with hydropower). I propose that this narrative is of special anthropological interest because not only is it about the relevance of modern technology but it has also flourished as a medium through which access to the alpine environment is facilitated. It also reveals cultural ideas about the nature and social life of technology. With it, the local narrative describes the idea of the "adjustment" of technology to nature; this "adjustment" could be to nature's physical qualities and resources (such as cold and water), or it could be institutionalized in ideas about nature (such as environmental sustainability). When talking with others about various infrastructure-building projects in the glacier ski resort or somewhere down in the valley, I usually heard that particular infrastructure plans would either "go along with nature, or not." For example, a technician explained to me that the photovoltaic power plant installed in the glacier resort during my fieldwork in 2015 would "go along perfectly with nature." In a similar vein, the two hydropower plants built a few years ago along the valley's Pitze River were described as "going along with nature."[8] In contrast, the glacier company's plan to build a further large pool in the cryosphere environment for collecting more water for snowmaking was "not going along with nature." This plan was rejected by the relevant environmental assessment authorities. Likewise, while snowmaking by means of snow cannons requires cold temperatures, a special snowmaking machine called the All Weather Snowmaker (see below), installed to counter the effects of climate change, is able to produce snow even at very warm temperatures, and hence some workers considered it as "going against nature." Heinrich in particular explained that this machine "goes against nature because it is unnatural to have snow at warm temperatures." According to him, "Technology should always go along with nature and not against it" (interview, 14 November 2014). For him, both natural snowfall and making it snow require cold and not heat, and this dictates which technology is "good." Quite in contrast to him, his colleague Markus, a technician, reaffirmed his conviction that "any technology available in the market should be used for guaranteeing snow" (interview, 2 November 2016). According to him, anything possible should be done to ensure winter tourism and with it the future of the Pitztal. But what are the current techniques for providing "snow reliability"? The next section engages with them in detail.

Challenges in the Alpine Cryosphere: Facing Vanishing Glaciers and Melting Permafrost

Although glaciers in general have been shrinking since the end of the Little Ice Age in the nineteenth century, from the mid-1980s glaciers worldwide have undergone a more or less dramatic retreat (Bender et al. 2011: 407). As shown in a number of anthropological studies, glacial loss and lack of snow cover not only affect the ecological balance in locations around the world but also directly impact people living in the vicinity, affecting their local economies, regional and global tourism, modes of perception, and senses of place (Cruikshank 2005; Orlove, Wiegandt, and Luckman 2008; Wiegandt and Lugon 2008; Dunbar et al. 2012). The retreat of the glaciers in the Ötztaler Alps, which are (still) home to the largest end-to-end glaciated area in the Eastern Alps in Europe, is particularly drastic (Fischer 2017).[9]

"Our existence is literally melting away," stated Reinhold (field note, 8 August 2015), the chief technical manager of the Pitztal glacier ski resort, thus voicing his concerns about the retreating glaciers and snow cover in the ski resort in one of our conversations. Standing with me at around twenty-nine hundred meters in summer 2015, this local man in his mid-fifties described to me how much farther the glaciers had reached when he had begun working in the resort in the 1980s. Like him, several other staff members described the drastic retreat of the glaciers that they had witnessed during the last few decades. Nevertheless, glacier ski areas are marketed to tourists as providing "true snow reliability" when compared with resorts at lower elevations, particularly as climate research models are projecting that ski areas in Europe located below twelve hundred meters, in contrast to those in higher areas, will disappear toward the end of the century due to the ongoing temperature rises in the European and Austrian Alps (APCC 2014: 16, 25; Marty et al. 2017). However, a recent study has revealed that by then there may be a decrease in snow depth of about 50 percent even for elevations above three thousand meters

(Marty et al. 2017). This forecast echoes Reinhold, who finds it difficult to imagine their glacier ski area as having reliable snow in the long-term future: "We'll have to make even greater efforts and fight even harder on the glacier in ten or twenty years. We'll need more equipment and more staff" (interview, 8 August 2015).[10] His colleague Heinrich was convinced that there will still be skiing on "their" glacier for the next few decades at least. The high altitude combined with snow depots (huge hills of harvested snow—see below) and snowmaking would continue to make it possible (interview, 6 August 2015).

Coping with "Atypical Dangers" and Climate Change

Similar to ski resorts at lower elevations, glacier resorts assemble a whole range of legal provisions and adopt a multitude of standard operating procedures as a prerequisite for their establishment and maintenance as well as for safety reasons. In high-altitude resorts, many of these regulations relate to their particular weather, ice, and snow conditions. Familiarity with these regulations is very important when carrying out field research in order to fully comprehend the diverse tasks and narratives of the workers. In particular, so-called "atypical dangers" such as avalanches, a piste entirely freezing, crevasses that cannot be filled in, and ablation (i.e., the melting of snow and ice cover over large areas) must be constantly safeguarded against and/or eliminated.[11] If this is impossible, then pistes must be closed (Amt der Tiroler Landesregierung n.d.: 23). According to these regulations, the workers fill in crevasses with snow and ice for safety reasons. Another task relates to the natural movement of glaciers: as they move, they destabilize the towers of the T-bar lift. As a result, each year workers must adjust and relocate the lift towers. Ablation poses another challenge, as I saw during my fieldwork. Bare rocks and debris or permafrost soil appear in ablated areas, which must then be adapted into a "piste-friendly" base by flattening the ground. To do so, workers use drilling and blasting technologies to break up the rocks, and vast amounts of stones must be removed in trucks. "Pay attention and keep away for the next few minutes!" workers repeatedly warned me for security reasons.

While such tasks are mainly defined by provincial piste security regulations, other practices, widely described as snow management, have emerged more from the economic and competitive imperative of "snow reliability" and its current interlinkage with the changing global climate. These are regulated by environmental assessment legislation. However, both categories of practices intersect with one another to a certain extent insofar as they share the safety of tourists and piste security as their prime concerns.

The issue of "snow reliability" in particular has received greater attention in studies of climate change in ski resorts since the turn of the millennium (Mayer, Steiger, and Trawöger 2007) because of the impact the changing climate has on the cryosphere environment. Among other effects, glacial retreat and the permafrost degradation of alpine rocks, debris, and soil are considered to be major hazards in alpine regions. They cause the break-up of rocky slopes and rock falls that endanger the built environment and infrastructure and cause casualties (Krautblatter and Leith 2015: 147). To Heinrich, melting permafrost poses the most serious problem, as the following quote illustrates:

> Really my biggest concern up here is the permafrost. . . . This is very dangerous because you never know when it will start falling apart. But once it is melting, the rocks break apart, and rockslides then become a big danger to our guests and all of us. I know when and how to trigger avalanches for security reasons. But we don't know how to deal with the melting of permafrost except by covering some areas with textiles. (Interview, 14 November 2014)

Making Snow and Glaciers "Reliable" in Times of Climate Change

Against the backdrop of challenges, problems, and dangers facing the company's workers, it is pertinent to describe the three most important techniques for making snow "reliable" and the pistes secure within the rapidly changing cryosphere. These involve making snow depots and covering them with geotextiles,[12] covering sensitive and dangerous zones of glacier and permafrost with the textiles, and making snow by means of technology. While these practices are mainly linked to the economic value of snow and the touristic adaptation narrative, they also have social significance and value to the workers.

The most crucial practice in providing "snow reliability" is making pistes out of the snow stored in huge outdoor snow depots. At the beginning of the season in September. The workers make the depots either during or at the end of the season in May. They drive snow groomers to collect the snow and later distribute it into thick (fifty-centimeter) white stripes running across the rocky brown-gray landscape. This method has been in use for more than a decade. The harvested snow is composed of two-thirds natural and one-third technologically produced snow, which in 2015 together amounted to approximately three to four thousand cubic meters. While these hills of collected snow have an essential economic value, workers also attribute a social significance to them, as illustrated by Heinrich, who emphatically stated, "These depots stand for my future job security!" (field note, 29 July 2015). For him, the snow depots are both powerful performers in the landscape and represent his future occupational security.

Once collection is complete, after several weeks the workers cover the hills with large white geotextiles in order to prevent the snow from melting too soon. During my fieldwork stay in 2015, there were seven such huge hills of snow awaiting distribution and 7.5 hectares of textiles protecting them. Usually, the textiles can be reused for three seasons. As Heinrich explained to me, depending on the natural snow cover on the glacier, some covered snow depots could stand there for as long as three season's without melting. However, the snow conditions were different in 2015, when Heinrich lamented, "There was nearly no snowfall this summer. It might be that there won't be enough snow in the depots to open in September. However, if so, then we'll have to go along with nature and not against it." "And what does that mean?" I asked him. "It means that we have to open the season later," he responded (field note, 29 July 2015). Opening later, however, would cause the company to make a financial loss, he added. Later he admitted: "Even though humans can control most of nature by technology, there will of course always be some part that humans can't control" (field note, 29 July 2015).

In addition to the snow depots, certain glacier and permafrost areas are covered with textiles during the summer. The aim is to prevent the ice and permafrost from rapidly melting and to keep the pistes safe. The specific "sensitive" and dangerous zones protected in this way are around the ski lift towers, the rocky outcrops on the glaciers, and retreating and collapsing glacier terminuses (Mayer et al. 2007: 165; Olefs 2009: 35). The covering method has been exploited at various lower glaciers in Europe since the mid- to late 1990s (Mayer et al. 2007: 165n12), but the fleece textiles were used for the first time in Austria, including at the Pitztal glacier resort, in the early 2000s as a consequence of the extremely hot summer in 2003 (field note, 29 July 2015; "Ein Pflästerli für die Gletscher" 2006: 8). This sort of covering took place even before glaciologists from the University of Innsbruck had experimented with different textiles in glacier ski areas in Austria, including the Pitztal, between 2004 and 2008. Their research results have since shown that the covering method results in a 60 percent decrease in ice and snow ablation (Fischer, Olefs, and Abermann 2011: 95). The scientific findings were echoed in the narratives of several glacier workers when they attributed a social and ecological value to the snow management practices: "We are sometimes blamed by environmentalists for destroying the glacier," Heinrich emphasized, with strong feeling; he continued, "But the opposite is the case: we're protecting the glacier and caring for it!" (field note, 29 July 2015). However, as he and others stated

Figure 9.2. A worker uses a digger for removing some textiles in late summer 2015. © Herta Nöbauer.

in various conversations, it is impossible to cover the whole glacier and completely stop it from melting.

In contrast to the research cited above, ecologists and environmentalists have criticized the use of textiles. According to their critique, textiles mar the appearance of the landscape and have a negative impact on the microorganisms of the snow and ice (BMWFW 2017). These distinct competing scientific standpoints are also reproduced in the differing national-regional environmental regimes regarding the use of textiles. In contrast to Switzerland, for instance, where covering with textiles requires approval by the respective canton's spatial planning administration ("Die Gletscher sind wieder abgedeckt" 2013), in Austria it is defined by environmental regulations as part of the maintenance of glacier ski areas (BMWFW 2017).

The history of snowmaking by means of "modern" technology reaches as far back as the 1930s (Nöbauer 2017, 2018). To produce snow, water is pumped through pipes and snow cannons under high pressure, and the droplets sprayed out need to freeze. Thus, the temperature of the air must be between minus four and zero degrees Celsius, and the water around zero degrees Celsius. Low relative humidity is also essential, as the drier the air, the more snow can be produced. Large amounts of water and energy are also required for this process. The water is taken from communal water sources, collected in large, specially constructed pools, and then pumped through an extensive network of pipes to the snow cannons. Snowmaking can be activated by a fully automated computer system or manually by workers. The quantity and quality of water used are strictly regulated by provincial and national legislation in Austria. In contrast to countries such as Switzerland, Canada, and the United States, which permit the use of chemical or bacterial additives in snowmaking, the "water purity rule" in Austria prohibits this. In the Pitztal glacier resort, glacier water and some spring water are used for snowmaking. Two-thirds of the electrical energy required to drive the process is taken from the Tyrolean power grid (widely based on hydropower) and one-third from the glacier company's own photovoltaic solar power plant.

In Austria, around 70 percent of all ski slopes are currently supplied with technically produced snow (WKO 2018a). This high percentage puts Austria among the leading countries in Europe in terms of human-made snow in alpine ski tourism. Ski lift companies consider snowmaking indispensable for securing winter tourism. They invest approximately €120 million annually into snowmaking infrastructure (WKO 2018b). The snowmaking technology industry meanwhile is participating in the global economy, with an annual turnover in the billions of euros. In the province of Tyrol, which has by far the most ski areas in Austria, nearly all slopes are supplied with technically made snow (Steiger and Abegg 2015: 323).

Snowmaking had already started to be used on a few glacier terminuses in the 1980s. Its further employment since the mid-1990s was intended to enable the ski resort companies open early in the season when there is insufficient snow to cover the rocky base (Mayer et al. 2007: 161–63). The subsequent development in the early 2000s to making snow in even huge glaciated areas has been identified by scientists as a clear and exclusive effect of global warming (Mayer et al. 2007: 161–62).

In the Pitztal glacier ski resort, around 15 percent of its eighty-five hectares of slopes are supplied with technically produced snow. Although this proportion of human-made snow is still rather small compared with the Austrian average and with other (glacier) ski areas,[13] it is surprising for such a high elevation. The first snow cannons were introduced on the Pitztal glacier in 1991. However, in their accounts to me, a couple of workers already retrospectively associated this early installation with "climate change." The retreat of glaciers back then had indeed begun to disrupt the course of slopes. Facing these changes and the decrease in numbers of skiers during the summer, the company decided to reduce the skiing season from twelve to eight months.

Beside the snow cannons, the All Weather Snowmaker, a unique machine otherwise used only in Switzerland, was adopted at the glacier resort in autumn 2009. It is a huge, multiton machine affixed into a building specially constructed for it at an elevation of around twenty-nine hundred meters. As its name indicates, the machine, which made a long and challenging journey from Israel to the Pitztal glacier, works independently of the weather and is capable of producing snow even at ambient temperatures as high as thirty degrees Celsius. The desire for such weather-independent technology on a glacier attracted my interest from the very beginning of my research on snow and, in fact, prompted my decision to conduct fieldwork in the Pitztal in particular. Its operation requires a very high amount of energy, which provoked some glacier workers to criticize the snowmaker. Moreover, as already mentioned, Heinrich asserted that making snow at warm temperatures "goes against nature." In contrast to the snow cannons, the snowmaker is based on the vacuum ice principle. The same cryogenic method has been applied in various extreme environments and at varying heights and depths, such as for desalinating sea water and cooling gold mines in different regions of the world (for details see Nöbauer 2017, 2018). After countless problems (some of which still have to be resolved) due to its emplacement at high altitudes, in 2009 its operation began for the first time in the alpine cryosphere environment (Nöbauer 2017, 2018).

Snowmaking has attracted a significantly broader and stronger critique of its ecological impact than has the use of textiles. Against the backdrop of a whole range of scientific and technological projects established in Austria and elsewhere to reduce the amount of energy and water used, the criticism has primarily been due to the machines' high consumption of energy and water (de Jong 2013; Gross and Winiwarter 2015). Conflicting environmentalist standpoints are echoed by equally conflicting scientific discourses on snowmaking. All in all, however, the different environmental legislative frameworks related to the use of water and the diverse energy sources (fossil fuels, hydro and solar power) are often not taken into account in the controversy about the ecological impacts of snowmaking.

Conclusion

This chapter has engaged with the "anthroposcene" in the "occupationscape" of the Pitztal glacier ski resort in Austria. By focusing on the engagement of the local men working in this particular alpine cryosphere with the dramatic environmental changes occurring there, I explored the multilayered impacts that the globalized tourist economy and climate change have on the environment and workers. My exploration of their and wider local narratives about human-environmental relations along with their practices of snow and ice management shows that, against the backdrop of the highly competitive tourist economy, the workers and the glacier company are under increasing pressure. Primarily, glaciers receding at unprecedented speed, retreating snow cover, and degrading permafrost dominate their daily work, feelings, and thoughts regarding future prospects. These drastic environmental effects and the techniques to counter them obviously impact the workers' sense of individual and regional identity and, in particular, of job security and regional economic stability. However, as demonstrated by their polyphonic explanations of their actions and thinking, workers do make social and cultural sense of such changes. In fact, the predominant local narratives about human-environmental relations, which I have traced, reveal a rich, complex, and sometimes conflicting arrangement of explanations. Placing these distinct narratives about adaptation to their environment within the political economy of past and current alpine livelihoods, I argue that they provide important "cultural sets of practices and ideas" (Wolf 1982: 391) that local people are using and manipulating in order to make sense of ongoing changes. Beside the significance of local narratives of adaptation, workers also incorporate particular scientific discourses and knowledge about "countering climate change" in the alpine cryosphere into their cultural set so they can empower themselves and legitimize their practices. Sciences such as glaciology thus are given a special place in local explanations and, so I propose, have a particular responsibility as a result.

While the modern business of snow and glaciers is clearly created by a local and global political economy (cf. Wolf 1996), it is also very much shaped by a political ecology. In particular, the three adaptation narratives point to the different and often conflicting approaches to land and landscape and the ownership, use, and consumption of it. As I have shown, the widespread and highly complex local narrative of "going along with nature, or against it" implies distinct manifestations of "adaptation." They range from using and cultivating agricultural land and caring for the landscape, as represented in the first "traditional agricultural adaptation narrative"; and extend to modifying, controlling, and, indeed, dominating nature for touristic ends, as illustrated in the "touristic" and "technological adaptation narratives." However, the limits of domination over nature notwithstanding, "capitalist nature" and "technonature" have also become evident.

A closer analysis of the narratives and techniques reveals a major and perhaps surprising idea: the deep belief that the effects of climate change can be countered by means of modern technology. This belief invites us to learn more about hegemonic cultural ideas of adaptation to climate change. While technology is clearly matter of culture, such an exclusive focus on "technological fixes" (Rosner 2004) should make anthropologists (and others such as environmentalists) skeptical. Rather, it also seems increasingly important to critically consider other dimensions of cultural practices, such as consumption and production styles or eventually other cultural ideas of a noncommodified "nature."

To conclude, this ethnographic study has demonstrated how people in a particular alpine cryosphere location are attempting to make sense *of* and *in* an "overheated" world (Eriksen 2016) that is exerting increasing pressure on them.

Acknowledgments

I wish to thank Julene Knox for her diligent English editing and the Department of Social and Cultural Anthropology of the University of Vienna for funding the editing.

Dr. Herta Nöbauer is senior lecturer at the Department of Social and Cultural Anthropology at the University of Vienna. Her research and teaching interests include various topics associated with the anthropology of the environment, technology, and emotions. In her current research, she focuses on the anthropology of snow and vertical globalization in the European Alps. Among others, she has published on human-snow-technology relations, weather issues and climate change in the alpine cryosphere, and identity and work issues related to the alpine environment.

Notes

1. Paul Sillitoe uses the term "anthroposcene" as a contraction of "anthropological scene," particularly the anthropological scene with respect to environmental change due to human activities, to which the now familiar term "Anthropocene" refers. It signifies a play on words to indicate that anthropology has something important to contribute to the study of the Anthropocene (email communication with Paul Sillitoe on 8 June 2019).
2. I use "vertical globalization" to describe the increasing flows of people, ideas, infrastructure, communication technology, trade, and finance oriented toward high mountain areas and the sky. While this orientation is geographically directed upward, vertical globalization may also be directed downward beneath the surface of the earth (e.g., toward the maritime areas or the extraction of diverse resources).

3. German original: "Wir legen unsere Existenz, unsere Zukunft und all unsere Kraft in den Tourismus" (Hochzeiger Bergbahnen 2009: 6).
4. I adopt my subchapter's expression "skiing into modernity" from Andrew Denning's (2015) book title.
5. The first line of Austria's anthem starts with "Country of mountains. . ." (in German "Land der Berge").
6. The significant rise in the costs associated with ski holidays is another reason why skiing is increasingly contested. This popular leisure activity is indeed on track to become an elitist activity.
7. The local meaning of "rewilding" is nature or land getting wild. In contrast to approaches to conservation (Carey 2016) which mainly focus on attempts of reconstructing damaged ecosystems and landscapes, the vernacular use in the Pitztal has negative connotations. Letting land go wild means having no access to the land for agricultural and touristic use.
8. For illuminating details about the tensions between economic and ecological conditions of hydropower in the Pitztal, see Tina Wimmer's master's thesis (2019).
9. On average the glaciers in the Pitztal region retreated more than 24 m in 2016; the year before, they shrank even more, with the average at nearly 66 m (Fischer 2017: 23).
10. During one of my later visits, in December 2018, Reinhold explained that the retreat of glaciers and the degrading of permafrost were accelerating faster than they had expected. "Climate change is quicker than politics and bureaucracy," he affirmed. He anticipated that the areas of bare rock would rapidly enlarge, necessitating new "rock management" practices to flatten them. Such practices, however, have been strictly prohibited by environmental politics until now. Therefore, according to Reinhold, the company and politics would need to react appropriately within the next five years in order to maintain the operation of the ski resort (Reinhold, pers. comm., 4 December 2018).
11. For very recent details, see endnote 10.
12. Some Austrian glaciologists (Olefs 2009; Fischer, Olefs, and Abermann 2011) have experimented with various materials (including different colors and thicknesses) for covering, such as membranes, biodegradable textiles, and nonwoven fabric. Their results have shown that white-colored geotextiles made from nonwoven fabric that is breathable and permeable are the most effective. The Pitztal glacier resort had applied exactly these latter textiles even prior to the glaciologists' experiments. After three seasons of use on the glacier, the textiles are being returned to the construction industry, where they are reused.
13. In the neighboring Ötztaler glacier ski areas, 77 percent of the 111 kilometers of slopes are covered with technically made snow (https://www.soelden.com/schneeanlagen).

References

Amt der Tiroler Landesregierung, Abteilung Sport. n.d. *Tiroler Pisten-Gütesiegel*. Innsbruck. Retrieved 12 July 2019 from https://www.tirol.gv.at/fileadmin/themen/sport/berg-und-ski/downloads_berg_und_ski/piste.pdf.

APCC (Austrian Panel on Climate Change). 2014. *Österreichischer Sachstandsbericht Klimawandel 2014* (AAR14). Vienna: Verlag der Österreichischen Akademie der Wissenschaften.

Bätzing, Werner. 2003. *Die Alpen: Geschichte und Zukunft einer europäischen Kulturlandschaft*. C. H. Beck Verlag: Munich.

Bender, Oliver, Axel Borsdorf, Andrea Fischer, and Johann Stötter. 2011. "Mountains under Climate and Global Change Conditions—Research Results in the Alps." In *Climate Change—Geophysical Foundations and Ecological Effects,* edited by Juan Blanco and Houshang Kheradman, 403–22. Rijeka: InTech.

Beniston, Martin, Daniel Farinotti, Markus Stoffel, Liss M. Andreassen, Erika Coppola, Nicolas Eckert, Adriano Fantini, Florie Giacona, and Christian Hauck, et al. 2018. "The European Mountain Cryosphere: A Review of Its Current State, Trends and Future Challenges." *Cryosphere* 12: 759–94.

Berthoud, Gérald. 2001. "The 'Spirit of the Alps' and the Making of Political and Economic Modernity in Switzerland." *Social Anthropology* 9(1): 81–94.

BMWFW (Bundesministerium für Wissenschaft, Forschung und Wirtschaft). 2017. Sparkling Science. Research project COVER.UP summary. May 2017. Retrieved 13 July 2019 from https://www.sparklingscience.at/_Resources/Persistent/f30348ce96b3059c36b778dcb1803be6532d2033/SpSc%204S%2004-025_UNTERWEGS_R%C3%BCckbl_WEB.pdf.

Carey, John. 2016. "Rewilding." *Proceedings of the National Academy of Sciences of the United States of America* (PNAS) 113(4): 806–8. Retrieved 6 April 2020 from https://www.pnas.org/content/113/4/806.

Carey, Mark. 2007. "The History of Ice: How Glaciers Became an Endangered Species." *Environmental History* 12(3): 497–527.

Cruikshank, Julie. 2005. *Do Glaciers Listen? Local Knowledge, Colonial Encounters, and Social Imagination*. Canadian Studies Series. Vancouver: University of British Columbia Press.

de Jong, Carmen. 2013. "(Über)Nutzung des Wassers in den Alpen." *Jahrbuch des Vereins zum Schutz der Bergwelt* 78: 19–44.

Denning, Andrew. *2015. Skiing into Modernity: A Cultural and Environmental History.* Oakland: University of California Press.

"Die Gletscher sind wieder abgedeckt." 2013. *Südostschweiz*. 1 October. Retrieved 10 July 2019 from https://www.suedostschweiz.ch/zeitung/die-gletscher-sind-wieder-abgedeckt.

Dunbar, K. W., Julie Brugger, Christine Jurt, and Ben Orlove. 2012. "Comparing Knowledge of and Experiences with Climate Change across Three Glaciated Mountain Regions." In *Climate Change and Threatened Communities: Vulnerability, Capacity and Action,* edited by Dan Taylor, David W. Brokensha, and Alfonso Peter Castro, 93–106. Rugby: Practical Action Publishing.

"Ein Pflästerli für die Gletscher." 2006. *CIPRA INFO* 81, Deutsche Ausgabe, p. 8. Retrieved 5 May 2019 from http://www.cipra.org/de/publikationen/2773/459_de/inline-download.

Elixhauser, Sophie. 2015. "Climate Change Uncertainties in a Mountain Community in South Tyrol." In *Averting a Global Environmental Collapse: The Role of Anthropology and Local Knowledge,* edited by Thomas Reuter, 45–64. Cambridge: Cambridge Scholars Publishing.

Eriksen, Thomas Hylland. 2016. *Overheating: An Anthropology of Accelerated Change.* London: Pluto.

Escobar, Arturo. 1999. "After Nature: Steps to an Antiessentialist Political Ecology." *Current Anthropology* 40(1): 1–30.

Fischer, Andrea. 2017. "Gletscherbericht 2015/16: Sammelbericht über die Gletschermessungen des Österreichischen Alpenvereins im Jahre 2016." *Bergauf* 71(141): 18–25.

Fischer, Andrea, Marc Olefs, and Jakob Abermann. 2011. "Glaciers, Snow and Ski Tourism in Austria's Changing Climate." *Annals of Glaciology* 52(58): 89–96.

Gross, Robert and Verena Winiwarter. 2015. "Commodifying Snow, Taming the Waters: Socio-ecological Niche Construction in an Alpine Village." *Water History* 7: 489–509.

Hochzeiger Bergbahnen. 2009. *Hochzeiger Chronik*. Jerzens, Tirol.

Horak, Roman, and Georg Spitaler. 2003. "Sport Space and National Identity: Soccer and Skiing as Formative Forces: On the Austrian Example." *American Behavioral Scientist* 46(11): 1506–18.

Hudson, Mark J., Mami Aoyama, Mark C. Diab, and Hiroshi Aoyama. 2011. "The South Tyrol as Occupationscape: Occupation, Landscape, and Ethnicity in a European Border Zone." *Journal of Occupational Science* 18(1): 21–35.

Huggel, Christian, Mark Carey, John C. Clague, and Andreas Kääb, eds. 2015. *The High Mountain Cryosphere: Environmental Changes and Human Risks*. Cambridge: Cambridge University Press.

Ingold, Tim. 1993. "The Temporality of the Landscape." *World Archeology: Conceptions of Time and Ancient Society* 25(2): 152–74.

IPCC (Intergovernmental Panel on Climate Change). 2013. *Climate Change 2013: The Physical Science Basis; Working Group I Contribution to the Fifth Assessment Report (AR5) of the Intergovernmental Panel on Climate Change*. New York: Cambridge University Press.

Johler, Reinhard. 2002. "Is There an Alpine Identity? Some Ethnological Observations." In *MESS—Mediterranean Ethnological Summer School (Piran/Pirano, Slovenia 1999 and 2000)*, edited by Bojan Baskar and Irena Weber, 4:101–13. Ljubljana: University of Ljubljana.

Krautblatter, Michael, and Kerry Leith. 2015. "Glacier- and Permafrost-Related Slope Instabilities." In *The High-Mountain Cryosphere: Environmental Changes and Human Risks*, edited by Christian Huggel, Mark Carey, John C. Clague, and Andreas Kääb, 147–65. Cambridge: Cambridge University Press.

Marty, Christoph, Sebastian Schlögl, Mathias Bavay, and Michael Lehning. 2017. "How Much Can We Save? Impact of Different Emission Scenarios on Future Snow Cover in the Alps." *Cryosphere* 11: 517–29.

Mayer, Marius, Robert Steiger, and Lisa Trawöger. 2007. "Technischer Schnee rieselt vom touristischen Machbarkeitshimmel—Schneesicherheit und technische Beschneiung in westösterreichischen Skidestinationen vor dem Hintergrund klimatischer Wandlungsprozesse." *Mitteilungen der Österreichischen Geographischen Gesellschaft* 149: 157–80.

Müllner, Rudolf. 2013. "The Importance of Skiing in Austria." *International Journal of the History of Sport* 30(6): 659–73.

———. 2017. "Self-Improvement in and through Sports—Cultural Historical Perspectives." *International Journal of the History of Sport* 33(14): 1592–605.

Nöbauer, Herta. 2017. "Die multidimensionale Reise technischer Schneeerzeugung: Rekonfigurationen von maskuliner Technik, Umwelt und Ökonomie." *Blätter für Technikgeschichte* 78–79: 41–61.

———. 2018. "Von der Goldmine zum Gletscher: All Weather Snow als multiples Frontier-Phänomen." *Zeitschrift für Technikgeschichte* 85(1): 3–38.

———. 2021. "Weather, Agency and Values at Work in a Glacier Ski Resort in Austria." In *The Anthroposcene of Weather and Climate: Ethnographic Contributions to the Climate Change Debate*, edited by Paul Sillitoe, 124–45. New York: Berghahn Books.

Olefs, Marc. 2009. "Intentionally Modified Mass Balance of Snow and Ice." PhD diss., Leopold-Franzens University Innsbruck, Austria.

Orlove, Ben, Ellen Wiegandt, and Brian H. Luckman, eds. 2008. *Darkening Peaks: Glacier Retreat, Science, and Society*. Berkeley: University of California Press.

Orlove, Ben, Heather Lazrus, Grete K. Hovelsrud, and Alessandra Giannini. 2014. "Recognitions and Responsibilities: On the Origins and Consequences of the Uneven Attention to Climate Change around the World." *Current Anthropology* 55(3): 249–75.

Pechtl, Willi., ed. 2005. *Abbilder des Erhabenen: Photographische Annäherungen an die Ötztaler Alpen*. Oetz: Turmmuseum Oetz.

———. 2015. *Im Tal leben: Das Pitztal längs und quer*. Innsbruck: Studia Verlag.

Rosner, Lisa., ed. 2004. *The Technological Fix. How People Use Technology to Create and Solve Problems*. New York: Routledge.

Sillitoe, Paul., ed. 2021. *The Anthroposcene of Weather and Climate: Ethnographic Contributions to the Climate Change Debate*. New York: Berghahn Books.

Steiger, Robert, and Bruno Abegg. 2015. "Klimawandel und Konkurrenzfähigkeit der Skigebiete in den Ostalpen." In *Tourismus und mobile Freizeit: Lebensformen, Trends, Herausforderungen*, edited Roman Egger and Kurt Luger, 319–32. Norderstedt: BoD—Books on Demand.

Stoddart, Mark C. J. 2012. *Making Meaning out of Mountains: The Political Ecology of Skiing*. Vancouver: UBC Press.

Strauss, Sarah. 2009. "Global Models, Local Risks: Responding to Climate Change in the Swiss Alps." In *Anthropology and Climate Change: From Encounters to Action*, edited by Susan Crate and Mark Nuttall, 166–74. Walnut Creek, CA: Left Coast Press.

Tschofen, Bernhard. 1999. *Berg, Kultur, Moderne: Volkskundliches aus den Alpen*. Vienna: Sonderzahl.

———. 2004. "Tourismus als Modernisierungsagentur und Identitätsressource: Das Fallbeispiel des Skilaufs in den österreichischen Alpen." *Histoire des Alpes Storia delle Alpi Geschichte der Alpen* 9: 265–82. (Tourisme et changements culturels. Tourismus und kultureller Wandel).

Wiegandt, Ellen, and Ralph Lugon. 2008. "Challenges of Living with Glaciers in the Swiss Alps, Past and Present." In *Mountains: Sources of Water, Sources of Knowledge*, edited by Ellen Wiegandt, 33–48. Dordrecht: Springer.

Wimmer, Tina. 2019. "Das Restwasser als Schnittpunkt zwischen Ökonomie und Ökologie: Nachhaltiges Wassermanagement am Beispiel zweier Pitztaler Kleinwasserkraftwerke" [The residual water as the point of intersection between economy and ecology: Sustainable water management exemplified by small hydropower plants in Pitztal]. MA thesis, University of Vienna, Austria.

WKO (Wirtschaftskammer Österreich). 2018a. *Factsheet—Technische Beschneiung in Österreich*. Stand Oktober 2018. Retrieved 12 December 2018 from https://www.wko.at/branchen/transport-verkehr/seilbahnen/Factsheet-Beschneiung.pdf.

———. 2018b. *Factsheet—Die Seilbahnen Österreichs*. Stand November 2018. Retrieved 12 December 2018 from https://www.wko.at/branchen/transport-verkehr/seilbahnen/Infoblatt-Die-Seilbahnen-in-Zahlen.pdf.

Wolf, Eric R. 1982. *Europe and the People without History*. Berkeley: University of California Press.

———. 1996. "Global Perspectives in Anthropology: Problems and Prospects." In *The Cultural Dimensions of Global Change: An Anthropological Approach*, edited by Lourdes Arizpe, 31–44. Paris: UNESCO Publishing.

Part III

Politics, Policies, and Contestation

Philosophers and social scientists of all kinds are replicating two similar ideas on their writings about climate change and its repercussions on Western societies: there is an instated *mal-être,* and it is impossible to imagine the future, at least a bright one. Candidly, things are not alright now and do not look better ahead. Meanwhile, around the world many people with other cosmogonies and/or worries do not care about what those writing from Paris, Oxford, or Boston are declaring. Meanwhile, in London, Sidney, or Vancouver, others are trying hard to imagine the future and slow the warming down by doing organic agriculture, organizing communities around sharing economies, creating new currencies based on trust, advocating the love for other animals, and promoting notions of well-being rather than amassing coin in order to be happy. Philosophers, social scientists, and tree lovers do not exclude one another, though. Actually, in many instances they are the same person. The chapters in section III of *Cooling Down* are a good illustration of how diverse social movements, politics, public policies, practices, and ideas can coexist, be paradoxical, and be very liquid.

The writings now travel through places like the Elbe River Valley, around Dresden, Germany; Tribal Nations of the North American Southwest; the surrounding mountains of Guarda, Portugal; the Isthmus of Tehuantepec in Oaxaca, Mexico; or Boulder, Colorado, and Isle de Jean Charles, Louisiana, both in the United States.

Kristoffer Albris takes us back to the 2002 and 2013 floods of the Elbe River in Dresden, to "describe how individuals and families have recovered and rebuilt their homes ... and how they reflect on the future as being uncertain." In spite of the recurring floods and the conviction of more in the near future, residents "defiantly but also ambivalently" usually choose to stay. Issues of local policies, risk management, and political economy but also of emotions and feelings are at play. Albris follows two families to produce an ethnography about how people adapt to risk according to their memory, personal experience, solidarity, and feelings of belonging.

Julie Maldonado and Beth Rose Middleton enquire how traditional knowledge of the various tribes of the Southwest United States "offers innovation, guidance, and place-based, time-tested knowledge on how to address climate change from a holistic framework that foregrounds equity and justice." There are 182 federally recognized tribes in the Southwest and many other tribes and communities not recognized by the federal government. The authors ponder indigenous populations with traumatic pasts and agitated contemporary lives, full of uncertainty and "marginalized forces." These populations are facing climate change effects while observing others, in particular Europeans, whom they consider the first perpetuators of nature exhaustion, leading dubious forms of mitigation and adaptation to climate change. Once again, this chapter reports contested knowledges and powers but also contested pasts and memories.

Guilherme José da Silva e Sá addresses a process of rewilding in Portugal. The northeast of Portugal is a mountainous area that was until recently considered among the poorest regions of Europe. Its continental climate with very cold winters, hot summers, and relatively low rates of precipitation combined with depopulation and inexpensive land contribute to forms of reimaging wilderness. Consequently, international organizations are introducing a rewilding project to recreate what they believe to be a lost pristine nature and to mitigate climate change. The views on what the past nature used to be, on land ownership, local development policies, and access to pastures are being disputed.

Amanda Leppert and Roberto E. Barrios examine how public-private partnerships between state agencies and renewable energy companies are organized as a response to anthropogenic climate change in the Isthmus of Tehuantepec in Oaxaca, Mexico. Following energy renewal policies and developments in the region, Leppert and Barrios scrutinize issues such as modernization and consequent risk perception and management as well as environmentalist cosmopolitism and socioeconomic inequity. Foucault and Beck's "hopeful hypotheses concerning cosmopolitan environmentalisms and the vanishing and emergence of epistemic objects" serve as the theoretical framework. The authors emphasize epistemology, development and environmental policies, and international environmental movements but also the potentialities of ethnography.

The last chapter, written by Susanna Hoffman, may be understood as an epitome of *Cooling Down*. The author, though emphasizing that there are multiple factors intervening in and to disasters, states, "One contemporary driver, however, is contributing far more than any other to the recent increased frequency and magnitude of disasters. It is global warming." Two case studies, "Both [coming] from the seemingly impervious United States where, despite the warnings of scientists, significant climate change denial continues to prevail," are introduced as examples of how disaster and climate change are interconnected. The first, a detailed description of an unusual extreme weather phenomenon in Boulder, Colorado, and the second, about a more acknowledged massive seawater inundation and violent hurricanes in the State of Louisiana that are generating "America's First Climate Refugees," not only introduce acumen of how climate change and disaster are linked but also highlight different ways of reacting and adapting in response.

Chapter 10

Where Floods Are Allowed

Climate Adaptation as Defiant Acceptance in the Elbe River Valley

Kristoffer Albris

Introduction

In this chapter, I examine how residents adapt to living in flood risk inundation zones in the Elbe River Valley, around the Saxon capital of Dresden. I describe how individuals and families have recovered and rebuilt their homes after the most recent flood events in 2002 and 2013, and how they reflect on the future as being uncertain. I outline the difficult choices they face in considering whether to move away from an area they feel attached to, or to stay, in spite of the growing realization that more floods will come in the future as a result of changing climatic patterns in this region of Europe. Importantly, the ethnography shows that people defiantly but also ambivalently accept the uncertain conditions they face by staying. Moreover, the ethnography describes how such defiant forms of risk acceptance need to be understood within a political economy of flood risk management, whereby some areas are seen as worthy of protection by the authorities while others, for various reasons, are not.

I take the ethnographic examples of two families in the Elbe River Valley as starting points to discuss how defiantly accepting to stay in a zone of risk despite being aware of the hazards faced can be seen as a form of adaptation. As other anthropological accounts have described in recent years (Marino 2015; Simpson 2013; Ullberg 2013), which I will echo in this

Notes for this chapter begin on page 266.

chapter, people are compelled to act and adapt to risks as a result of factors related to both memory, personal experience, the ability to forget, collective solidarity, community and place attachment, and, not least, political and economic forces that limit their choices. As such, adaptive actions cannot be reduced to single-variable explanations of risk perceptions—a point also raised in a recent review of the risk perceptions and disaster preparedness (Wachinger et al. 2013)—but must be viewed and understood in the context of political and economic forces at work in the given social and cultural environment.

The idea of defiant acceptance that I present in this chapter is intended to conceptualize how people at risk of a natural hazard might be well aware of the dangers they face, based both on personal experience and the possibilities they have of preventing harm to themselves and their family (i.e., to move away). Yet, because of different factors—such as attachment to place, community solidarity, and political-economic configurations of risk management plans—they choose to accept the circumstances in a spirit of defiance. Thereby they exert a sense of agency in relation to the predicaments in which they find themselves. As such, defiant acceptance can be seen as a form of adaptation in relation to climate change and disaster risk and perhaps even as a form of adaptive resilience (Tierney 2014: 197), although resilience as a term carries with it a host of analytical and political issues (Barrios 2014; 2016; Hastrup 2009) that I will not address in detail in this chapter.

The ethnography research underpinning this chapter involved eleven months of fieldwork in the Elbe River Valley area in Saxony, Germany, between 2014 and 2016. The term "Elbe River Valley" has traditionally been used to designate various areas of different scales in the Elbe catchment area in Eastern Germany. Here I take the term to refer to the culturally specific area between the towns of Meissen and Pirna, with its gravitational center being the Saxon state capital of Dresden, and in which the majority of the fieldwork was also conducted. The majority of the ethnographic data stems from participant observation of community life in four periurban settlements along the Elbe River and of interviews with flood-affected residents, exploring how they have attempted to adapt to recurring floods by remodeling their homes or by whether they have considered moving away. I first describe the overall context of the flood events in the Elbe River Valley and how these events have prompted both citizens and local authorities to pursue different climate and flood risk strategies. Following this, I present two ethnographic cases of families living in different parts of the region. Before concluding, I discuss these two cases in light of my proposed arguments as outlined in the introduction, highlighting how existing theories in disaster and climate risk perception need to consider the paradoxical aspects of human action and thought, of which the idea of defiant acceptance is an example.

Floods and Climate Change in the Elbe River Valley

The Central European floods of 2002 and 2013 are the specific disaster events in focus in this chapter. Both events caused massive damage and affected hundreds of thousands of people across the region. The economic costs of the 2013 floods numbered €11.7 billion in total (Munich Re 2014). The even more damaging 2002 floods were one of the costliest disasters in European history. Hundreds of towns and settlements across Central Europe were flooded, and twenty-one people died in Germany alone. In Dresden, the costs of the 2002 floods have been estimated at €1.36 billion (Dresden Brand- und Katastrophenschutzamt 2013). Tens of thousands of people were evacuated. Hundreds of homes were destroyed, ruined, or made uninhabitable for months or years. Stories of people losing their homes or livelihoods and seeing their insurance premiums rise were widespread in the media and in official reports (Dresden Umweltamt 2012). The 2013 floods, although almost as massive in terms of the height of the water levels, caused comparatively less damage than the 2002 event, with costs amounting to "only" €137.1 million in Dresden (Freistaat Sachsen 2013: 12). Some structural flood protective measures, increased cooporation with the Czech authorities upstream in the Elbe catchment area, and a heightened sense of risk awareness among the population are some of the factors officially explaining the lower costs of the 2013 floods (Freistaat Sachsen 2013: 12). Similar cost-reducing measures have been documented across other sites in Germany that suffered from the same flood events (Thieken et al. 2016). An official evaluation report argued that the reduced impact of the 2013 flood was directly due to the massive investments in structural flood protection that followed the 2002 event (Kirchbach, Popp, and Schröder 2013). In Dresden, the local administration calculated that its investment of €26 million in flood management plans between 2003 and 2013 had been directly responsible for the damages being almost a tenth of what they were in 2002 (Dresden Brand- und Katastrophenschutzamt 2013). Still, hundreds of houses were flooded, thousands of people were evacuated, and just as many people were made homeless for a considerable period. Indeed, for many people in Dresden, the 2013 floods were equally as bad as, or much worse than, the 2002 event. Since insurance plans had become less comprehensive, relief aid from state and civil society sources was less generous, and reconstruction firms sought higher profits with poorer-quality work. Most importantly, few residents of Dresden had expected that a flood on a par with the 2002 event would occur within such a short time frame. As one affected riverside resident in Dresden noted while reflecting on the recurrence of the disaster: "I thought I had already had my flood." Suddenly, people in the Elbe River Valley—and the flood experts producing statistical projections of recurring events—have had to reevaluate what a hundred-year flood might mean.

In almost all parts of Europe, river and coastal flooding are the most frequent types of natural hazards, and they have shaped and impacted European societies for as long as there have been human settlements (Mauelshagen 2009; Wanner et al. 2004). The history of Saxony, Dresden, and the Elbe River Valley has been shaped in particular by the recurring presence of riverine floods (Fügner 2002; Korndörfer 2001). In addition, flood disasters have been on the rise across Europe in recent years. The European Environment Agency (2013) has, however, pointed out that it cannot yet be assumed that climate change is driving the increasing number of flood events and damage costs. Nonetheless, more severe and unpredictable weather patterns, as expected for most parts of Europe, would exacerbate the impact of existing cycles of flood events in the region. Across many parts of Europe, it seems, floods have become more frequent and, one could argue, a more regular aspect of life for millions across the continent. One could also argue that, in some places, although they were once thought to be highly exceptional, floods are now increasingly perceived to be the rule rather than the exception.

That residents of the Elbe River Valley region now view floods as a frequent part of life rather than as exceptional events is supported by survey studies conducted in the area. In one study, Kreibich et al. (2011) analyzed the flood preparedness of households and businesses after the 2002 and 2006 floods in the Elbe catchment area. They found that before the 2002 event, 30 percent of households and 54 percent of businesses reported that they had taken no precautionary measures, and only 26 percent of households had the appropriate knowledge of how to respond to floods. These numbers were quite different after the 2006 event, which was not only far less severe in terms of flood levels but also less surprising. Following this event, only 10 percent of households and 26 percent of business reported being unprepared. In a study that surveyed households in Dresden after the same floods, Kreibich and Thieken (2009) concluded that just 13 percent of households had undertaken precautionary measures before the 2002 floods. However, the number of households with some degree of preparedness rose to 67 percent before the 2006 event. The reason for the decrease in general unpreparedness seems to be a rather commonsense one: when you have experienced an event in the recent past, you will be more aware and, thus, better prepared for a similar event. This is supported by my ethnographic data in which people articulated concern about not being able to sustain the kind of knowledge required to prepare for floods. In this sense, risk perception and adapting to risks is a matter of seeing risk as memory bumped forward, as Sheila Jasanoff (2010) has put it.

Such a reasoning resonates with findings from other anthropological accounts of climate change and disasters. Elizabeth Marino's (2015) ethnography of climate change and disasters in Shishmaref, Alaska, portrays in great detail how it is becoming increasingly difficult to demarcate climate change from disaster events in the arctic region, with one exacerbating the effects of the other. Indeed, many recent anthropological publications on climate change show how environmental change and adaptation are being tied to environmental crisis and rupture, including disasters and catastrophes (Hastrup and Rubow 2014). It might not even be that climatic changes are directly felt, as is the case in Shishmaref. But the existence of the perception that it will come to influence or already influences climate and weather variability frames a sense of anxiety about the future as uncertain. The same alarm, I argue, is also the case in the Elbe River Valley, and the concern with climate change is an integral part of such a perception across Germany (Kachelmann 2002) and locally in and around Dresden (Korndörfer et al. 2011).

While local politicians call for private insurance schemes to bear the burden of flood damages, parts of the river basin have been declared as inundation and retention zones, effectively meaning that floods in these areas have to be allowed in order to safeguard others. Many who live in these peripheral flood-prone areas, where structural protection is impossible or unwanted for one reason or another, face a dilemma: either adjust to the flood risk or move away. Unfortunately, for some, houses are hard to sell, while insurance is often difficult to obtain.

Adjusting to floods has consequently been for many a matter of trying to remodel, refit, or restructure homes and houses in such a way that they can withstand at least minor flood events. The owners often choose to make small adjustments rather than wait for the authorities to come up with a comprehensive solution such as a wall or a dike. To add further complexity to the matter, in some areas of the valley, locals protest plans to build structural protection measures, as these would change the landscape so dramatically as to undermine the reason for living in those areas. As Ernst Fischer,[1] a resident of the riverine settlement Laubegast, which is prone to flooding and where locals have resisted the building of a floodwall, explains:

> *Ernst*: I am no hydrologist or building engineer. We do not know what to do to prevent the flood problem. So, people have to be prepared. We try to organize ourselves. We must be smart and find the relevant knowledge for how to deal with it.

Despite protests against walls and dikes in the Elbe River Valley—a phenomenon more pronounced than in any other area of Germany (Otto, Hornberg, and Thieken 2018)—most people would like to see some form of collective flood prevention measures and, preferably, sustainable solutions such as water retention basins or widening the riverbed. Some locals have also argued that a collective insurance fund for flood damages should be created, where everyone living close to the river should pay a monthly or annual contribution. This idea has, however, not materialized as of yet. In the absence of such solutions, people devise small-scale plans for how they can continue everyday life when the ground floor of their house is under water, when their neighborhood has been isolated from the rest of Dresden, or when basic infrastructure utilities such as electricity or gas are cut off. Pictures of people in canoes or kayaks circulating online or in local news media outlets are a common illustration of the predicament, as locals use alternative means of transportation to get to dry land in order to catch a bus or a tram to work. Many have also moved their kitchens above the ground floor, thereby minimizing the risk of essential items such as refrigerators and stoves being damaged or destroyed. Others have readied temporary solutions to ensure continued water and power supply to their house. Yet, despite such creative and microscale attempts to live with flood hazards, many constantly confront themselves with critical and unnerving questions, such as whether to relocate up to higher ground or simply move far away from the Elbe.

As I will show, however, it is not necessarily clear what it would mean for people to move away or whether that is even possible. Understanding the dilemmas that flood-affected individuals and families face means exploring the subjective and social mechanisms at play for why people choose to remain in flood inundation zones while being perfectly aware of the risks they face, even, as I came across, reporting a sense of pride about the fact that they chose to stay. I will describe this through the ethnographic cases presented in the following sections of the chapter.

Coming to Terms with Uncertainty

The Krüger family lives in Meusslitz, at the far eastern end of the Dresden municipality. The father, Andreas, works as an office clerk (*Bürokaufmann*), while the mother, Petra, works in childcare (*Kinderbetreuerin*). They have lived in their house since 2001 and have two children between the ages of ten and fifteen.

As I knock on their door on a very warm summer day in July 2015, Andreas invites me in and leads me into their comfortable shady garden, complete with a garage, tool shed, and trampoline for the kids. The garden is big, the house has three stories, and just behind the hedge that goes around the entire plot you can catch a glimpse of the Old Elbe Arm, which is essentially a long strip of idle land. This was where the water that flooded their house in 2002 and 2013 came from.

When they bought their house in 2001, it was a complete ruin, and they needed to do a whole lot of work to get it into shape. Growing up, Andreas's father taught him how to use tools, and he became quite handy, so they took it upon themselves to restore the house and turn it into their dream home. With the help of friends, they spent a whole year restoring the house, clearing it of asbestos and taking care of other wear and tear that it had endured over the years. "We probably would not have bought the house if we had known just how bad a shape it was in," Andreas reflects. In part because they have spent so much energy trying to design the house according to their visions of a good life, Andreas tells me, they really do not want to move away.

Like many of the other people I interviewed in flood-prone areas, Andreas and Petra bought their house just before the 2002 floods. Since there had not been a flood for many years, they thought it would be fine. "We did not really think about it," Petra says, and explains their story from the beginning:

> *Petra*: We had no idea how high the water could get. No one had told us. There was nothing in the media about what to expect. We probably did not even know whether there would be floods here. A neighbor remembered a flood back in 1941. So, we thought, "No worries!" But then the water came. We took what we could, but we were not quite finished building the house at the time. We tried to protect it with sandbags. Everyone in the area did that. Everyone. Today we know that you cannot stack sandbags particularly high, so it really does not make that much of a difference, but back then we did it. The water level was above the ground floor of the house, almost to the first floor. Everything in the garden floated around.
>
> *Andreas*: The worst thing about cleaning up and repairing damages after the 2002 floods was the dried mud and sludge [*Schlamm*]. The sludge coated the house in a very thick layer, and it had even sealed the door so completely that we could not get inside. The water was gone, but the sludge remained. The house is old, and we had not quite finished the restoration work, so we spent the whole time during that first phase trying to get it clean once again after having bought it just one year earlier.

In the aftermath of the floods, while it was summer and while the process of drying the walls was underway, they set up a kitchen and bathroom in the garden and slept upstairs on the first and second floors, which had not been as severely affected. But, as autumn and winter approached, they had to rent a house through some of their friends. They took whatever furniture they could along with whatever they needed to live a decent, temporary life. They used foldout beds and camping equipment. In the meantime, Petra had become pregnant with their second child.

Andreas and Petra have no objection to living with family or friends when the first floor of their house is underwater, including during the extensive period after the floods when restoration and repair work is ongoing. Some of their neighbors are more stubborn, refusing to leave their homes, and instead use small boats and auxiliary power generators to enable everyday life to continue despite the water. Some neighbors even fished from their balconies and terraces for food. This is something that Andreas and Petra laugh about now, of course, as it was not really a matter of survival but about showing a kind of spirit of refusal on the part of their neighbors to let the floods disrupt the ability to stay where they were.

People who live in areas that have experienced floods explained to me that it has taken more than one flood event for them to ascertain exactly what needs to be done to prepare for the many blows to one's home and psyche: "One learns to make one's house ready and stay updated through the media," as Andreas phrased it. Andreas and Petra have obtained a great deal of knowledge in dealing with floods over the years, including how to be flexible and prepared and, not least, to have foresight about when and how fast the water rises:

> *Andreas*: We now estimate that it takes about ten hours from the first official warning before the water starts to enter our house. We know exactly how much time we have when the water is at a certain level at the measuring station in Dresden. In 2013, this prediction fit pretty well. In 2002, we had no idea what awaited us. And now here in 2013, we knew what was coming—but even though we knew that we had to wait, it was still stressful.

They had no flood insurance in 2002, but they received many private donations to help rebuild. They also invested their own savings. Still, much of the financial support came from the Development Bank of Saxony (Sächsische Aufbaubank, SAB). In 2002, they did not know how much damage the house had sustained or what needed to be repaired. The SAB sent an appraiser who was also an architect and who helped them make plans for how to repair the damage to the house. At that time, the old eighteenth-century house was categorized as a heritage site, which added to the complexity of the process. It was not clear to them how everything was decided upon, but over the next thirteen months, they repaired as much as they could with the help of the architect from the SAB so that they would not have to do anything more for several decades.

What is interesting to note here is what Andreas and Petra think about their future: do they want, or will they be forced, to move?

> *Andreas*: Petra has considered it, but it is not realistic. No one will buy the house. They know that the house has been completely under water at least twice and almost a third time in 2006. We also have a mortgage to pay to the bank, and every time we consider moving, we feel this double burden hinders our plans. It is also very beautiful here, and we do not really want to move. Maybe after the next flood. But selling the house right now makes little sense because we cannot get enough for it to buy something that suits our needs. And, I must be honest, I will not spend the rest of my life in a home where I am unhappy. When the next flood comes, maybe things will be different.

As is the case for many of the people in living in flood-prone areas in Dresden, the Krügers settled down here around the time when they were having their first child because it is far from both the noise and the pollution of the city but still in proximity to the public infrastructure system of buses and trams. Andreas tells me that they have talked about moving on several occasions, especially after the 2013 flood. But it was mostly Petra that wanted to move and he that wants to stay. That is, he corrects himself, if another major flood comes in the next few years, he might reconsider whether to move. For the time being, he likes living here. There is easy transport to the city center, many green spaces, and a close-knit community feeling, both in relation to their surrounding neighbors and to the local community association that was established after the 2002 floods.

What about structural flood protection? I mention the issue of floodwalls to Andreas and Petra. The city decided in 2005 that this part of Dresden is officially classified as a flood inundation zone, and there are no plans for protection. For Andreas and Petra, this means they have been left behind by the city:

> *Petra*: So, they let us consciously be flooded with water because then they can save other areas of the city.
>
> *Andreas*: We have heard that the city is considering building some of the major roads in this part of Dresden higher up, which would make it easier to get to and from these areas that become islands when there is flooding. The city is also considering different bridge configurations so the water can flow easily away and does not get pushed back to us. But nothing has been done so far.

On the one hand, Andreas and Petra think it is understandable from a rational point of view that the city does not intend to safeguard them from floods. There are only a couple of dozen houses in their area of the city and not many stores or significant industry. On the other hand, says Petra, their very existence is at stake. Andreas is more skeptical, asking rhetorically if the city decided to protect this area against floods, how much would the water level rise in the rest of the city? "Maybe you cannot even measure it in centimeters. If that is the case, the only logical course is to build some form of protection here." The question, of course, adds Andreas, echoing the debate in Laubegast, is that some of the locals will have to decide whether they would prefer to lose the beautiful view from their ground floor or be protected from flooding. Petra jumps in and adds:

> *Petra:* The people who decide these things on the city council or the Saxony parliament do not live here themselves, and they probably have never experienced any flooding. They have probably never been affected by water masses themselves.

While the Krüger family's case highlights in detail the worries on the part of families in flood-prone areas of feeling left behind by the local authorities, the case I describe in the following section delves into the question of insurance and the tedious troublesome work around doing repairs to one's home.

Insurance and the Economy of Flood Repairs

On the opposite end of the Dresden municipality, the peri-urban settlements of Cossebaude and Gohlis are like Meusslitz: areas that are partly used as retention spaces for flood management.

I visited the Schneider family in July 2015 on a warm summer night. As I arrive, Dieter Schneider opened his garden gate for me. On the opposite side of the garden lay vast fields of barley. "Look!" Dieter exclaimed as he threw his arms wide, trying to grasp the wideness of the dark blue evening sky with the setting sun in the background. "Who wouldn't want this? You have everything. The fields, the sunshine, the calmness, the trees, the fresh air, and the river not far away. But, just remember, if you can't see the Elbe, then you're fine. Most of the time."

Dieter lives with his wife Helma and their two children on a street that connects Cossebaude and Gohlis on the western end of the Dresden municipality. There are a number of similar houses in this small settlement, with mostly middle-class families that moved out of the Dresden city area to raise children and have space and calm to pursue a comfortable life. All the houses were built between 2000 and 2001. Dieter and Helma are both from the Johannstadt, a neighborhood close to the city center. They moved to Cossebaude in 2000 because they had friends that lived in the area that they wanted to live closer to.

When I visited them, they offered me coffee in their backyard, which was meticulously well kept with beds, shrubberies, water pools for fish, and a finely carved-out terrace. As I complimented the garden, Dieter uttered a long sigh and explained how they have had to repair and rebuild parts of their house over and over again. For them, it seems like their life has been nothing but repair work since the time they moved here.

Their story of the 2002 emergency is strikingly similar to the Krüger family's. Compared to the Krügers, however, they have experienced more frustration during the reconstruction phases. Interestingly, although the 2002 floods were a greater shock than those of 2013, the latter event caused a great deal more trouble. In 2002, they managed the best they could:

> *Helma*: My parents were on vacation, and we lived there for three weeks, and then we found a place to live. Our son had just been born; he was only three months old. And, then we got a home with kitchen, and another one just beside it with a couple of rooms for sleeping. And, for the two months it lasted, we camped out there. We got a couple of pieces of furniture from our damaged house. It was hard. Really hard, with two small children, the small one was three months and the old one was five years. The childcare center was also flooded, so we had to look after them. I was home, at that time I did not work. I had to constantly bring the baby to my parents', go to the house to clean up, then go breastfeed, and then go back again to clean the house. It was a very stressful period.

As with the Krüger family, not long after Dieter and Helma bought the plot and built the house it was designated a flood inundation zone by a 2000 city council decision. That means that the city allows a certain amount of inundation in that area, using the land as a retention space.

> *Dieter*: In the contract and in the law, it says that we had to build in distinct ways, higher than normal. We have the highest house in the area, and we built on a concrete foundation. We were among the first, a lot of the other houses in the area were built after us.

They bought the plot of land, and then the building plans needed to be revised according to flood risk rules. By then, they had already invested in the house. As Dieter explained to me, that they had already invested in the house is crucial for the insurance policies they have held and to the kind of compensation they are entitled to when their house is flooded. In 2002, they had flood insurance with Allianz; the insurance company had taken over their policy from a public flood insurance scheme from the time of the German Democratic Republic (GDR) that many people in flood-prone areas across the former East Germany have had or still have. Although most of the damages that people sustained to their houses and belongings are fairly manageable compared to other disaster contexts, the costs quoted sometimes took me by surprise:

Dieter: The damage was around 100,000 euros in total. The costliest damages were to the walls, piping, electricity, and other basic services that a house needs to function. But everything was very well organized on the part of the insurance company. At that time, very few people here were insured; we were among the only ones. The insurance compensation came very quickly, it functioned very effectively, and they sent a contractor to help us who was very effective. The floor tiles were like sheets of ice. You could just break off pieces of them. Then fungus started to form, and we had to prevent that somehow.

Kristoffer: And that was covered by the insurance?

Dieter: Yes, apart from the 5,000-euro deductible that we had to pay ourselves. The process lasted from August to November, and then it was over. In comparison with 2013, it was a very quick process.

Kristoffer: Did you think after the 2002 floods that you would move?

Helga: No, not at that time. We didn't think it would come again so quickly, and never that it would be as high as it was in 2002. It was a hundred-year flood. So, you thought that was it for the next hundred years . . .

Dieter and Helga were spared flooding during the minor 2006 event, but not long after came the 2013 floods. The aftermath of this event turned out to be more complicated for the Schneiders than the Krügers, as will be evident from the following dialogue:

Dieter: What happened then was horrible. Before the 2013 floods, we had moved our insurance policy from Allianz to a new company. They sent a flood damage expert and an insurance representative, and we signed the contract. And we agreed on what was going to happen with the repairs. But then it developed in a bizarre manner. A water sanitation company came with a group of cheap foreign workers. These workers caused more damage than repairs because they were not skilled enough to know what ought to be removed and what should not, and what had been affected by the water. The company estimated that they would have to spend 20,000 euros to dry the interior walls. And that is not normal. That is much too expensive. And then it continued. Many, many emails with photos and all the information we could possibly send, these tiles, this piece of furniture, and so on. They just did the same thing for all the houses in the area, although there are big differences between the houses in terms of building materials.

Helga: It was bad. It lasted forever. The companies did not need the floods back in 2002. In 2013, the economy was riper for companies to do work on houses after the floods. In 2013, they needed the contracts because suddenly there was an industry around it, so they tried to do it as cheaply as possible.

The case of the Schneider family points to several issues, one of which is the problematic encounter with insurance company policies and insurance laws from state and national legislators, as such policies and laws tend to be altered when events such as floods recur on a regular basis. Another issue is that of renovating and construction companies that offer their services to flood-affected families through the insurance companies. As has been noted in much more extreme examples in the United States (Klein 2007; Adams 2013) and elsewhere in the world (Gunewardena and Schuller 2008), market actors can capitalize on destruction and calamity in ways that rarely end up benefitting those affected by disasters. Yet, the case also echoes points raised earlier in connection with the Krüger family, namely that the Schneider family insists on staying. Although, as Dieter told me as he escorted me out of their garden, each time the floods have engulfed their house, he has considered moving. As time goes by, he reverts stubbornly back to the opinion that they can manage the inundation and that things will become better in the future.

Defiant Acceptance and the Politics of Flood Protection

As is evident from the two cases I have described above, some parts of Dresden are defined as areas where floods are allowed. That some areas are allowed to be flooded is not only due to a lack of solutions or funds, it is sometimes also an integral and necessary part of flood management. In this case, its purpose is to keep other parts of Dresden that are of higher value dry. A kind of center-periphery problem thus occurs when most of the flood protection and management investments are made in the city center (*Altstadt*), where few inhabited buildings are at risk but where buildings of great symbolic value and significance to the city are concentrated.

If complete and total flood protection in all areas of a city like Dresden is impossible, then it is up to those living in flood risk zones to withstand future inundation by bearing more of the financial burden themselves. The Free State of Saxony tries to help and to compensate flood-affected citizens through the SAB, but as we have seen, the process of receiving this financial aid is complicated and lengthy. Moreover, for property owners to receive full compensation, they must carry private insurance as well, but for many it is becoming increasingly hard to keep these insurance policies. As the case of the Krüger family shows, because of falling house prices, it is not always possible for people to move, and in the case of the Schneider family we observe that things do not necessarily get handled more efficiently the second or third time around.

As is also clear from the cases of the Schneider and Krüger families above, one of the questions I asked most often was whether people had considered moving away. There are no official statistics that can shed definitive light on this question in any comprehensive manner. Official demographic statistics on the different districts of Dresden show no clear indicators of this occurrence (Landeshauptstadt Dresden 2016). Any attempt to conclude that people have moved for one reason or the other makes this question complex to answer in a quantifiable way. In Gohlis, I was told that four to five families moved after the 2002 floods, a very small proportion of the total population given that the entire village was flooded. Locals even told me that there was a building boom after 2002, because land and house prices fell. The floods, thus, not only made people leave but also attracted newcomers, and just as companies sought to profit from flood damages, so too did the falling home prices seem attractive to some.

During my many interviews and conversations with flood-affected people, I observed their interesting tendency to talk about the floods as one event. Collapsing such discrete events into one ontological category initially seemed to me to be an indication that people ascribe the existence of floods as being a more or less unavoidable fact in the future. However, as the ethnographic cases above illustrate, when prompted to expand on their experiences, my interlocutors easily singled out and analyzed the different flood events according to their similarities and differences. In other words, even though flood events are collapsed into a single category in everyday language, people do not necessarily take for granted that it means that floods of a similar severity will come in the near future, or that they are predetermined and unavoidable, but neither does it mean that floods will not come. It means, rather, that people living in areas that are allowed to be flooded are beginning to a certain extent to embrace contingency and uncertainty as a condition of life. Rather than being either a lack of risk awareness or a defeatist and apathic resignation, I argue that these dispositions against recurring floods can be viewed as a form of defiant acceptance of the circumstances of a changing environment and climate in a political economy where floods are prevented in some places and allowed in others.

Defiant acceptance should not be interpreted to mean that people sit back and do nothing about their houses being flooded, which ought to be clear from the cases I have described in this chapter. It does mean that, especially after the 2013 event, flood-affected citizens of Dresden do not take expert estimates of statistical return periods or predictions of the future at face value. Rather than using the past to know the present or predict the future, the past casts the future in uncertain terms. The future is to some degree knowable although not predictable, and living with floods is either a fact of life or a period of their lives that is already over—that is, if there are no more floods in the coming decades.

The idea of defiant acceptance as I propose in this chapter also points toward the fact that social capital and social cohesion in a local community—indeed a postdisaster solidarity (Oliver-Smith 1999)—plays an important role in keeping the community together and discouraging people to move. As political scientist Daniel Aldrich has pointed out with respect to postdisaster communities in Japan, the commitment to stay and rebuild, recover, and adapt after a disaster is fueled by a sense of belonging to that place (Aldrich 2012; Tierney 2014: 221). Seeing environmental adaptation as a form of acceptance thus also involves a process of finding a form of peace with nature, or at least to understand it. In this case it is the Elbe River, as Stefan Reuter who lives in Gohlis remarks:

> *Stefan*: You of course thought, floods happen, you live by a river. Everyone thinks by themselves that it doesn't start a fire by oneself, but there is always the likelihood. And without this basic faith and trust in nature, you cannot live. You would live with fear and anxiety every single day. "A new flood will come!" Two years after the major 2002 floods, in 2004, I was badly sick. And a year after that, the roof of the barn flew away. If I was afraid of every single thing that could happen, well then . . . you have to live with it.

The notion of living in an area where floods are allowed to happen, and in which individuals and families have to live with them, captures the situation and stance on part of many citizens in the peripheral urban areas of the Elbe River Valley as a form of defiant acceptance. The following remark by Günther Koch is emblematic:

> *Günther*: In my view, there are only two possibilities: either you protect yourself from the floods or you live with the floods. People living in mountains have other problems with snow or landslides. Coastal people have their problem with storms and tsunamis. We have floods from the river. That is how it is.

Such a perception of living with environmental risk and change—whether perceived to be a result of climatic changes or not—echoes much of the recent literature in the anthropology of climate and environment (Hastrup and Rubow 2014; Crate 2011). Adaptation understood as defiant acceptance might appear to be a paradox, as it suggests that people surrender to the circumstances of risk and choose not to prepare or reduce risks, yet such a narrow understanding of adaptation and risk perception (Grothmann and Reusswig 2006) overlooks the inherent ambiguity and ambivalence that people report in postdisaster situations with respect to how they should respond to future risks. The complexities of risk awareness and perception among local communities continues to be a challenge also for policymakers across Europe (Albris, Lauta, and Raju 2020). As Wachinger et al. (2013) have shown convincingly in a thorough review of risk perceptions across a number of studies, factors such as age, gender, media coverage, education, income, or social status play a minimal role in understanding the relationship between risk perceptions and levels of preparedness. Instead, the personal experience of past hazards and degrees of trust in local authorities and/or one's local community are vital factors explaining the relationship between risk perception and preparedness (Wachinger et al. 2013: 1059).

Although such a perspective on preparedness and risk might seem initially a matter of common sense, it risks overlooking the central importance of political and economic decisions regarding what is prioritized and what is not in the name of risk management. For instance, in the Elbe River Valley, what are termed risk inundation zones might at first seem like an objective risk assessment indicator, specifying which areas are at risk of flooding. In fact, these labels are applied to certain geographical areas in order to legitimize an absence of preventive measures on part of the local and regional authorities. In other words, these are areas where floods are accepted for the time being and where people stay *in spite* of them. As Hoffman (2002: 140) notes, it is not always easy to explain why it is that people return to the place where calamity has happened, although some simply have no choice. Yet, of the people whose stories of calamity I have presented in this chapter, all, in fact, are able to move, although it would most likely put them in dire economic positions. Nonetheless, they do in fact have the possibility. As such, staying in a zone of risk, as Hoffman (2002) notes with reference to disaster symbolism, can also be seen as a form of ownership created in conjunction with an attachment to the places they inhabit in the Elbe River Valley. Having had to cope with several flood disaster events seems to have prompted a kind of routinization complete with dealing with renovation, reconstruction, insurance policies, and, not least, empty promises of flood protection from the local government that never materialize. As part of this routinization and temporary acceptance of risk, I argue, we can also a detect a kind of proud defiance of being able to cope with recurring floods without the aid of structural flood protection schemes from the local authorities.

Conclusion

In this chapter, I have discussed how the process of adapting to recurring floods and perceived climatic changes can be seen as a form of defiant acceptance. I have presented two cases of individuals and families in the

Elbe River Valley who experienced the three flood events, discussing how they have struggled to rebuild and repair their homes, the different ways they experienced the flood emergencies, and their views of the future. I presented these cases in order to show the complexity of how people are trying to adjust their lives to the perceived fact of recurring floods by paying attention to how they see this adaptation and adjustment as not necessarily having an end goal, other than for things to return to normal as quickly as possible after their houses are flooded. I also showed how even the question of whether to move from or stay in a flood-prone area is never a decision that families or individuals take in isolation but one made in dialogue with the presence of floods as a matter of public concern and of a political economy underlying flood protection and risk management. Indeed, as I have aimed to describe throughout this chapter, adaptation or normalization of extraordinary events is never merely a private matter; it is also a social and political one just as much as it is an economic one.

I have argued that the notion of defiant acceptance points to a stance in which staying put and performing minimal amounts of adaptive actions are not irrational or based on erroneous risk perceptions. Rather, such defiance is a reaction to both a general sense of uncertainty with the future of climate risks and also to a political economy of flood protection whereby some parts of the riverine landscape are valued by local authorities and thus prioritized as worthy of protection, to the detriment and sacrifice of other areas. As some citizens see floods as an inescapable although uncertain fact of life, they choose to stay. In doing so, they practice a form of defiant acceptance that is both a collective demonstration of the pride and capability to cope with calamitous events in their local communities as much as it is a critique of the local authorities.

Acknowledgments

The research underpinning this chapter was done as part of the Changing Disasters research project, which was funded by a grant from the 2016 University of Copenhagen Excellence Programme for Interdisciplinary Research.

Kristoffer Albris is assistant professor at the Copenhagen Center for Social Data Science (SODAS) and the Department of Anthropology, University of Copenhagen. His main research interests include people's responses to natural hazards (primarily floods and storm surges), the politics of climate risk adaptation projects, and the relationship between extreme events and the everyday. He is currently working on his first monograph, investigating how people in the Elbe River Valley are coming to terms with recurring floods.

Note

1. Names of interlocutors appearing in this chapter have been pseudonymized.

References

Adams, Vincanne. 2013. *Markets of Sorrow, Labors of Faith: New Orleans in the Wake of Katrina*. Durham, NC: Duke University Press.

Albris, Kristoffer, Kristian Cedervall Lauta, and Emmanuel Raju. 2020. "Disaster Knowledge Gaps: Exploring the Interface between Science and Policy for Disaster Risk Reduction in Europe." *International Journal of Disaster Risk Science* 11: 1–12.

Aldrich, Daniel P. 2012. *Building Resilience: Social Capital in Post-disaster Recovery*. Chicago: Chicago University Press.

Barrios, Roberto. 2014. "'Here, I'm Not at Ease': Anthropological Perspectives on Community Resilience." *Disasters* 38(2): 329–50.

———. 2016. "Resilience: A Commentary from the Vantage Point of Anthropology." *Annals of Anthropological Practice* 40(1): 28–38.

Crate, Susan. 2011. "Climate and Culture: Anthropology in the Era of Contemporary Climate Change." *Annual Review of Anthropology* 40(1): 175–94.

Dresden Brand- und Katastrophenschutzamt. 2013. *Hochwasser Juni 2013: TELKurzbericht zum Hochwasser der Elbe 03.—10.06.2013 in der Landeshauptstadt Dresden*. Dresden: Landeshauptstadt Dresden.

Dresden Umweltamt. 2012. *Umweltbericht 2012: Dresden—10 Jahren nach den Hochwassern 2002*. Dresden: Landeshauptstadt Dresden.

European Environment Agency. 2013. *Flood Risk in Europe: The Long-Term Outlook*. Retrieved 20 April 2020 from www.eea.europa.eu/highlights/flood-risk-in-europe-2013.

Freistaat Sachsen. 2013. *Der Wiederaufbau im Freistaat Sachsen nach dem Hochwasser im Juni 2013: Freistaat Sachsen; Sächsische Staatskanzlei*. Retrieved 20 April 2020 from https://publikationen.sachsen.de/bdb/artikel/20553.

Fügner, Dieter. 2002. *Hochwasserkatastrophen in Sachsen*. Taucha: Tauchaer Verlag.

Grothmann, Torsten, and Fritz Reusswig. 2006. "People at Risk of Flooding: Why Some Residents Take Precautionary Action While Others Do Not." *Natural Hazards* 38(1): 101–20.

Gunewardena, Nandini, and Mark Schuller, eds. 2008. *Capitalizing on Catastrophe: Neoliberal Strategies in Disaster Reconstruction*. Lanham, MD: AltaMira Press.

Hastrup, Kirsten. 2009. "Waterworlds: Framing the Question of Social Resilience." In *The Question of Resilience: Social Responses to Climate Change*, edited by Kirsten Hastrup, 11–32. Copenhagen: The Royal Danish Academy of Sciences and Letters.

Hastrup, Kirsten and Cecilie Rubow, eds. 2014. *Living with Environmental Change: Waterworlds*. London: Routledge.
Hoffman, Susanna. 2002. "The Monster and the Mother: The Symbolism of Disaster." In *Catastrophe and Culture: The Anthropology of Disasters*, edited by Susanna Hoffman and Anthony Oliver-Smith, 113–42. Santa Fe: School of American Research Press.
Jasanoff, Sheila. 2010. "Beyond Calculation: A Democratic Response to Risk." In *Disaster and the Politics of Intervention*, edited by Andrew Lakoff, 14–41. New York: Columbia University Press.
Kachelmann, Jörg, ed. 2002. *Die Grosse Flut: Unser Klima, Unsere Umwelt, Unsere Zukunft*. Hamburg: Rowohlt Verlag.
Kirchbach, Hans-Peter von, Thomas Popp, and Jörg Schröder. 2013. *Bericht: der Kommission der Sächsischen Staatsregierung zur Untersuchung der Flutkatastrophe 2013*. Freistaat Sachsen: Sächsische Staatskanzlei. Retrieved 20 April 2020 from: https://publikationen.sachsen.de/bdb/artikel/20534.
Klein, Naomi. 2007. *The Shock Doctrine: The Rise of Disaster Capitalism*. New York: Henry Holt.
Korndörfer, Christian. 2001. "Die Dresdner Elbauen—Hochwasserschutz und Refugium für Menschen und Natur." *Dresdner Hefte* 67(3/01): 22–29.
Korndörfer, Christian, Peter Teichmann, and Frank Frenzel. 2011. "Integrated Climate Adaptation in Dresden: Insights from Flood Prevention." In *Resilient Cities: Local Sustainability*, edited by Konrad Otto-Zimmermann, 1:291–98. Dordrecht: Springer.
Kreibich, Heidi, Isabel Seifert, Annegret H. Thieken, Eric Lindquist, Klaus Wagner, Bruno Merz. 2011. "Recent Changes in Flood Preparedness of Private Households and Businesses in Germany." *Regional Environmental Change* 11: 59–71.
Kreibich, Heidi, and Annegret H. Thieken. 2009. "Coping with Floods in the City of Dresden, Germany." *Natural Hazards* 51: 423–36.
Landeshauptstadt Dresden. 2016. *Statistische Mitteilungen: Bevölkerung und Haushalte 2015*. Dresden: Büro des Oberbürgermeister. Retrieved 20 April 2020 from http://www.dresden.de/de/leben/stadtportrait/statistik/publikationen/stadtteilkatalog.php.
Marino, Elizabeth. 2015. *Fierce Climate, Sacred Ground: An Ethnography of Climate Change*. Fairbanks: University of Alaska Press.
Mauelshagen, Franz. 2009. "Disaster and Political Culture in Germany since 1500." In *Natural Disasters, Cultural Responses: Case Studies toward a Global Environmental History*, edited by Christoph Mauch and Christian Pfister, 41–76. Lanham, MD: Lexington Books.
Munich Re. 2014. "Overall Picture of Natural Catastrophes in 2013 Dominated by Weather Extremes in Europe and Supertyphoon Haiyan." Press release. Retrieved 20 April 2020 from https://www.munichre.com/en/media-relations/publications/press-releases/2014/2014–01–07-press-release/index.html.
Oliver-Smith, Anthony. 1999. "The Brotherhood of Pain: Theoretical and Applied Perspectives on Post-disaster Solidarity." In *The Angry Earth: Disaster in Anthropological Perspective*, edited by Anthony Oliver-Smith and Susanna Hoffman, 156–72. New York: Routledge.
Otto, Antje, Anja Hornberg, and Annegret H. Thieken. 2018. "Local Controversies of Flood Risk Reduction Measures in Germany: An Explorative Overview and Recent Insights." *Journal of Flood Risk Management* 11(S1): S382–94.
Simpson, Edward. 2013. *The Political Biography of an Earthquake, Aftermath and Amnesia in Gujarat, India*. Oxford: Oxford University Press.

Thieken, Annegret H., Tina Bessel, Sarah Kienzler, Heidi Kreibich, Meike Müller, Sebastian Pisi, and Kai Schröter. 2016. "The Flood of June 2013 in Germany: How Much Do We Know about Its Impacts?" *Natural Hazards and Earth System Sciences* 16(6): 1519–40.

Tierney, Kathleen. 2014. *The Social Roots of Risk: Producing Disasters, Promoting Resilience.* Stanford, CA: Stanford University Press.

Ullberg, Susann. 2013. *Watermarks: Urban Flooding and Memoryscape in Argentina.* Stockholm Studies in Social Anthropology N.S. 8. Stockholm: Acta Universitatis Stockholmiensis.

Wachinger, Gisela, Ortwin Renn, Chloe Begg, and Christian Kuhlicke. 2013. "The Risk Perception Paradox—Implications for Governance and Communication of Natural Hazards." *Risk Analysis* 33(6): 1049–65.

Wanner, Heinz, Christoph Beck, Rudolf Brázdil, Carlo Casty, Mathias Deutsch, Rüdiger Glaser, Jucundus Jacobeit, Jürg Luterbacher, Christian Pfister, Stefan Pohl, Katrin Sturm, Peter C. Werner, and Eleni Xoplaki. 2004. "Dynamic and Socioeconomic Aspects of Historical Floods in Central Europe." *Erdkunde* 58(1): 1–16.

Chapter 11

Climate Resilience through Equity and Justice

Holistic Leadership by Tribal Nations and Indigenous Communities in the Southwestern United States

Julie Maldonado and Beth Rose Middleton

Climate-induced impacts are adversely affecting the health, livelihoods, cultural resources, and spiritual well-being of Indigenous communities in the southwestern United States. Hotter temperatures, intensified drought, more flooding, vast tree mortality, and increased wildfires are disrupting the ecosystems upon which Indigenous people depend. At the same time, many Indigenous communities and Tribal nations in the region are leading actions to mitigate against and adapt to the effects of the climate crisis, and to secure a sustainable future for generations to come. Indigenous and Traditional Knowledges are increasingly recognized by Western scientists as necessary and valuable to inform and guide climate adaptation (CTKW 2014; IPCC 2014). Indigenous leadership in the face of global planetary change offers innovation, guidance, and place-based, time-tested knowledge on how to address climate change from a holistic framework that foregrounds equity and justice. This chapter draws largely from the research conducted by the authors for the 2018 Fourth United States National Climate Assessment (NCA), Southwest Chapter, Key Message 4: Indigenous Peoples (Gonzalez et al. 2018). In the NCA, the Southwest region includes Arizona, California, Colorado, Nevada, New Mexico, and Utah. Thus, what is termed the "southwestern US" in this chapter covers the geographic and political boundaries spanning vast

Notes for this chapter begin on page 282.

and diverse landscapes and climates—from the arid desert to the coastal rainforest.

Background

The US Southwest has the largest population of Indigenous peoples of any region in the United States. There are currently 182 federally recognized tribes in the Southwest (Federal Register 2019; National Conference of State Legislatures 2020), as well as numerous non–federally recognized tribes and Indigenous communities, many of whom are already experiencing the impacts of a changing climate. Cumulative climate stressors and responses cannot be viewed in an isolated, static context. These stressors are layered on top of and exacerbate existing challenges, historical legacies and traumas, and marginalizing forces, including the creation of energy sacrifice zones and the establishment of hydroelectric and water storage infrastructure without Indigenous consent. Indeed, many Indigenous people articulate that climate change began with European contact and the shift in associated land management "from carrying capacity sustainability to a resource extraction model" (Goode et al. 2018). It is critical to take a long perspective, both toward the past and the future, in order to understand the present effects, projected impacts, and current and proposed actions to mitigate and adapt to climate change.

To begin with, the present and emerging climate change impacts on Indigenous peoples, lands, and resources in the Southwest must be placed within the context of continued settler colonization and forced removals (Wildcat 2009; Goldtooth and Awanyankapi 2010; Whyte 2016, 2017), such as the Long Walk endured by the Diné (Navajo) (Denetdale 2009) and the march of Concow Maidu people to Round Valley in northern California (Bauer 2016). Historical traumas, socioeconomic and political pressures, and extractive infrastructure all combine to impact Indigenous peoples' adaptive capacity and exacerbate current and projected climate impacts (Maynard 2014; Whyte 2013, 2017).

The colonizers forced Native populations across the Southwest onto marginal lands within their territories that non-Native settlers found inhospitable due to the limited water quantity and poor soil quality. For example, as settlers moved into the Southwest, the Diné underwent a series of colonial-driven policies and initiatives that left them impoverished and marginalized both physically and in other ways. They were forced in the reservation movement to less fertile territory replete with rocky dirt, sandstone outcroppings, and lack of water. To compensate, the US government provided them with sheep, and then when the sheep population grew beyond what the government deemed viable on the land, the sheep were slaughtered. In consequence, and with few resources, many Diné were further impoverished. The original and abiding arid conditions converged with poverty and economic and political marginalization to convert a prolonged drought across the Navajo Nation into a widespread disaster (Redsteer et al. 2011) when extreme storms in 2013 resulted in flooding that displaced dozens of Diné families (ICMN 2013).

In addition, what colonial powers once saw as wasteland turned out to be rich in natural resources, including fossil fuel energy resources. The legacies of colonialism and forced removals established the twenty-first-century pathway for the extreme levels of extraction by the petroleum industry (Powell 2018), which saw the area as one to exploit as an energy sacrifice zone—"a place where human lives are valued less than the natural resources that can be extracted from the region" (Buckley and Allen 2011: 171). The Four Corners region of the Southwest has over forty thousand existing natural gas wells and associated infrastructure, as well as two large coal plants (Four Corners Power Plant and San Juan Generating Station). The Navajo Generating Station coal-fired power plant provides power to pump water from the Colorado River to urban centers such as Phoenix, while Diné communities adjacent to the station lack running water (Powell and Maldonado 2017).

Diné are not alone in facing intertwined impacts of coloniality and climate change. In the face of such challenges, many tribes in the Southwest have limited resources for risk reduction, mitigation planning, and adaptation. Only a few Southwestern tribes have completed specific drought plans (National Drought Mitigation Center 2021), and many tribes have reported that they are not able to comprehensively monitor environmental shifts (Ferguson and Crimmins 2009; Redsteer et al. 2013). Additionally, the limited funding that is available to tribes is only available in the form of short-term grants, which are insufficient to deal with long-ranging mitigation planning and adaptation responses (Ferguson et al. 2010).

Current Climate Change Effects on Tribes in the US Southwest

Perhaps in its most serious fallout, drought exacerbated by human-caused climate change across the Southwest (Diffenbaugh, Swain, and Touma 2015; Udall and Overpeck 2017) has contributed to the degradation of traditional foods of the region's tribes. These consist, in part, of food sources such as corn, squash, acorn, and pronghorn antelope herds, staples of distinct Southwest Indigenous populations' agricultural and stewardship/subsistence legacy. Increasing temperatures have further caused a decline in Southwest pine nut populations, an essential food for some tribes (Lanner 1981; Washoe Tribe 2009; Redmond, Forcella, and Barger 2012; Redmond et al. 2017). Diné elders have observed that intensifying dryness has not only contributed to declines in corn and other culturally significant crops but also wild vegetation and animal populations, even avian species such as eagles, as well as the crucial flow of specific water springs (Redsteer et al. 2015). The loss of traditional foods accompanies increasing rates of diabetes and heart disease across Indigenous populations (Norgaard 2005; Lynn et al. 2013), resulting in cumulative physical, mental, and spiritual health effects.

Some tribes in the Southwest are also noting and experiencing extreme fluctuations in the region's prevailing natural state, once quite stable. The Navajo Nation has experienced prolonged drought since the 1990s, after which it underwent, as mentioned above, extreme storms that resulted in flooding that displaced many Diné families (ICMN 2013). Higher temperatures and evaporation, diminished soil moisture (Seager and Hoerling 2014; Diffenbaugh et al. 2015; Seager et al. 2015; Williams et al. 2015; Cheng et al. 2016), and impacts to snow-related parameters (Pierce and Cayan 2013) have further resulted in the decline of surface vegetation and an increase in sand dune mobility on some Navajo and Hopi lands, putting rangeland productivity at risk along with damaged infrastructure and loss of native plants (Redsteer et al. 2011; Redsteer 2012).

Climate disruption is contributing to another dire circumstance in the Southwest, namely increasing wildfires (Littell et al. 2009; Abatzoglou and Williams 2016). Affecting many tribes and their lands, subsistence, and community and cultural infrastructure, the fires have expanded in number, severity, and location. In addition to human-caused climate change, the fires are also, in part, fueled by underlying historical colonial-driven policies, including mismanagement of the land and fire suppression (Norton-Smith et al. 2016), extractive logging activities, monoculture replanting, and postlogging effects on water and certain fish, wildlife, and plants, such as tanoaks and beargrass, which some tribes in the greater southwestern region rely on for food and cultural purposes (Karuk Tribe 2010; Voggesser et al. 2013). Due in part to the suppression of traditional tribal burning leading to the increased density of other vegetation, the ability of oaks to withstand climate-related stress has decreased (Long et al. 2017). Declining surface soil moisture and higher temperatures and evaporation (Seager et al. 2014; Diffenbaugh et al. 2015; Williams et al. 2015; Cheng et al. 2016) converge with oak trees' decreased resilience, diminished acorn production, and fire and pest threat to reduce the availability and quality of acorns for tribal food consumption and cultural purposes (Voggesser et al. 2013).

For coastal tribes in the northern Pacific coast of the region, their homelands were once essentially a cold climate rainforest, abundant in coastal shellfish, freshwater fish in streams, ocean fish, and forest wildlife and edibles. One of the main foods sustaining the people was salmon. Today, increasing ocean and coastal water temperatures, in addition to nonclimate factors such as dam infrastructure, habitat degradation, and overharvesting by non-Natives, are affecting species such as salmon, on which northern Pacific Coast tribes rely for subsistence and cultural perpetuation (Jenni et al. 2014; Sloan and Hostler 2014; Hutto et al. 2015). These changes also affect mental and spiritual health, disrupting cultural connections to plant and animal relatives and to related place-based identity and practices (Donatuto et al. 2014; Rising Voices 2014).

Projected Future Climate Change Effects on Tribes in the Southwestern United States

It is projected that climate changes in the southwestern United States will worsen over the next several decades. The effects will no doubt cause further impacts to traditional foods, medicinal plants, and cultural resources (Lynn et al. 2013; Sloan and Hostler 2014). In the southwestern US, increased variability in water supply due to what will likely become consistent droughts may lead to heightened insecurity in settlements located on or reliant upon rivers and other water sources. For example, the 2004 Gila River Indian Community Water Rights Settlement Act allows the Gila River Indian Community to lease their unused water supplies. While such leases are a possible new source of economic development, the diversion of water allocations could place communities at risk of losing their own necessary water supply if there is then not enough water to go around, yet they are committed to provide purchased water to other entities per a lease agreement. Murphy cites National Climatic Data Center statistics on increasing drought in Arizona and warns that water supplies will decrease while demand continues to increase, resulting in a situation in which "water sold must be delivered, regardless of the condition of the selling reservation . . . it is possible for a group to oversell its appropriated water. In this worst-case scenario, the Community will have to breach its contracts for the survival of its people" (Murphy 2003–4: 185).

Increasing drought also leads to increased food-security and food-sovereignty risks. Projected reductions of runoff from decreasing and melting snowpack would increase the salinity of Pyramid Lake in Nevada. Diverse Paiute peoples live in a part of the southwestern region between the Rocky and Sierra Nevada Mountains characterized by isolated high, snowy mountain peaks, stretches of salt flats, land bearing little vegetation, hot summers, and cold winters. Traditional foods include small wildlife such as rabbits, as well as pine nuts and cattails, and fish if available. Climate change–induced decreasing snowpack and earlier snowmelt in this case will reduce fish biodiversity and, in particular, affect the spawning and sustenance of the endangered cui-ui fish, which is of central importance to the Pyramid Lake Paiute Tribe (Gautam, Chief, and Smith Jr. 2013).[1]

During colonization, some tribal economies throughout the southwestern United States became dependent on livestock, largely sheep but also cattle, which continues today (e.g., at Navajo Nation, Pyramid Lake Paiute Tribe, Southern Ute Tribe, and tribes at the US-Mexico border, e.g., at Tohono O'odham Nation Pascua Yaqui Tribe). These tribal populations are at risk of climate change–related stresses that are unbalancing the rangelands' ecosystems and, as a result, impacting their economic resources and livelihoods (Cozzetto et al. 2013; Redsteer et al. 2013; Nania et al. 2014; Redsteer et al. 2015; Norton-Smith et al. 2016). In addition, much of the southwestern US is at high risk of exposure to the expansion of invasive plant and animal species under new climate conditions (Early et al. 2016), namely, invasive cheatgrass, leafy spurge, and other species that reduce forage for livestock.

Many California tribes (for example, Yurok, Paiute, Miwok, Western Mono) are concerned about loss of acorns, a highly nutritional traditional food also used for medicinal purposes and basketry (Ortiz 2008; Long et al. 2017, 2016), due to "sudden oak death," which is spread by a pathogen that is known to increase with shifts in humidity and temperature (Guo, Kelly, and Graham 2005; Liu et al. 2007; Redsteer et al. 2013; Norton-Smith et al. 2016). Projected climate change effects may continue to shift bark beetles up in elevation (Sidder et al. 2016), which have already devastated much mountain fir forest in their rapid and fierce climate-driven expansion, potentially causing more losses to other traditional tribal food resources.

Especially relevant for coastal northern California tribes, projected climate changes increase the risks to salmon (Dittmer 2013; Jenni et al. 2014; Montag et al. 2014). Increased sea level rise and ocean temperatures, along with ocean acidification, increase risks of inundation of shellfish beds (Lynn et al. 2013), pathogens that cause shellfish poisoning (Cozzetto et al. 2013; Sloan and Hostler 2014), and damage to shellfish populations, which can cause cascading effects in food and ecological systems upon which some tribes depend (Feely et al. 2012; Dalton, Mote, and Snover 2013; Lynn et al. 2013).

In addition to warming, projected extreme shifts in weather include colder periods during the winter. Some tribes, such as some pueblos and tribes of New Mexico, have related human health concerns, as colder winter periods mean increasing use of wood-burning stoves for heating, resulting in worsening air quality, increased exposure to particulate matter, and accompanying higher incidences of asthma (National Tribal Air Association 2009). For many nations, extreme heat is also a concern. In California, the number of extreme heat days is expected to rise following current trends, creating especially dangerous conditions for low-income elders who may not be able to afford air conditioning or may not have access to "cool zones" (Gaughen in Goode et al. 2018; Climate Central 2016).

Emerging "Indigenuity" Responses

Although historical traumas, socioeconomic and political pressures, and extractive infrastructure have already reduced many Indigenous peoples' adaptive capacity to current and projected climate impacts (Maynard 2014; Whyte 2013, 2017), Indigenous peoples, nations, and pueblos across the Southwest and other areas are among those leading the way in innovative adaptation and mitigation actions. They are employing "indigenuity" in devising responses rooted in traditional knowledge (Wildcat 2009, 2013). Traditional knowledge embodies ways to "live well with" human and nonhuman beings within a framework that recognizes and honors interspecies relationality (Todd 2018). Kathy Sanchez (San Ildefonso Pueblo) with Tewa Women United calls for a "relational culture," one in which "everyone is Indigenous to the land from where they are from, in which we know how to be centered in that land, and in which we work together based on a 'relational culture'" (Powell and Maldonado 2017). Melanie Yazzie (Diné) and Cutcha Risling Baldy (Hupa, Yurok, Karuk) foreground a "radical relationality," which centers Indigenous understandings of relationship/relatedness/reciprocity as part of a radical shift away from colonial ways of thinking and treating the land and water (see Yazzie and Risling Baldy 2018). Indigenous approaches to addressing settler colonial disruptions and the ways in which such disruptions have and continue to exacerbate, extend, and amplify the impacts of climate change are rooted in Indigenous epistemologies and Indigenous resilience.

Indigenous and Traditional Knowledges[2] (IK/TKs), which often include knowledge about interrelationships between species and interconnectivity within an ecosystem, are increasingly recognized as necessary and valuable to inform and guide climate adaptation (CTKW 2014; IPCC 2014). Traditional Ecological Knowledge (TEK) about traditional plant species and habitat composition can provide early warning or detection of invasive species and support ecological restoration (ITEP 2012). Some tribes are also using TEK to reintegrate traditional foods into their diets, such as the Tesuque Pueblo of New Mexico, who are reviving their Indigenous agricultural techniques (Viles 2011). Others, such as the Karuk Tribe (2010), the North Fork Mono (Long et al. 2017; Goode et al 2018), and the Mountain Maidu in California (Middleton 2012), use TEK to guide resource management efforts.

Building on generations of accrued knowledge, some tribes use fire to resist the impacts of climate change, increase ecosystem resilience, manage crops, and enhance productivity of significant traditional food sources and culturally important species (Voggesser et al. 2013; Norgaard 2014a, 2014b; Vinyeta and Lynn 2013). Fire is traditionally used as a central tool in social, cultural, and spiritual practices. For example, fire enhances three-quarters of Karuk traditional food and culturally important species (Norgaard et al. 2016). Over the past hundred years, US policy approached wildfire mitigation through wildfire suppression and exclusion (Norton-Smith et al. 2016). In the Southwest, tribes, pueblos, agencies, and organizations are reinvigorating Indigenous burning practices to mitigate the increasing threat of fire under current conditions and projected climate changes.

Traditional burning practices restore natural habitats, reduce hazardous forest fire fuel loads, and protect culturally important plants such as hazel and beargrass for basketweaving, medicinal plants, traditional food sources, and animal species (Middleton 2012; Lake and Long 2014; Norgaard 2014a, 2014b; Yurok Today 2014a; Long et al. 2017, 2016; Goode et al. 2018). Such practices not only reduce the threat of major fires by thinning densely stocked forests, but also increase grasslands, thereby preserving groundwater that would be taken up by the succession of woody plants. While the impacts of floral changes will vary depending upon other climatic factors, such as changes in precipitation, practicing traditional land stewardship may contribute to offsetting these impacts by stabilizing or increasing the amount of groundwater available by maintaining meadows (Scott et al. 2014).

Utilizing Indigenous and Tribal knowledge and management principles to guide the regular use of fire on the landscape can both mitigate against the damaging risks of spreading wildfires and protect public and tribal trust resources as the impacts of climate change expand (Norgaard 2014b; Norgaard et al. 2016). The Yurok tribal and community members in Northwest California, for instance, have formed the Cultural Fire Management Council (CFMC) to bring fire back to the landscape as a form of restoration (Yurok Today 2014a). Through the CFMC, the Yurok Tribe, in partnership with the Nature Conservancy Fire Learning Network, Firestorm Inc., Yurok Forestry/Wildland Fire, the Northern California Indian Development Council, and the US Forest Service, is reinvigorating their cultural burning practices. The collaboration is also designed to train the Yurok Wildland Fire crew through the Prescribed Fire Training Exchange (TREX), a training exchange program between firefighters from federal and state agencies, nongovernmental organizations, and the tribe's fire crew (FLN 2014; Yurok Today 2014a; TREX 2017). "Restoration of the land means restoration of the people," said Margo Robbins, the CFMC president. "Returning fire to the land enables us to continue the traditions of our ancestors" (Yurok Today 2014a).

Some tribes in the Southwest have also developed climate adaptation plans. The Yurok Tribe and the Gila River Indian Community have, for example, partnered with the Institute for Tribal Environmental Professionals, and the Tohono O'odham Nation collaborated with the University of Arizona's Center for Climate Adaptation Science and Solutions, becoming among the first tribes in the region to develop climate adaptation and resiliency plans. The Navajo Nation has collaborated with outside scientists to develop the Navajo Nation Assessment (Nania et al. 2014) and used projected climate changes to inform their drought contingency plan (Bierbaum et al. 2013) and their hazard mitigation and State Wildlife Action planning processes (Navajo Nation Department of Water Resources 2003).[3] However, a notable gap is the available capacity and funds to move from planning to implementation (Black et al. 2015).

Drawing on multiple knowledge systems to inform ecological restoration within their ancestral territory, the Amah Mutsun Tribe of the central California coast has developed a land trust that is engaged in a collaborative "eco-archaeology" project with archaeologists and ecologists at the University of California–Berkeley. The partners are triangulating multiple methods to identify the historical ecology and associated management practices at the Quiroste Valley Cultural Preserve (Lightfoot and Lopez 2013). Youth members of the Amah Mutsun Native Stewardship Corps are working to remove invasive plants and restore the landscapes so that culturally important species (and the human practices associated with them) can thrive (Hannibal 2016).

A number of tribes are also working in the legal arena to protect culturally/ecologically important resources. Whether applied to mining, agriculture, or residential development, western water-using groups have become increasingly dependent on groundwater, and "with the effects of climate change looming, it is highly likely that tribes will [also] become increasingly dependent on groundwater" (Irwinsky 2014–15: 565). Until the 2015 and 2017 rulings in favor of tribes in *Agua Caliente Band of Cahuilla Indians v. Coachella Valley Water District*, the courts were largely silent on whether tribes had rights to groundwater that were the same as the surface rights confirmed in *Winters v. US* (1908) (Irwinsky 2014–15). Following *Agua Caliente,* tribes with *Winters* rights also have affirmed groundwater rights, including rights to ensure adequate quality and quantity of groundwater. Such a clarification of tribal jurisdiction is increasingly important in an era of uncertain water availability due to climatic change. Tribes also continue to challenge established water doctrines that have long ignored both tribes' responsibilities (Wildcat 2009; Tonino 2016) and rights to water. In California, the North Coast region (Region 1) of the State Water Resources Control Board (CA-SWRCB) was the first region to recognize Tribal Cultural and Subsistence uses of water as formal beneficial uses (see Reed, Middleton Manning, and Martinez 2020), and tribal representatives and allies are working to expand that category to SWRCB regions statewide.

Meanwhile tribal leaders also continue to advocate for a human right to water and actualizing the principles of the United Nations Declaration on the Rights of Indigenous Peoples. Winnemem Wintu Chief and Spiritual Leader Caleen Sisk is working in northern California to fight the proposed raising of Shasta Dam and to encourage a sustainable solution for the return of salmon to the McCloud River watershed above the dam, citing international recognition of Indigenous peoples' rights to cultural perpetuation, which requires healthy homelands (Marcus 2015; Winnemem Wintu, n.d.). The Yurok Tribe recently recognized the rights of the Klamath River (Yurok Tribe, 2019; Yurok Today 2019), and the Yurok, Karuk, and other partners continue to work toward the removal of dams on the lower Klamath River, which would be the largest dam removal project in US history (Flaccus 2020; see also KRRC 2018). Water throughout the southwestern United States has long been central to the conveyance and processing of minerals and fossil fuels within Indian territory. While policies such as the Indian Mineral Leasing Act have encouraged various tribes' dependence on fossil fuel resources (Voggesser 2010), and extraction has created energy sacrifice zones across southwestern tribal territories in particular (Smith and Frehner 2010), tribes across the southwestern US especially are leading the way in using renewable energy sources (Powell and Long 2010) and resisting the imposed limitations of western water law to reinvigorate Indigenous water stewardship (Curley 2019; Nolan 2019).

A number of tribes are employing solar, wind, geothermal, and biomass to meet their energy needs (EIA 2016). The Tonto Apache Tribe in Arizona has undertaken a renewable energy initiative to build a 249-kilowatt solar photovoltaic system for the tribe's Mazatzal Hotel on the tribe's reservation (Office of Indian Energy Policy and Programs 2015b). Yoeme at the US-Mexico border are working to develop the second largest solar array on a reservation (Black et al. 2015). The Southern Ute Tribe in Colorado has established a facility to make fuel from algae, which is grown adjacent to a natural gas–processing facility on the reservation (ICMN 2010). In California, the Ramona Band of Cahuilla has established a microgrid to utilize renewable resources to meet all of their energy needs, becoming the first reservation in the US to be completely off grid. Also in California, the Campo Kumeyaay Nation was the first tribe in the United States to develop a utility-scale wind project on land leased from the tribe (EIA 2016).[4] In New Mexico, Santo Domingo Tribe received a grant in 2015 to install a 115-kilowatt solar PV system, which will power the tribe's water pump and water treatment facility (USDOE 2015). Jemez Pueblo in New Mexico has initiated renewable energy projects (including solar and geothermal) and energy efficiency and planning throughout the Pueblo (ITEP 2012). Fort McDowell Yavapai Nation in Arizona is initiating solar power projects (ITEP 2013a). Santa Ynez Band of Chumash Indians in Southern California has developed projects to work toward energy independence based on renewable energy sources, and the Chumash Casino is focused on developing green initiatives such as a roof liner on the casino to absorb heat and reduce the energy required to maintain a comfortable temperature in the building (ITEP 2013b). NativeSun (Hopi Nation) has installed hundreds of solar units around the region; Native American Photovoltaics has installed dozens of solar systems for homes previously without electricity on the Navajo Reservation; the Campo and Viejas Bands of Kumeyaay people in Southern California, in partnership with Superior Energy LLC, established a reservation-based commercial wind farm. The Blue Lake Rancheria Tribe in California has established a "first-of-a-kind biogas fuel cell system," fueled by wood waste from timber harvesting (EIA 2016). Further, the tribe was named a climate action champion in 2015–16 for implementing innovative climate actions such as an "all-of-the-above renewable strategy of transportation, residential, and municipal renewable energy projects" (Office of Indian Energy Policy and Programs 2015a).

Some tribes have also signed agreements to sell energy to outside entities to help meet their energy demands. Moapa Paiute's 250-megawatt solar project on the Moapa River Indian Reservation in Nevada has a power purchase agreement to sell solar power to the Los Angeles Department of Water and Power (Lott 2014). However, within the context of nontribal (both private as well as federal) ownership of tribal homelands, inadequate cultural resource protection laws, and short-term economic incentives, tribes are also at risk of being negatively affected by renewable energy projects. There are instances, in southern California and Nevada in particular, of tribal lands being encroached upon and desecrated by solar and wind projects (Maynard 2014; Sahagun 2014).

In addition to finding culturally appropriate ways to produce energy as climate change effects encroach upon tribal homelands, some tribes are also developing projects to sequester emissions from greenhouse gases, often in collaboration with surrounding states. States including California have recognized the importance of keeping carbon sequestered in high-carbon environments, such as temperate mixed coniferous and redwood forests (McKittrick 2014; Middleton Manning and Reed 2019). One mechanism to incentivize carbon sequestration is the carbon credit market, which is either voluntary or mandatory and operates on regional, national, and international scales. On the international level, REDD (Reducing Emissions from Deforestation and Forest Degradation) is a mechanism that provides countries financial incentives to reduce emissions through forest management options. Unfortunately, implementing these incentives often involves privatized land grabs (see Indigenous Environmental Network n.d.; Lohmann 2006). Market-based carbon sequestration regimes can be problematic for Indigenous peoples if their land rights are not recognized or if they are not centrally involved in project design, planning, development, and implementation (Campbell 2015–16). "Because [these] initiatives essentially turn forests into commodities by drawing financial resources into developing countries, land tenure rights are crucial to determining who has the authority to accept or reject [carbon credit] projects, who can manage the forests, and who is ultimately the financial beneficiary of the program" (Campbell 2015–16: 206).

Two tribes in California (Yurok and Round Valley) have developed carbon-offset projects on their forestlands under California's robust cap-and-trade program. These nations have asserted their sovereignty and developed these projects to support tribally led restoration and stewardship. As a result of their carbon-offset projects, the Yurok Tribe has placed over fifty thousand acres of ancestral land back into tribal ownership after nearly one hundred years (Middleton and Reed 2019). Former Yurok Tribal Chairman Thomas O'Rourke described why the tribe has chosen to participate in the carbon market: "To not only do our part with global warming, but to preserve our way of life so that our future generations can see the pristine forest that our parents' grandparents saw" (Barboza 2014; see also Yurok Today 2014b).

Conclusion

Experiencing impacts to their health, livelihoods, economies, subsistence, cultural resources, lifeways, and spiritual well-being, Indigenous communities, nations, and peoples in the southwestern United States are leading resiliency efforts and innovative adaptation and mitigation actions to address the climate crisis for current generations and those to come. Indigenous leadership in climate resilience is changing not only adaptation processes and implementation but also the very definitions of related concepts like food security. For example, an expanded definition of food security in a context of ongoing colonial disruption and climatic change accounts for the importance of the availability of traditional foods:

> Food security means more than simply whether or not sufficient food is being produced or harvested in a "one-size-fits-all" food-to-nutrition relationship, and expands to include all of the various ways in which a food system supports health in the biophysical, social, and ecological dimensions. These include the importance of culturally preferred foods. . . . wild fish and game . . . are important for food security . . . because they are important to the preservation and transmission of traditions and cultural practices, for the maintenance of social networks and interpersonal relationships, and for supporting individual and community sense of self-worth and identity. (Gerlach and Loring 2013)

As such, achieving food security means securing Indigenous land and water rights and responsibilities, as both land and water are transforming in unforeseen ways.

To address the climate crisis in a way that is attentive to environmental and climate justice and Indigenous rights, it is vital to bring multiple knowledges and knowledge holders together. Collaborations and partnerships between Indigenous knowledge holders and Indigenous and non-Indigenous climate and earth scientists, while ensuring the safeguarding of sensitive information (Maldonado et al. 2016; Tonino 2016; Indigenous Science Statement 2017; Rising Voices 2017), can focus beyond short-term economic gain and toward long-term solutions (Smith Jr. et al. 2014). Recognizing tribal land and water rights and centering Indigenous leadership in project design and implementation are essential to devising mitigation and adaptation projects that support Indigenous lifeways and livelihoods (Campbell 2015–16). Climate adaptation collaborations are most effective when they incorporate spiritual and cultural perspectives on mitigation and adaptation pathways (Powell and Maldonado 2017). It is beyond time to listen, learn, and act in response to the climate crisis, guided by the deep place-based knowledges of those who, despite generations of trauma and centuries of violence, have survived, adapted, and continue to be resilient.

Julie Maldonado is associate director for the Livelihoods Knowledge Exchange Network (LiKEN), a nonprofit link-tank for policy-relevant research toward postcarbon livelihoods and communities. In this capacity, she serves as codirector of the Rising Voices Center for Indigenous and

Earth Sciences, which facilitates intercultural, relational-based approaches for understanding and adapting to extreme weather and climate change. Dr. Maldonado is also a lecturer in the University of California-Santa Barbara's Environmental Studies Program. She previously worked for the US Global Change Research Program and is an author on the 3rd, 4th, and 5th US National Climate Assessments. As a public anthropologist, her work focuses on climate adaptation, disasters, displacement, resettlement, and environmental and climate justice. Her recent book, *Seeking Justice in an Energy Sacrifice Zone: Standing on Vanishing Land in Coastal Louisiana*, emerged from years of collaborative work with tribal communities in coastal Louisiana experiencing and responding to repeat disasters and climate chaos. The book was released shortly before the release of her coedited volume, *Challenging the Prevailing Paradigm of Displacement and Resettlement: Risks, Impoverishment, Legacies, Solutions*.

Beth Rose Middleton is a professor at the Department of Native American Studies, University of California, Davis. Her research centers on Native environmental policy and Native activism for site protection using conservation tools, and her broader research interests include intergenerational trauma and healing, rural environmental justice, digital humanities, and Indigenous analyses of climate change. She received her BA in Nature and Culture from UC Davis and her PhD in Environmental Science, Policy, and Management from UC Berkeley. She has written two books: *Trust in the Land: New Directions in Tribal Conservation* (University of Arizona Press, 2011), on Native applications of conservation easements, and *Upstream: Trust Lands and Power on the Feather River* (University of Arizona Press, 2018), on the history of Indian allotment lands at the headwaters of the California State Water Project.

Notes

1. The Pyramid Lake Paiute Tribe's "Paiute name is *Kuyuidokado*, or cui-ui eaters, named after the Pyramid Lake sucker fish" (Gautam et al. 2013).
2. Following the "Guidelines for Considering Traditional Knowledges in Climate Change Initiatives" (CTKW 2014), we use the plural "knowledges."
3. For a list of US Bureau of Indian Affairs–funded Tribal resilience programs across the Southwest, see https://biamaps.doi.gov/tribalresilience/.
4. For further information, see Map: The US Department of Energy Office of Indian Energy supported energy efficiency and renewable energy projects, https://energy.gov/indianenergy/maps/tribal-energy-projects-database.

References

Abatzoglou, John T., and A. Park Williams. 2016. "Impact of Anthropogenic Climate Change on Wildfire across Western US Forests." *Proceedings of the National Academy of Sciences of the USA* 113: 11770–75.

Barboza, Tony. 2014. "Yurok Tribe Hopes California's Cap-and-Trade Can Save a Way of Life." *Los Angeles Times*, 16 December.

Bauer, William J., Jr. 2016. *California through Native Eyes: Reclaiming History*. Seattle: University of Washington Press.

Bierbaum, Rosina, Joel B. Smith, Arthur Lee, Maria Blair, Lynne Carter, F. Stuart Chapin III, Paul Fleming, Susan Ruffo, Missy Stults, Shannon McNeeley, Emily Wasley, and Laura Verduzco. 2013. "A Comprehensive Review of Climate Adaptation in the United States: More than Before, but Less than Needed." *Mitigation and Adaptation Strategies for Global Change* 18: 361–406.

Black, Mary, Karletta Chief, Katharine Jacobs, Schuyler Chew, and Lynn Rae. 2015. *Tribal Leaders Summit on Climate Change: A Focus on Climate Adaptation Planning and Mitigation*. Workshop Report. Native Nations Climate Adaptation Program, University of Arizona, Tucson. 12–13 November. Retrieved 20 January 2021 from http://www.nncap.arizona.edu/sites/default/files/NNCAPSummitFinal.pdf.

Bryan, Michelle. 2017. "Valuing Sacred Tribal Waters within Prior Appropriation." *Natural Resources Journal* 57 (Winter): 139–81.

Buckley, Geoffrey L., and Laura Allen. 2011. "Stories about Mountaintop Removal in the Appalachian Coalfields." In *Mountains of Injustice: Social and Environmental Justice in Appalachia*, edited by Michele Morrone, Geoffrey L. Buckley, and Jedidiah Purdy, 161–80. Athens: Ohio University Press.

California Air Resources Board. 2013. *California Air Resources Board's Process for the Review and Approval of Compliance Offset Protocols in Support of the Cap-and-Trade Regulation*. May.

Campbell, Cindy. 2015–16. "Implementing a Greener REDD+ in Black and White: Preserving Wounaan Lands and Culture in Panama with Indigenous-Sensitive Modifications to REDD+" *American Indian Law Review* 40(2): 193–232.

Cheng, L., M. Hoerling, A. AghaKouchak, B. Livneh, X. W. Quan, and J. Eischeid. 2016. "How Has Human-Induced Climate Change Affected California Drought Risk?" *Journal of Climate* 29: 111–120.

Climate Central. 2016. "US Faces Dramatic Rise in Extreme Heat, Humidity." 13 July. Retrieved 28 June 2021 from https://cires.colorado.edu/outreach/sites/default/files/2020-03/Climate%20Resiliency%20HS2%20-%20Climate%20Central%20Report%20Denver%20Focused.pdf.

Cozzetto, Karen, Karletta Chief, Kyle Dittmer, Mike Brubaker, Bob Gough, Kalani Souza, Frank Ettawageshik, Sue Wotkyns, Sarah Opitz-Stapleton, Sabre Duren, and Prithviraj Chavan. 2013. "Climate Change Impacts on the Water Resources of American Indians and Alaska Natives in the US." *Climatic Change* 3: 569–84.

CTKW (Climate and Traditional Knowledges Workgroup). 2014. "Guidelines for Considering Traditional Knowledges in Climate Change Initiatives." Retrieved 11 July 2021 from http://climatetkw.wordpress.com/.

Curley, Andrew. 2019. "Unsettling Indian Water Settlements: The Little Colorado River, the San Juan River, and Colonial Enclosures." *Antipode* (15 April).

Dalton, Meghan M., Philip W. Mote, and Amy K. Snover. 2013. *Climate Change in the Northwest: Implications for Our Landscapes, Waters, and Communities*. Washington, DC: Island Press.

Denetdale, Jennifer. 2009. *The Long Walk: The Forced Navajo Exile*. New York: Chelsea House Publishers.

Diffenbaugh, Noah S., Daniel L. Swain, and Danielle Touma. 2015. "Anthropogenic Warming Has Increased Drought Risk in California." *Proceedings of the National Academy of Sciences of the USA* 112: 3931–36.

Dittmer, Kyle. 2013. "Changing Streamflow on Columbia Basin Tribal Lands—Climate Change and Salmon." *Climatic Change* 120: 627–41. doi:10.1007/s10584–013–0745–0.

Donatuto, Jamie, Eric E. Grossman, John Konovsky, Sarah Grossman, and Larry Campbell. 2014. "Indigenous Community Health and Climate Change: Integrating Biophysical and Social Science Indicators." *Coastal Management* 42(4): 355–73. doi:10.1080/0892 0753.2014.923140.

Early R., B. Bradley, J. Dukes, J. Lawler, J. Olden, D. Blumenthal, P. Gonzalez, E. D. Grosholz, I. Ibanez, L. P. Miller, C. J. B. Sorte, and A. Tatem. 2016. "Global Threats from Invasive Alien Species in the Twenty-First Century and National Response Capacities." *Nature Communications* 7 (12485).

EIA (Energy Information Administration). 2016. "California: Profile Analysis." Retrieved 11 July 2021 from https://www.eia.gov/state/analysis.php?sid=CA.

Federal Register. 2019. "Indian Entities Recognized and Eligible to Receive Services from the United States Bureau of Indian Affairs." 84(22): 1200–1205. Retrieved 11 July 2021from https://www.federalregister.gov/documents/2019/02/01/2019-00897/indian-entities-recognized-by-and-eligible-to-receive-services-from-the-united-states-bureau-of.

Feely R. A., T. Klinger, J. A. Newton, and M. Chadsey. 2012. *Scientific Summary of Ocean Acidification in Washington State Marine Waters: National Oceanic and Atmospheric Administration (NOAA) OAR Special Report*. Contribution No. 3934 from NOAA/Pacific Marine Environmental Laboratory.

Ferguson, Daniel, and Michael Crimmins. 2009. "Who's Paying Attention to the Drought on the Colorado Plateau?" *Southwest Climate Outlook* 8(7): 3–6. https://climas.arizona.edu/sites/default/files/pdf2009juldroughtcoplateau.pdf.

Ferguson, Daniel, Christina Alvord, Michael Crimmins, Margaret Redsteer, Michael Hayes, Chad McNutt, Roger Pulwarty, and Mark Svoboda. 2010. *Drought Preparedness for Tribes in the Four Corners Region Workshop Report*. April. Flagstaff, AZ. https://www.drought.gov/sites/default/files/2020-06/wksp_Drought%20Preparedness%20for%20Tribes%20in%20the%20Four%20Corners%20Region_Workshop%20Report_04-08-10_0.pdf.

FLN (Fire Learning Network). 2014. *Yurok Prescribed Fire Training Exchange*. Weitchpec, CA, May/June.

Flaccus, Gillian. 2020. "Largest US Dam Removal Stirs Debate over Coveted West Water." Associated Press, 29 March. Retrieved 9 June 2021 from https://apnews.com/article/dams-us-news-ap-top-news-wa-state-wire-ca-state-wire-c3d6e788310a359 1210dc4433793d919.

Gaughen, Shasta. 2018. "Human Health Impacts from Climate Change." In *Summary Report from Tribal and Indigenous Communities within California*, edited by Ron Goode et al. *California's Fourth Climate Change Assessment*, State of California Governor's Office of Planning and Research, State of California Energy Commission, and California Natural Resources Agency. Retrieved 7 July 2021 from https://www.energy.ca.gov/sites/default/files/2019-11/Statewide_Reports-SUM-CCCA4-2018-010_TribalCommunitySummary_ADA.pdf.

Gautam, Mahesh, Karletta Chief, and Wiliam James Smith Jr. 2013. "Climate Change in Arid Lands and Native American Socioeconomic Vulnerability: The Case of the Pyramid Lake Paiute Tribe." *Climatic Change* 3: 585–99.

Gerlach, S. Craig, and Philip A. Loring. 2013. "Rebuilding Northern Foodsheds, Sustainable Food Systems, Community Well-Being, and Food Security." *International Journal of Circumpolar Health* 72: 21560.
Goldtooth, Tom, and Mato Awanyankapi. 2010. "Earth Mother, Pinons, and Apple Pie." *Wicazo Sa Review* 25(2): 11–28.
Gonzalez, Patrick, Gregg M. Garfin, David D. Breshears, Keely M. Brooks, Heidi E. Brown, Emile H. Elias, Amrith Gunasekara, Nancy Huntly, Julie K. Maldonado, Nathan J. Mantua, Helene G. Margolis, Skyli McAfee, Beth Rose Middleton, and Bradley H. Udall. 2018. "Southwest." In *Impacts, Risks, and Adaptation in the United States: Fourth National Climate Assessment*, edited by D. R. Reidmiller, C. W. Avery, D. R. Easterling, K. E. Kunkel, K. L. M. Lewis, T. K. Maycock, and B. C. Stewart, 2:1101–84. Washington DC: US Global Change Research Program. doi: 10.7930/NCA4.2018.CH25.
Goode, Ron W., S. Gaughen, M. Fiero, D. Hankins, K. Johnson-Reyes, B. R. Middleton, T. Red Owl, and R. Yonemura. 2018. "Summary Report from Tribal and Indigenous Communities within California." *California's Fourth Climate Change Assessment*. Retrieved 7 July 2021 from https://www.energy.ca.gov/sites/default/files/2019-11/Statewide_Reports-SUM-CCCA4-2018-010_TribalCommunitySummary_ADA.pdf.
Guo Qinghua, Maggi Kelly, and Catherine H. Graham. 2005. "Support Vector Machines for Predicting Distribution of Sudden Oak Death in California." *Ecological Modeling* 182(1): 75–90.
Hannibal, Mary Ellen. 2016. "New Paradigms for Stewardship." *Bay Nature*, 31 March.
Hutto, S. V., K. D. Higgason, J. M. Kershner, W. A. Reynier, and D. S. Gregg. 2015. *Climate Change Vulnerability Assessment for the North-Central California Coast and Ocean. Marine Sanctuaries Conservation Series ONMS-15–02*. US Department of Commerce, National Oceanic, and Atmospheric Administration, Office of National Marine Sanctuaries, Silver Spring, MD.
ICMN (Indian Country Media Network). 2010. "Southern Ute Tribe Invests in Algae Fuel." *Indian Country Today*, 16 December. Retrieved 7 July 2021 from https://indiancountrytoday.com/archive/southern-ute-tribe-invests-in-algae-fuel.
———. 2013. "Flash Flooding on Navajo Nation Displaces Scores, Wrecks Homes with Mold and Mud." *Indian Country Today*, 14 September. Retrieved 7 July 2021 from https://indiancountrytoday.com/archive/flash-flooding-on-navajo-nation-displaces-scores-wrecks-homes-with-mold-and-mud.
Indigenous Environmental Network. n.d. *REDD: Reaping Profits from Evictions, Land Grabs, Deforestation, and Destruction of Biodiversity*. Retrieved 7 July 2021 from http://ienearth.org/REDD/.
Indigenous Science Statement for the March for Science. 2017. Indigenous Science. Retrieved 7 July 2021 from http://www.esf.edu/indigenous-science-letter/.
IPCC (Intergovernmental Panel on Climate Change). 2014. *Climate Change 2014: Synthesis Report; Contribution of Working Groups I, II and III to the Fifth Assessment Report of the Intergovernmental Panel on Climate Change*. Geneva, Switzerland.
Irwinsky, Mallory. 2014–15. "Coalbed Methane Development in Wyoming and Montana: The Potential Impacts of *Montana v. Wyoming*, Coalbed Methane Development, and Water Quality on the Tribes of the Powder River and Wind River Basins." *American Indian Law Review* 39(2): 553–83.
ITEP (Institute for Tribal Environmental Professionals). 2012. "Climate Change and Invasive Species: What It Means to Tribes and How We Can Adapt." Retrieved 7 July 2021 from http://www7.nau.edu/itep/main/tcc/docs/resources/om_InvasiveSpeciesFactSheet_081512.pdf.

———. 2011. "Jemez Pueblo of Jemez: 'Leading the Way to a Renewable Future.'" Tribal Climate Change Profile: Pueblo of Jemez. April.

———. 2013a. "Fort McDowell Yavapai Nation: Harnessing Solar Power for Energy Independence and Utilities Savings." Tribal Climate Change Profile: Fort McDowell Yavapai. January.

———. 2013b. "Santa Ynez Band of Chumash Indians: 'Climate Change and Environmental Management Programs.'" Tribal Climate Change Profile. July.

Jenni, K., D. Graves, J. Hardiman, J. Hatten, M. Mastin, M. Mesa, J. Montag, T. Nieman, F. Voss, and A. Maule. 2014. "Identifying Stakeholder-Relevant Climate Change Impacts: A Case Study in the Yakima River Basin, Washington, USA." *Climatic Change* 124: 371–84.

Karuk Tribe. 2010. *Department of Natural Resources Eco-cultural Resource Management Plan*. Karuk Tribe of California, Department of Natural Resources. http://www.karuk.us/karuk2/images/docs/dnr/ ECRMP_6-15-10_doc.pdf.

KRRC (Klamath River Renewal Corporation). 2018. Definite Plan for the Lower Klamath River. Retrieved 7 July 2021 from http://www.klamathrenewal.org/definite-plan/.

Lake, Frank K., and Jonathan W. Long. 2014. "Fire and Tribal Cultural Resources." In *Science Synthesis to Support Socioecological Resilience in the Sierra Nevada and Southern Cascade Range*, edited by J. W. Long, L. J. Quinn-Davidson, and C. N. Skinner, 173–86, chap. 4.2. General Technical Report PSW-GTR-247. Albany, CA: US Department of Agriculture, Forest Service, Pacific Southwest Research Station.

Lanner, R. M. 1981. *The Pinon Pine: a Natural and Cultural History*. Reno: University of Nevada Press.

Lightfoot, Kent, and Valentin Lopez. 2013. "The Study of Indigenous Management Practices in California: An Introduction." *California Archaeology* 5(2): 209–19.

Littell, Jeremy S., Donald McKenzie, David L. Peterson, and A. Leroy Westerling. 2009. "Climate and Wildfire Area Burned in Western US Ecoprovinces, 1916–2003." *Ecological Applications* 19: 1003–21.

Liu, Desheng, Maggi Kelly, Peng Gong, and Qinghua Guo. 2007. "Characterizing Spatial-Temporal Tree Mortality Patterns Associated with a New Forest Disease." *Forest Ecology and Management* 253: 220–31.

Lohmann, Larry. 2006. "Carbon Trading: A Critical Conversation on Climate Change, Privatization, and Power." *Development Dialogue* 48: 2–360.

Long, Jonathan W., Ron W. Goode, Raymond J. Gutierrez, Jessica J. Lackey, and M. Kat Anderson. 2017. "Managing California Black Oak for Tribal Ecocultural Restoration." *Journal of Forestry* 115(5): 426–434.

Long, Jonathan W., Lenya Quinn-Davidson, Ron W. Goode, Frank K. Lake, and Carl N. Skinner. 2016. *Restoring California Black Oak to Support Tribal Values and Wildlife*. General Technical Report PSW-GTR-251. Albany, CA: US Department of Agriculture, Forest Service, Pacific Southwest Research Station.

Lott, Melissa C. 2014. "First Utility-Scale Solar Project on Tribal Land Breaks Ground in Nevada." *Scientific American*, 5 April. Retrieved 7 July 2021 from https://blogs.scientificamerican.com/plugged-in/first-utility-scale-solar-project-on-tribal-land-breaks-ground-in-nevada/.

Lynn, Kathy, John Daigle, Jennie Hoffman, Frank Lake, Natalie Michelle, Darren Ranco, Carson Viles, Garrit Voggesser, and Paul Williams. 2013. "The Impacts of Climate Change on Tribal Traditional Foods." *Climatic Change* 3: 545–56.

Maldonado, Julie, Heather Lazrus, Bob Gough, Shiloh Bennett, Karletta Chief, Carla Dhillon, Linda Kruger, Jeff Morisette, Stefan Petrovic, and Kyle Whyte. 2016. "The Story of Rising Voices: Facilitating Collaboration between Indigenous and Western Ways of Knowing." In *Responses to Disasters and Climate Change: Understanding Vulnerability and Fostering Resilience*, edited by Michele Companion and Miriam Chaiken, 15–26. Boca Raton, FL: CRC Press.

Maloy, Stanley. 2016. "US-Mexico Border, Climate Change, and Infectious Diseases." Presentation. Retrieved 7 July 2021 from https://www.epa.gov/sites/production/files/2016-02/documents/us-mexico_border_climate_change_and_infectious_diseases.pdf.

Marcus, Diveena S. 2015. "Indigenous Activism beyond Borders." *South Atlantic Quarterly* 114(4): 892–906.

Maynard, Nancy G., ed. 2014. *Native Peoples–Native Homelands Climate Change Workshop II, Final Report: An Indigenous Response to Climate Change*. Washington, DC: US Global Change Research Program. Retrieved 8 July 2021 from https://earth.gsfc.nasa.gov/sites/default/files/NPNH-Report-No-Blanks.pdf.

McKittrick, Erin. 2014. "A Fossil Fuel Economy in a Climate Change Vulnerable State." *Environment* 56(3): 25–35.

Middleton, Beth Rose. 2012. "Fuels: Greenville Rancheria." *Smoke Signals* (Bureau of Indian Affairs, Branch of Wildland Fire Management) 24: 7–9.

Middleton Manning, Beth Rose, and Kaitlin Reed. 2019. "Returning the Yurok Forest to the Yurok Tribe: California's First Tribal Carbon Credit Project." *Stanford Environmental Law Review* 39(1): 71–124.

Montag, J. M., K. Swan, K. Jenni, T. Nieman, J. Hatten, M. Mesa, D. Graves, F. Voss, M. Mastin, J. Hardimann, and A. Maule. 2014. "Climate Change and Yakama Nation Tribal Well-Being." *Climatic Change* 124: 385–98.

Murphy, Lindsay. 2003–4. "Death of a Monster: Laws May Finally Kill Gila River Adjudication." *American Indian Law Review* 28: 173–87.

Nania, Julie, Karen Cozzetto, Nicole Gillet, Sabre Duren, Anne Marie Tapp, Michael Eitner, and Beth Baldwin. 2014. "Considerations for Climate Change and Variability Adaptation on the Navajo Nation." Research paper, University of Colorado–Boulder.

National Conference of State Legislatures. 2020. "Federal and State Recognized Tribes." Retrieved 8 July 2021 from https://www.ncsl.org/research/state-tribal-institute/list-of-federal-and-state-recognized-tribes.aspx#az.

National Drought Mitigation Center. 2021. "Tribal Drought Plans." Retrieved 28 June 2021 from https://drought.unl.edu/droughtplanning/Plans/TribalPlans.aspx.

National Tribal Air Association. 2009. "Impacts of Climate Change on Tribes in the United States." Retrieved 8 July 2021 from https://www.ntaatribalair.org/climate-change/.

Navajo Nation Department of Water Resources. 2003. "Navajo Nation Drought Contingency Plan." Retrieved 8 July 2021from http://drought.unl.edu/archive/plans/drought/tribal/NavajoNation_2003.pdf.

Nolan, Ruth. 2019. "Paiute Traditions Inform Water Management Practices in Once-Lush Owens Valley." KCET. Retrieved 8 July 2021 from https://www.kcet.org/shows/tending-nature/paiute-traditions-inform-water-management-practices-in-once-lush-owens-valley.

Norgaard, Kari M. 2005. "The Effects of Altered Diet on the Health of the Karuk People." Submitted to Federal Energy Regulatory Commission Docket # P-2082 on Behalf of the Karuk Tribe of California. Retrieved 8 July 2021 from https://pages.uoregon.edu/norgaard/pdf/Effects-Altered-Diet-Karuk-Norgaard-2005.pdf.

———. 2014a. "Retaining Knowledge Sovereignty: Expanding the Application of Tribal Traditional Knowledge on Forest Lands in the Face of Climate Change." Prepared for the Karuk Tribe Department of Natural Resources. Retrieved 8 July 2021 from https://pages.uoregon.edu/norgaard/pdf/Retaining-Knowledge-Sovereignty-Norgaard-2014.pdf.

———. 2014b. "The Politics of Fire and the Social Impacts of Fire Exclusion on the Klamath." *Humboldt Journal of Social Relations* 36(1): 73–97.

Norgaard, Kari Marie, Kirsten Vinyeta, Leaf Hillman, Bill Tripp, and Frank Lake. 2016. "Karuk Tribe Climate Vulnerability Assessment Assessing Vulnerabilities from the Increased Frequency of High Severity Fire." Karuk Tribe Department of Natural Resources. Retrieved 8 July 2021 from https://karuktribeclimatechangeprojects.files.wordpress.com/2016/11/final-karuk-climate-assessment1.pdf.

Norton-Smith, Kathryn, Kathy Lynn, Karletta Chief, Karen Cozzetto, Jamie Donatuto, Margaret Hiza Redsteer, Linda Kruger, Julie Maldonado, Carson Viles, and Kyle Whyte. 2016. *Climate Change and Indigenous Peoples: A Synthesis of Current Impacts and Experiences*. US Department of Agriculture Forest Service. Retrieved 8 July 2021 from https://un-declaration.narf.org/wp-content/uploads/fed.-climate-change-and-indigenous-.pdf.

Office of Indian Energy Policy and Programs. 2015a. "Blue Lake Rancheria—Forging a Path toward Climate Resiliency." 22 January. Retrieved 8 July 2021 from https://www.energy.gov/indianenergy/articles/blue-lake-rancheria-forging-path-toward-climate-resiliency.

———. 2015b. "Tonto Apache Tribe—2015 Project." Retrieved 8 July 2021 from https://energy.gov/indianenergy/tonto-apache-tribe-2015-project.

Ortiz Beverly. 2008. "Contemporary California Indians, Oaks and Sudden Oak Death (Phytophthora ramorum)." In *Proceedings of the Sixth California Oak Symposium: Today's Challenges, Tomorrow's Opportunities*, edited by Adina Merenlender, Douglas McCreary, and Kathryn L. Purcell, 39–56. General Technical Report. Rep. PSW-GTR-217. Albany, CA: US Department of Agriculture, Forest Service, Pacific Southwest Research Station.

Pierce, David, and Dan Cayan. 2013. "The Uneven Response of Different Snow Measures to Human-Induced Climate Warming." *Journal of Climate* 26(12): 4148–4167.

Powell, Dana. 2018. *Landscapes of Power: Politics of Energy in the Navajo Nation*. Durham, NC: Duke University Press.

Powell, Dana, and Dailan Long. 2010. "Landscapes of Power: Renewable Energy Activism in Diné Bikéyah." In *Indians & Energy: Exploitation and Opportunity in the American Southwest*, edited by Sherry L. Smith and Brian Frehner, 231–262. Santa Fe: School of Advanced Research Press.

Powell, Dana, and Julie Maldonado, eds. 2017. *Just Environmental and Climate Pathways: Knowledge Exchange among Community Organizers, Scholar-Activists, Citizen-Scientists and Artists*. Workshop report. Society for Applied Anthropology Annual Meeting, Santa Fe, New Mexico. 28 March. Retrieved 8 July 2021 from https://likenknowledge.org/wp-content/uploads/2021/05/Climate-Pathways-Workshop-Report_Santa-Fe_March-2017.docx.pdf.

Redmond, Miranda D., Frank Forcella, and Nichole N. Barger. 2012. "Declines in Pinyon Pine Cone Production Associated with Regional Warming." *Ecosphere* 3(12): 120. http://dx.doi.org/10.1890/ES12-00306.1.

Redmond, Miranda D., Katherine C. Kelsey, Alexandra K. Urza, and Nichole N. Barger. 2017. "Interacting Effects of Climate and Landscape Physiography on Pinon Pine Growth Using an Individual-Based Approach." *Ecosphere* 8(3): e01681.

Redsteer, Margaret Hiza. 2012. "Disaster Risk Assessment Case Study: Recent Drought on the Navajo Nation." Climate Adaptation Futures International Conference, Tucson, AZ, 29–31 May.

Redsteer, Margaret Hiza, Kirk Bemis, Karletta Chief, Mahesh Gautam, Beth Rose Middleton, Rebecca Tsosie. 2013. "Unique Challenges Facing Southwestern Tribes: Impacts, Adaptation and Mitigation." In *Assessment of Climate Change in the Southwest United States: A Technical Report Prepared for the US National Climate Assessment*, edited by Gregg Garfin, Angela Jardine, Robert Merideth, Mary Black, and Sarah Leroy, 385–404. Washington, DC: Island Press. Retrieved 8 July 2021from https://swccar.org/sites/all/themes/files/SW-NCA-color-FINALweb.pdf.

Redsteer, Margaret Hiza, Rian C. Bogle, and John M. Vogel. 2011. "Monitoring and Analysis of Sand Dune Movement and Growth on the Navajo Nation, Southwestern United States." US Geological Survey Fact Sheet 2011–3085. Retrieved 8 July 2021 from http://pubs.usgs.gov/fs/2011/3085/.

Redsteer, Margaret Hiza, K. B. Kelley, H. Francis, and D. Block. 2015. "Accounts from Tribal Elders: Increasing Vulnerability of the Navajo People to Drought and Climate Change in the Southwestern United States." In *Indigenous Knowledge for Climate Change Assessment and Adaptation, Part III: Facing Extreme Events*, edited by Douglas Nakashima, Igor Krupnik, and Jennifer Rubis, 171–87. New York: Cambridge University Press.

Reed, Kaitlin, Beth Rose Middleton Manning, and Deniss Martinez. 2020. "Becoming Storms: Indigenous Water Protectors Fight for the Future." In *Lessons in Environmental Justice*, edited by Michael Mascarhenas. Sage.

Rising Voices. 2014. *Adaptation to Climate Change and Variability: Bringing Together Science and Indigenous Ways of Knowing to Create Positive Solutions*. Workshop report. National Center for Atmospheric Research, Boulder, CO. 30 June–2 July. Retrieved 8 July 2021 from https://risingvoices.ucar.edu/sites/default/files/rv2_full_workshop_report_2014.pdf.

Sahagun, Louis. 2014. "Native Americans Challenge Construction of Mojave Desert Solar Plant." *Los Angeles Times*, 12 December.

Scott, Russell L., Travis E. Huxman, Greg A. Barron-Gafford, G. Darrel Jenerette, Jessica M. Young, and Erik P. Hamerlynck. 2014. "When Vegetation Change Alters Ecosystem Water Availability." *Global Change Biology* 20: 2198–210.

Seager, Richard, and Martin Hoerling. 2014. "Atmosphere and Ocean Origins of North American Droughts." *Journal of Climate* 27(12): 4581–606. http://dx.doi.org/10.1175/jcli-d-13-00329.1.

Seager, Richard, Martin Hoerling, Siegfried Schubert, Hailan Wang, Bradfield Lyon, Arun Kumar, Jennifer Nakamura, and Naomi Henderson. 2015. "Causes of the 2011–14 California Drought." *Journal of Climate* 28(18): 6997–7024. http://dx.doi.org/10.1175/ JCLI-D-14-00860.1.

Sidder, Aaron M., Sunil Kumar, Melinda Laituri, and Jason S. Sibold. 2016. "Using Spatiotemporal Correlative Niche Models for Evaluating the Effects of Climate Change on Mountain Pine Beetle." *Ecography* 7: 1–22.

Sloan, Kathleen, and Joe Hostler. 2014. *Utilizing Yurok Traditional Ecological Knowledge to Inform Climate Change Priorities*. Final report. Yurok Tribe Environmental Program. Submitted to the North Pacific Landscape Conservation Cooperative and US Fish and Wildlife Service.

Smith, Sherry Lynn, and Brian Frehner, eds. 2010. *Indians and Energy: Exploitation and Opportunity in the American Southwest*. Santa Fe: School of Advanced Research Press.

Smith, William James, Jr., Zhongwei Liu, Ahmad Saleh Safi, and Karletta Chief. 2014. "Climate Change Perception, Observation and Policy Support in Rural Nevada: A Comparative Analysis of Native Americans, Non-Native Ranchers and Farmers and Mainstream America." *Environmental Science and Policy* 42: 101–22.

State Water Resources Control Board. 2016. "Beneficial Uses Development: Tribal Traditional and Cultural, Tribal Subsistence Fishing and Subsistence Fishing Beneficial Uses; Stakeholder Outreach Document." Retrieved 8 July 2021 from https://www.waterboards.ca.gov/about_us/public_participation/tribal_affairs/docs/bu_outreach.pdf.

Todd, Zoe. 2018. "Refracting the State through Human-Fish Relations: Fishing, Indigenous Legal Orders and Colonialism in North/Western Canada." *Decolonization: Indigeneity, Education & Society* 7(1): 60–75.

Tonino, Leah. 2016. "Two Ways of Knowing: Robin Wall Kimmerer on Scientific and Native American Views of the Natural World." *Sun Magazine* (April): 4–14.

TREX (Prescribed Fire Training Exchange). 2017. Prescribed Fire Training Exchanges and Cooperative Burning. February. Retrieved 8 July 2021 from https://www.conservationgateway.org/ConservationPractices/FireLandscapes/FireLearningNetwork/Documents/PERFACT-Feb2017-TREX.pdf.

Udall, Bradley, and Jonathan Overpeck. 2017. "The Twenty-First Century Colorado River Hot Drought and Implications for the Future." *Water Resources Research* 53: 1–14.

USDOE (US Department of Energy). 2015. "Energy Department Selects 11 Tribal Communities to Deploy Energy Efficiency and Renewable Energy Technologies." Press release, 18 March.

Viles, Carson. 2011. "First Foods and Climate Change." Tribal Climate Change Profile: First Foods and Climate Change. Institute for Tribal Environmental Professionals.

Vinyeta, Kirsten, and Kathy Lynn. 2013. "Exploring the Role of Traditional Ecological Knowledge in Climate Change Initiatives." General Technical Report PNW-GTR-879. Portland, OR: US Department of Agriculture, Forest Service, Pacific Northwest Research Station.

Voggesser, Garrit. 2010. "The Evolution of Federal Energy Policy for Tribal Lands and the Renewable Energy Future." In *Indians and Energy: Exploitation and Opportunity in the American Southwest*, edited by Sherry L. Smith and Brian Frehner. Santa Fe: School of Advanced Research Press.

Voggesser, Garrit, Kathy Lynn, John Daigle, Frank Lake, and Darren Ranco. 2013. "Cultural Impacts to Tribes from Climate Change Influences on Forests." *Climatic Change* 120: 615–26.

Washoe Tribe of Nevada and California. 2009. "Wa She Shu: 'The Washoe People,' Past and Present." Retrieved 8 July 2021 from https://www.fs.usda.gov/Internet/FSE_DOCUMENTS/stelprdb5251066.pdf.

Western Rivers Conservancy. 2017. "Klamath-Blue Creek: Conserving One of the World's Great Salmon Runs." Retrieved 8 July 2021 from http://www.westernrivers.org/projectatlas/blue-creek/.

Williams, A. P., R. Seager, J. T. Abatzoglou, B. I. Cook, J. E. Smerdon, and E. R. Cook. 2015. "Contribution of Anthropogenic Warming to California Drought during 2012–2014." *Geophysical Research Letters* 42(16): 6819–28.

Whyte, Kyle Powys. 2013. "Justice Forward: Tribes, Climate Adaptation and Responsibility." *Climatic Change* 3: 517–30.

———. 2016. "Indigenous Peoples, Climate Change Loss and Damage, and the Responsibility of Settler States." East Lansing, MI: Michigan State University. Retrieved 8 July 2021 from https://papers.ssrn.com/sol3/papers.cfm?abstract_id=2770085.

Whyte, Kyle. 2017. "Indigenous Climate Change Studies: Indigenizing Futures, Decolonizing the Anthropocene." *English Language Notes* 55(1–2): 153–62.
Wildcat, Daniel. 2009. *Red Alert! Saving the Planet with Indigenous Knowledge*. Golden, CO: Fulcrum Publishing.
———. 2013. "Introduction: Climate Change and Indigenous Peoples of the USA." *Climatic Change* 120(3): 509–15.
Winnemem Wintu. n.d. "Our Traditional Leader: Caleen Sisk." Retrieved 8 July 2021 from http://www.winnememwintu.us/caleen-sisk/.
Yazzie, Melanie, and Cutcha Risling Baldy. 2018. "Introduction: Indigenous Peoples and the Politics of Water." *Decolonization: Indigeneity, Education & Society* 7(1): 60–75.
Yurok Today. 2014a. "Fire Council Ignites Long-Term Burn Plan: Yurok Fire Crew Trains to Conduct Cultural Burns on Tribal Lands." *Yurok Today* (June): 2–4.
———. 2014b. "Yurok Tribe Manages Carbon Program." *Yurok Today* (May): 8–9.
———. 2019. "Tribe Passes Powerful Resolution: Tribal Declaration Establishes Legal Rights for the Klamath River." *Yurok Today* (May): 3.
Yurok Tribal Council. 2019. "Resolution Establishing the Rights of the Klamath River." Resolution 19–40. 9 May.
Yurok Tribe Environmental Program. 2012. "Nue-ne-pueh Resource Health Report, Results from Tissue Sampling: 2010–2012." Yurok Tribe. Retrieved 8 July 2021 from https://kbifrm.psmfc.org/wp-content/uploads/2018/02/2014_0441_STAR-Outreach-Brochure.pdf.

Chapter 12

The Return to What Has Never Been

A View on the Animal Presence in Future Natures

Guilherme José da Silva e Sá

This chapter is based on data collected during ethnographic research initiated in 2014 in the Faia Brava Reserve, considered by the Institute for Nature and Biodiversity Conservation/Institute for Nature Conservation and Forests (ICNB/ICNF)[1] to be the first Private Protected Area in Portugal.[2] A specific characteristic of the Faia Brava Reserve is its purpose of ecological restoration, which has been promoted by the association that manages it—the Transhumance and Nature Association (ATN). This designation made the reserve the starting point of an ambitious project to renaturalize the western region of the Iberian Peninsula, one that foresees the reintroduction of large animal species in Portuguese territory through its integration into the Rewilding Europe network.[3] The motivations behind the rewilding initiative stem from the broad discussion on the impact of climate change on the planet, and especially in Europe. More than encouraging isolated reflections, the rewilding agenda gives rise to the possibility of directly intervening in these processes of climate change. This particular direct action, based on ecological restoration, aims to interrupt cycles of forest fires, which are lethal to several species in the affected areas and also secrete large amounts of CO_2 into the atmosphere. In addition, the regeneration of food chains involving large predators, herbivores, and necrophagous birds allows the dispersion of nutrients in the soil that are essential for the growth of local vegetation and for the future of recovered forests. Forests that are properly managed and in good condition can contribute to the capture of atmospheric carbon.

Notes for this chapter begin on page 311.

An Ethnographic Approach to the Field

I woke early to catch a bus that would take me to the bus station in Guarda,[4] a city where I would later rent a car to reach Figueira de Castelo Rodrigo. The trip to Guarda took around five hours, the last of which I spent admiring the contours of the Serra da Estrela, the highest mountain range in continental Portugal, situated in the middle west region of the country. On my way to the car rental store, I came across a public market that gave off the strong scent of aged handmade cheese, as well as some "Chinese" stores.[5] There was a Brazilian working at one of the stores who retained his accent from the countryside region of the state of São Paulo, even though he had lived in Portugal for fourteen years. I ate something in one of the twenty-four-hour gas station's convenience stores, where I bought a weekly newspaper of the region called *Terras da Beira* (which freely translates into English as "lands of the edge").

The news published in that issue of 28 August 2014 was particularly interesting, because it offered clues of what I would come across later. The main headline read: "The City of Figueira de Castelo Rodrigo Joins the 'New Populators' Program and Seia Shows Interest in Joining It as Well." "New Populators" was the name given to the rural repopulation program that offers assistance in the implementation or transfer of company projects into the Portuguese countryside, created in 2007 as an outcome of a chat between neighbors, one a sociologist and the other a technician from a local development association. This dynamic project aims to register "new populator" families for their later establishment in areas compatible with their profiles. This way, the concession given to each family also depends on the identification, made by a technical team, of the business potential for each region. In the case of Figueira de Castelo Rodrigo, the plan is to establish five families that fulfill the immediate need for people willing to work in cattle breeding and granite processing.

The territory's repopulation shared the front page's space with the news story, "Wildland Fires in the District: Less Scorched Area than Last Year." The wildfires that spread all over the region at that time of the year (the end of summer) are triggered by the low levels of rainfall and the constant change of wind direction. The forest firefighters are forced to work on many fronts to cover the large area of rocky terrain and ground vegetation. Even though the news reported an annual reduction of the scope of burned areas to that date, over the following days the TV news showed the rapid spread of the fire outbreaks. As I was told later, the wildland fires are one of the main concerns of the Faia Brava Reserve's managers. In order to prevent them, the perimeter is monitored daily by a watcher who looks for possible outbreaks that could threaten the reserve. On the very first day of research, I was able to accompany one of those night watches, and we observed a great arc of fire spreading with the wind over a region close to the Côa River Valley.

The theme of wildfires recurs in the memories and motivations that are part of the work of the Transhumance and Nature Association's general manager and the Portugal coordinator of the Rewilding Europe initiative; for now, however, it is important to discuss a short piece of news published in the weekly column of the newspaper: "The National Republican Guard (GNR) Identifies Suspect of Arson Fire in the Corujeira Area."

The fire that destroyed approximately sixty to eighty hectares of the Serra da Estrela Natural Park was caused, according to a source from the Territorial Unit of the GNR, by a man who was motivated by vengeance against his siblings concerning family heirlooms, "since the arson fire started next to the suspect's home and all around was burned, except for his property" (*Terras da Beira*, 22 August 2014). This notable event seems to be deeply connected to people's lives in this region. Such personal instigation as described in the newspaper seems to point to an inextricable presence of people in each place, in each route, and in each stone wall, which after centuries becomes mingled with the natural landscapes. There, life is all about the surrounding area.

In the same issue of the weekly paper, there appeared a column titled "Men and Wolves: A Summer Tale," written by the Wolf Group from the Animal Biology Department of the Faculty of Science at the University of Lisbon. It was a tale about Mondego, a sheepdog that would accompany a herd of cows alone:

> He basically stays there, watchful even if lying down; observing a dozen cows that went uphill with him and that soon will take him back down: his family. At least the only one Mondego ever knew; about his true origin, the siblings from his brood, no memory is left. He has found there, in the immensity where the herd wanders and grazes, his home, his freedom, and also his mission.

Years later, Mondego's owner would not tire of telling the story of what happened that night, now recounted with fanciful traces but still faithful to the core of what really occurred:

> When the cows went downhill, I noticed that a calf was missing. And Mondego also stayed behind on the hill. But there was a heavy fog, and I had to wait until the morning to go after them. When I came across the calf, it was lying down next to some rocks, to take shelter . . . with the dog leaning on him, as if they were two dogs. Mondego didn't want to leave the small being alone and spent the whole night watching over him. And I don't know if he had to defend him against some wolf . . .

The text finishes the heroic narrative:

> The one thing we know is that the story really happened, someplace in our fields. And if our hero was a Castro Laboreiro or a Serra da Estrela dog,[6] or any other, that is the least important thing. Because that is the life of many sheepdogs that accompany "their" cows, goats, and sheep every day, risking their lives in the face of the wolf, but also of the men, always ungrateful, with their traps, their cars, and their poisons. (*Terras da Beira*, 24 August 2014)

This is a story of adaptations and elements like everything in the everyday life of traditional Portuguese land use. Instead of a celebration of a nature that preserves insoluble borders, what is found is a message about a "re-nature," which survives through its compositions. Family is that to which an individual adapts once adopted. Dogs and cattle are no longer distinguished one from the other, each becoming antagonists of equally accepted enemies, the wolf and the man. Against the first—its agility and its pack—the dog's strong features (historically modeled by human hands) would not be enough. It is also necessary to refashion the anatomy of the dog, giving them thick collars full of spiky nails, a tool for protection against wolf bites. Against the "man," in a Hobbesian recombining inversion, the wolf becomes the man's wolf.

The return of the great predator to European terrain provides a new sense to old practices. The wolf reinvents the (once again purposeful) dog, which recreates the (again vibrant) pastures, with the purpose of giving a whole new sense to life in the countryside and to people's lives. Therefore, merely by renouncing their old (and new) machinations, human beings could rebuild this cycle. Instead of posing a tacit opposition between human activities—such as transhumance and regulated hunting—and the elements that are part of "nature" (fauna and flora), an agreement of coexistence is what emerges. In such agreement on mutual reinvention of "nature" and of human practices resides the hope of a future that, while it evokes a mythical past on the one hand, it is guided on the other by new terms capable of preventing the predatory actions of the past. In this manner, in yet another reported story, the news about the detention of two men for "hunting crimes"—one hunting with neither a firearms license nor a hunter's license and the other hunting nonauthorized species—the matter raised by the local newspaper is closed and becomes another beginning.

The research that I have been developing since 2014 mainly aims to follow up the implementing dynamics and practices of the several agents involved in a new kind of nature reserve that is adapted to the conditions established in Europe for the reconstruction and conservation of its environment. It also aims at the establishment of guidelines regarding "rewilding," its materialization, including natural parks that in the history of conservation biology are recent, no older than a decade. Along that line, it has been observed that one of the peculiarities of this sort of renaturalization program is its inextricable engagement with the proposition that human activities be guided by the idea of sustainable economic development. As far as they are opposed to the argument of preserving a "state of original nature," environmentalist partisans of renaturalization contend that nature should and could be recomposed through processes regarded as artificial. From this, it can be inferred that a vast range of interaction possibilities among species (human and nonhuman) is granted at the moment that the belief in a non-entropic nature is renounced.

I will start with the story of an encounter that, in its telling, identifies new possible areas for the expansion of the renaturalization project. This event, which took place when Rewilding Europe was celebrating its third year of existence in the region of Beira Alta Interior in Portugal, marked the beginning of a new stage of the renaturalization project in the western portion of the Iberian Peninsula. At that moment, after the establishment of the experiment at the Faia Brava Reserve, planning was initiated to extend the area along the valley of the Côa River.

First Act: In Search of the Void, Tracing the Course

Our encounter was around nine o'clock in the morning, at the crossroad of a small village along a Portuguese highway. The group—formed by two anthropologists (me and an intern from the reserve), two biologists who worked for the Faia Brava Reserve, and two directors of Rewilding Europe—went along a tortuous road, which soon became a narrow path of dirt and rocks, in a four-wheeler. After a steep ascent, we continued on foot to the top of a hill where it was possible to see a landscape that stretched for miles around us. Standing on a gigantic granite block, we looked through binoculars with one hand and pointed to the horizon with the other, as if with our fingertips we could scan the terrain.

The silence that is always present in that bucolic landscape was only interrupted by the enthusiastic conversations and the rushed steps of visitors. After initial surprise that a Brazilian anthropologist had just learned about the Rewilding Europe initiative, one of the directors started his explanation (being careful to be as didactic as possible in the presentation) of the aims of that field visit to Portugal. At one side of the valley, it was possible to find approximately "40 to 60 percent of human occupation;" at the hillside, rocky formations of granite, enthusiastically referred to as "the future of the rewilding project in Portugal," could be seen. He patiently explained that the aim of the Rewilding Europe initiative was to act in regions where human presence was scarce. As he spoke of the next stage of Portugal's renaturalization project, the terminology of percentages was used once again, since it would be implemented in territories that were "80 to 100 percent abandoned." According to the director, these demographic voids presented good conditions for nature to be reconstructed, and as would be stated later, in the future they could also represent a "good opportunity" for people that inhabit that region.

As we wandered along a path that crossed villages with a few dozen inhabitants—the majority of them elderly—rewilding action plans for the coming years began to take shape. The trails we covered led us to places where the ruins of stone walls prevailed, traces of century-old human occupation mingled with the originally rocky terrain, which resembled a mosaic of symmetrical shapes of green and gray. There were also traces of old windmills and irrigation canals, parts of a system that made the harvesting of vegetables viable due to the extraction of water from the creek that ran alongside. Many times, the director stopped and expressed his view on what the future landscape of these places would be like. Invariably, there were youngsters hitchhiking with their backpacks and riding mountain bikes along the valley trails. In the surrounding area, there would be herds of wild horses and bovines, as well as mountain goats balancing on the cliffs. Also composing the scenario, eagles and vultures would be flying in the sky. The focus on a viable future, as clarified here, is the main difference between the Rewilding Europe initiative and other identically named rewilding projects already carried out. While some of the proposals of renaturalization projects point to a return of the state of nature attested to in the past—as is suggested by the American "Pleistocene rewilding," Rewilding Europe concentrates its efforts on creating future interactive environments between human beings and the natural habitat. Within such logic, asking what the optimal point to be reached in renaturalization would be is no longer a relevant question. Renaturalization, according to Rewilding Europe's orientation, is primarily about what "nature" could become rather than what it was in the past.

The Genesis of the Rewilding Concept

Coined originally in the late 1990s (Soulé and Noss 1998), the term "rewilding" related to the idea of fomenting an alternative model to wildlife conservation reserves, mainly in North America. Also known as "Pleistocene rewilding," such proposals, formulated by a group of renowned specialists in conservation ecology, had the purpose of stipulating a concrete basis for the reintroduction of animal species, mainly herbivorous megafauna and great predators, in areas that were presently uninhabited. In evoking a past time, the reference to the Pleistocene was intended as an allusion to the environmental conditions found at the beginning of human habitation and expansion on the planet. Although they acknowledged the difficulty in "bringing back to life" animal species that had already been extinct for millennia, the proponents of this kind of "renaturalization" showed in two articles (which had significant repercussions in academic and environmentalist circles) what they considered to be the concrete basis for the recovering of these degraded ecosystems. The first of their principles argued that human beings have the moral authority, and even the ethical duty, to intervene in the natural environment, since their irresponsible actions directly or indirectly caused the extinction of several other species of animals and plants. Even so, according to Donlan (2005), human beings will continue to cause extinctions, to modify ecosystems, and to alter the course of evolution; this makes attempts to reach a solution political, although, and without denying human participation in the problem, also a highly desirable posture.

Such a perception seems to place the problem in a much broader arena of contemporary discussion: that is, the discourses relating to the imminence of a new geologic "era," widely known as the "Anthropocene." This latter is characterized as an event-moment that affirms the role of the human species as a new constant force of intervention in the planet's biophysical processes.

> However much we would wish otherwise, humans will continue to cause extinctions, change ecosystems and alter the course of evolution. . . . Our proposal is based on several observations. First, Earth is nowhere pristine; our economics, politics, demographics and technology pervade every ecosystem. . . . humans were probably at least partly responsible for the Late Pleistocene extinctions in North America, and our subsequent activities have curtailed the evolutionary potential of most remaining large vertebrates. We therefore bear an ethical responsibility to redress these problems. (Donlan 2005: 436)

> Far more than any other species in the history of life on Earth, humans alter their environments by eliminating species and changing ecosystem function. . . . Earth is now nowhere pristine, in the sense of being substantially free from human influence, and indeed, most major land masses have sustained many thousands of years of human occupancy and impacts. . . . Human-induced environmental impacts are now unprecedented in their magnitude and cosmopolitan in their distribution, and they show alarming signs of worsening. (Donlan et al. 2006: 660–61)

In a certain sense, the "naturalization" of human presence and action causes the rewilding model of conservationism to take on unique characteristics, because it attributes to humans some agency in the duty of returning the Earth to its old ecosystems, for which the proactive intervention in its dynamics and vital processes is necessary. Consequently, it makes sense that the notion of an untouched nature becomes detached from the vocabulary of the promoters of renaturalization strategy. Nature, therefore, would reserve within itself a great potential for artificialization, inasmuch as it could not be discussed in terms of the existence of isolated species but rather to a range of relations integrated into the actions of such species (which inevitably would include humans).

After overcoming the initial obstacle sustained by the myth of untouched nature, it is necessary to aim at restoring the functional "health" of ecosystems. For this, it is indispensable to adopt a proactive stance, or in the preferred terminology, an "optimistic" perspective toward twenty-first-century conservationism. Several possibilities for reconstructing certain ecosystems have been studied, identifying their functional processes of interaction and their trophic chains to evaluate the viability of species reintroduction, for instance, whether it is possible to relocate individuals from other areas or whether it will be necessary to use "proxy" species to fulfill the functional role passed on by those already extinct.

In this way, the program seeks not only to return independent species but also to favor the re-composition of functional interactions among them, and fundamentally to recreate their food chains. Such a characterization leads to the understanding that it would be necessary to prioritize the reintroduction of large predators or herbivores, or both. In this way, the recovery of the entire trophic chain from top to bottom would be ensured. It would entail that the reintroduction of a top-of-the-chain predator would demand appropriate conditions for its nutrition and survival. However, the reason for the highlighted entreaty for large animals transcends the organicists' explanations, even though it still preserves a certain pragmatism. According to the champions of rewilding, the emblematic animals of the megafauna are clearly those endowed with greater charisma, a fact that would mobilize interest, resources, and empathy more easily among human beings. This is a fundamental point considered throughout the renaturalization enterprise. Having the support of public opinion is vital in connecting sustainability and fundraising. Furthermore, the notion that the environment would recover to a state of economic sustainability becomes an outstanding strategy of persuasion regarding the viability and "rationality" of such enterprises, which at first sight may seem hardly reasonable.

Rewilding Europe

Although the proposal formulated by Donlan (2005; Donlan et al. 2006) has become a global reference point for the term "rewilding," it is far from being the only possible definition. Projects inspired by renaturalization are underway in different parts of the world, and all have their technical and ideological specificities. For example, the idea of a return to the Pleistocene is shared among North American and Russian initiatives, but it does not really represent the interests of the Rewilding Europe network, which focuses on what ecological niches could become in the future. The Rewilding Europe initiative contrasts with its corresponding programs even in terms of its viability of implementation; while projects aiming to return to the Pleistocene seem to exist only as marginal projections, Rewilding Europe's work has been underway since 2011.

With its headquarters based in Nijmegen, Holland, Rewilding Europe is composed of a network involving large and small conservationist NGOs, investors and banks that subsidize local projects, researchers linked to universities who provide the technical basis for the implementation of planned actions, rural landowners and agricultural producers, and tourists and volunteers who circulate around the eight rewilding model areas in Europe.

In the activities promoted by Rewilding Europe, the concept of "renaturalization" assumes a particular character that sees in the generation of social and economic opportunities a way of returning wildlife to Europe, and vice versa. It is therefore about committing to the planning of a future nature without perpetuating the old kind of ties inherited by natural history. Through the reappropriation and reoccupation of lands abandoned due to a historical process of rural exodus experienced in Europe during the twentieth century, renaturalization provides an ecologically viable model for the areas considered economically unproductive.

The effect generated by such intervention is the creation of private reserves in areas that are progressively purchased with the funds of small and big investors, who in turn become partners in the renaturalization enterprise. The reserves are generally managed by local NGOs that represent a broad range of shareholders. There are also alternative ways of integrating the rewilding project: for example, leasing land for the management of reintroduced natural resources, generally animal species, and establishing partnerships consisting of services related to ecotourism, primarily through rural hotels and small restaurants.

However, the extent of the model areas of renaturalization related to the Rewilding Europe initiative does not always coincide with the limits of the private reserves, and frequently transcends them. This occurs because the animal occupation areas may exceed the parks' borders. The renaturalization areas are conceived as large territorial extensions that must afford the animal and plant species' survival, whether they are reintroduced or recovered through management plans. The fact that a good proportion of the animals in question are migratory and, therefore, cannot be restricted to the reserves means that a renaturalization area must be understood as the occupation area of those species. The reserves themselves would function as future hot spots from which the animals could migrate, defining routes and ecological corridors that, with some human investment, would integrate the whole system.

While the rewilding areas in Europe are defined by the vital fluxes of the animals, they are also marked by their long records of anthropization. This element is regularly considered in the action plans of Rewilding Europe. At the same time that these zones are prepared to host animal reintroduction projects, the necessary conditions to ensure visits by tourists and researchers interested in wildlife are also put in place. An example of investment dedicated to this sort of visitor is the building of shelters inside the reserves, from which it is possible to observe and photograph the animals with the full discretion required. An effort is also made to improve the commercial activities involved in tourism around the protection areas, through training courses in the hospitality business, gastronomy, and sales of each region's traditional products.

In considering the reintroduction of species, there exists a prevalent consensus among the ecologists involved with Rewilding Europe, who understand that management of a reserve entails rural property. Wolves, bears, lynxes, equines, bovines, and goats in a wild state, as well as eagles and vultures in the sky, are some of the animal species envisioned in the project of the future repopulated European nature. Avoiding in particular the introduction of exogenous species, the intention is to recover native species, even through use of genetic research and direct environment intervention, when creating sanctuaries and food zones for the animal populations. The focus is to return them to the remodeled landscape in accordance with interests that combine environmentalism with sustainable economic development. To achieve these ends, there are a few restrictions on human intervention in the processes that are regarded as being "natural." It is common to hear that in areas historically abandoned by human occupation, "nature returns" not wholly at once but in a progressive manner to regain its space. Typically, this is how the process has been observed in some regions of Europe in the last decades. However, it is well known that the time required for autonomous recomposition of such alterations is reasonably long and that it is therefore beneficial for humans to provide an "initial boost." Nevertheless, the artificialization of nature is seen as a trigger rather than a substitute for nonhuman agents that will gradually drift to autonomy.

Second Act: A Good Trade, Strengthening the Strategy

After the new pathway for renaturalization in Portuguese territory had been defined, based on the exploratory excursions into the field during the technical visit paid for by Rewilding Europe's team, it was necessary to put the strategy into practice. Certain proceedings and measures were consequently carried out.

First of all, it was necessary to reexamine property maps and registers in order to precisely identify the overlay of the areas singled out for rewilding. Concurrently, meetings with the representatives from county and local entities were scheduled to inform them about the initiative that was underway. I had the opportunity to attend one of these events that was carried out at the Council Chamber of a county in Beira Alta Interior. The proposal was presented by the local coordinator of Rewilding Europe, who explained all the project's advantages: that is, the revaluation of the territory that had been abandoned for a long time due to the soil depletion that had rendered it no longer fruitful for conventional agricultural activities (i.e., farming and pasturing). The readjustment would be made possible by the redirection of economic activities toward ecotourism agriculture. As it was plausible to imagine that the replacement of one activity by another could result in an even greater depopulation of the fields, the coordinator explained that investing in nature could be a "good trade" that would even allow the resumption of some traditional activities, such as the artisan production of sweets, cheese, olive oil, jams, and various utensils for commercial objectives, given the presence of tourists.

The coordinator started to explain that the partnership system offered by the rewilding initiative would entail credit for the readjustment of herds, since the replacement of cattle and horses was considered exogenous to the actual reintroduction interest. The system would also include the possibility of land leasing and, finally, the implementation of small businesses connected to the rewilding enterprise. The development of a network of services, such as outdoor activity operators, photographic safaris, hotel businesses, and rural cuisine, needed to converge in a manner that would provide, in what is considered to be the motto of Rewilding Europe, an "experience with the wildlife" of European domain. After listening to the explanation in silence, the Council Chamber representative asked with some interest what was, in fact, required on their part. The coordinator replied that on that particular occasion he wanted only to notify them about Rewilding Europe's work in the region and to be able to count on the efforts of public representatives for the project's promotion. This request was successfully granted.

From that moment on, Rewilding Europe's strategy of action entered the next level of persuasion: the search for local supporters with the goal of expanding the renaturalization area. For this purpose, in the months following the exploratory survey, connections with the public and community representatives would be established to enable reliance on these groups as mediators between the organization and possible partners.

Rewilding Europe in Portugal

Originally, the area destined for the renaturalization in the western region of the Iberian Peninsula was also intended to be integrated with other conservation initiatives in Portugal and Spain. The territorial strip that stretches from the northeast of Portugal—the Guarda region—to the west of Spain—the Castilla y Leon region—has at its far reaches the Faia Brava Reserve (in Portugal) and the Campanarios de Azaba Reserve (in Spain). This borderland area presented common historical and geographical elements indicating a past of agricultural activities that slowly lost their relevance and interest among the new generations of inhabitants. This caused a progressive disinterest in the villages of the region. Some were even totally abandoned. Due to the migration of young people to big urban centers, such as Lisbon, Porto, Salamanca, and Madrid, and also to other countries, local and elderly people faced difficulties in maintaining their occupations, like pasturing and raising livestock in smallholdings.

After the first three years of action in the region, an evaluation of the outcomes up to that date was conducted, after which, in 2014, it became possible to start the renaturalization of the western Iberian Peninsula. While, on the one hand, the Portuguese initiative was highly praised for reaching its goals within the established deadlines, on the other, activities conducted in Spain did not achieve the desired results, and the partnership with Rewilding Europe was canceled in that country. It became necessary to rethink the organization's strategy for the subsequent years; Rewilding Europe in consequence revisited Portugal in order to explore new zones for the project's expansion, which, in light of the Spanish partner's withdrawal, would take a new course starting from the Faia Brava Reserve. Excursions were made along the course of the Côa River, a region that currently possesses a low demographic rate and few registers of agricultural and pasturing activities but that, in compensation, is composed of a terrain characterized by rocky cliffs and crystal-clear water.

Some Issues Rewilding May Bring to Anthropology

The transition from the twentieth century to the twenty-first has presented an ambiguous panorama: On the one hand, the depletion of ecosystems and the consequent threat to the survival of several animal and plant species has intensified, especially in the region between the tropics. On the other hand, some considerable advances have also been achieved in environmental legislation in the northern hemisphere and in public opinion mobilization concerning the need for an integrated ecological project for the planet.

The appearance of such ecological thought derived from the evident deforestation of huge areas along with the breakdown of cultivable regions, mainly in Europe. This situation has driven those of us participating to adopt two measures—with opposite moral footing—in the face of the decline of European agricultural production. The first measure expanded upon the exploratory nature of agricultural production, redirecting and creating new commercial and transnational pacts together with emerging economies from the southern hemisphere. This expansion of agricultural borders (on a global level) was equally responsible for the diffusion of deforestation problems on a global scale while taking advantage of local environmental legislation. Consequently, the ecological crisis has gone from being an easily located issue to a more systemic one, with global effects and causes propagated throughout history.

The rural exodus in Europe and the devaluation of parts of the traditionally cultivated lands caused the appearance of what Bernardina (2011) would call a "post-rural society," essentially, the resumption of a lifestyle determined by a certain defined "rurality" combined with an interest in providing viable conditions for accelerating local economies through rural and ecological tourism. The last dimension leads to the patrimonializing of customary field practices (such as local techniques, hunting, manufacturing, festivities, commensality, and cooperation) and the creation of natural parks (through reforestation and reintroduction of animals).

Both strategies aim to generate capital, though they diverge in focus through either the actions of multinational companies in Africa or Latin America or through those of small entrepreneurs who live in rural European regions. Such a duality, which is not redundant, helps to illuminate the original contexts of the programs discussed in this chapter. It is important to state here that rewilding projects are fundamentally oriented by the motivational principles that rule a capitalist system. The mobilization of resources, the way they communicate, their proposals, and their leeway to manage nature policies render these projects as another idyllic update of capitalism. However, once nature is no longer seen simply as a supplier of raw material but also as the product itself to be commercialized after certain transformations, a new setting for the production chain is presented. It is, thus, beholden to think of the business of environmental renaturalization as the main part of what is named the "Anthropocene" (or "Capitalocene") (Hache 2014). To add to the definition already suggested above, the "Anthropocene" is a term used by the biologist Eugene Stroemer and popularized since the 1980s by the famous chemist Paul Crutzen, who defends its use as follows: "It seems appropriate to apply the term 'Anthropocene' to the present days, a geological age which is dominated by the human kind in too many different ways" (Crutzen, cited in Kolbert 2015). Accordingly, the geologic age that we officially live in, the Holocene, would give way to a new context defined by the advance of human action as a geological force able to drastically interfere in those processes said to be "natural" for the planet. This concept evolved from the denomination "Capitalocene," which aims to make clear that the (peculiarly destructive) agency of human beings toward the planet is not an intrinsic characteristic of our species but rather a complicity with a certain manner of worldwide appropriation: capitalism. The term "Capitalocene" forms part of sociologist Jason Moore's (2017) perspective, in which, according to Danowski and Viveiros de Castro (2014: 28), "the Industrial Revolution initiated in the beginning of the 19th century is just a consequence of the social-economic mutation which generated capitalism in the 'long 17th' century, and therefore, the source of the crises is, ultimately, in the production relations, ahead of (and rather than) the productive forces, if we can express ourselves in such terms."

At first glance, it is possible to assume from preliminary ethnographic data, usually associated with the cataclysmic and destructive effects of human actions in nature, the clear logical deviation from the Anthropocene of forms of human intervention that intend instead to reconstruct nature. But what can appear to be the altruistic actions of so-called "Green Capitalism" can also reveal in certain cases (such as that of the Breakthrough Institute) a megalomaniac and technophile presumption.

> Certain relatives close to the Singularity people, however, have dedicated their attention to the problem by asking themselves about the immediate technological conditions to the survival of capitalism and its main achievements, freedom and security, in a scenario of increasing energetic consumption and persistent dependency on fossil fuels. Breakthrough Institute, an American think tank[7] (from California, as the Singularitarians[8]), with an uncertain political tendency, is maybe the most highlighted name among the defenders of that Green Capitalism which relies on centralized solutions able to implement ambitious techno-engineering projects through the Great Capital, with huge material investment, organically (if such adverb fits here) based on Big Science: the hydraulic frag-

> mentation of rocks to obtain fossil fuel, expansion and improvement of nuclear factories, great hydroelectric projects (barrages in the Amazon Basin, for example), generalization of transgenic vegetables monoculture, environmental Geo-engineering and so on. (Danowski and Viveiros de Castro 2014: 66–67)

If both motivations—negative and positive—seem to come from the same capitalist source, and this source determines the destruction of ecosystems, they also adapt and present themselves as a solution to its reconstruction. As Stengers states, "It is from Capitalism's nature to explore opportunities, that cannot be avoided. Through Capitalism's logical system, it is impossible not to identify the inclusion of the Earth with the appearing of a new field of opportunities" (Stengers 2015: 47).

However, I see the rewilding dream as being of a lesser utopian scale compared to the abovementioned Singularitarians, represented by the "Breakthrough Institute." The model of recovery and environmental management presented by Rewilding Europe intends to reorganize productive activities based on a sense of opportunity, in which investing in nature seems to be a good deal, but there is no idealistic notion that cutting-edge technology and large-scale projects may substitute the local—and deeply human—responsibility for "boosting" those processes recognized as vital.

In light of this it is essential to ask, exactly what issues are rewilding initiatives able to bring into the discussion in relation to the Anthropocene? To what extent do they enter into dialogue with the other initiatives of a collectivist nature that are receiving more visibility in this context?

Implications of the Reconstruction of a Natural Heritage

Taking into consideration some issues from the global level—at which Rewilding Europe is presented—and conversations about the Anthropocene, it can be concluded that the greatest contribution to this topic is to describe ethnographically the actions taken locally in partnership with the renaturalization project in Portugal. This leads us to a second dimension of this research, one that focuses on the processes of patrimonialization implicit in renaturalization programs. As previously mentioned in debates regarding the awareness of cultural heritage, the "nature patrimonialization" undertaken in a rewilding context no longer follows the standard parameters of inviolability and "authenticity" given to a specific natural landscape. Renaturalization theorists argue that the artificialization necessary to reconstruct these environments is part of the preservation of species and their own ecosystems. It is therefore said, according to

Gonçalves (1996), that if the authenticity and proximity to the past can be reevaluated regarding the attested cultural heritages, the same can be said about a natural heritage that was deliberately constructed by humans. In this latter case, it would be necessary to focus on the interactions of the species that allowed the reproduction of the functional roles that each one of them performed and not on the analysis of the species that inhabited a certain biome. As can be inferred from the following quotations, for those that support renaturalization, some features of the species disappear, and the interaction benchmarks remain.

> Such benchmarks would be defined not only by the presence or absence of species, but also by the presence or absence of species interactions—the true functional fabric of nature. (Estes, 2002, quoted in Donlan et al. 2006: 661)

> The focus of conservation biology is expanding to include not only species but species interactions. (Soulé et al., 2003, 2005, quoted in Donlan et al. 2006: 662)

In fact, the amount of intervention in the landscape through the reintroduction of animals and plants is measured by the evaluation and the technical capacity to put such measures into practice. Thus, frequent meetings take place with specialist ecologists about the reintroduction of wild animals, since, besides their expertise, thorough knowledge of the veterinary, sanitary, and legal requirements in each country is necessary. Only then is it possible to understand and manage a species reintroduction project and establish partnerships between organizations that foster renaturalization and public and private institutions (universities, research centers, regulatory organizations) that are in agreement with the consultants of the reintroduction processes. Those professionals, besides giving information about the ecology of the species in focus, have broad knowledge about the possible problems involved in the introduction of a specific animal or plant. Therefore, it is necessary that the choice of the species meets the viability criteria such as avoiding conflict with the human population, its trophic and territorial sustainability, the hunting legislation, the species reproductive cycles, and even the aesthetics efficiency with focus on public opinion sensitivity.

The relationship with the local inhabitants is of ultimate importance for the execution of a rewilding project. Not only are the adults constantly seen as possible partners but the new generation of children are also considered perfect mediators of the reappreciation of the natural landscape. To reach the children, the rewilding teams pays constant visits to schools. There are also social-environmental speeches and programs for the "adoption" of native seedlings that will later be planted in reforestation zones, including Faia Brava.

This fact leads to another issue regarding the way rewilding initiatives are locally implemented: the inclusive aspect of the human presence from the beginning development of activities in the reserve. If anthropization is not an epistemological problem, neither is it a practical barrier to the renaturalization actions. The consolidation of this model of natural reserve foresees the continuous mobilization of associations and local inhabitants, who will coexist with great predators such as wolves, Iberian lynxes, and birds of prey, as well as pigs, equines, bovines, shepherds, small local producers, and sometimes hunters, who will need to have their actions legalized by regulatory organizations.

However, there is some controversy between artificiality and authenticity in the way these natural parks are created. New landscapes arise as the environment is readjusted, and, therefore, human and nonhuman elements are responsible for the good functioning of their systems. If the elements that in the past ensured the subsistence of families—such as the unmeasured extraction and exploitation of natural resources—can no longer exist, it is necessary to replace them with new forms of interaction that will have a similar role. In this context, new "boosts," such as family hotel businesses and rural gastronomy, together with the farms, are considered to be proper sustainable methods to reestablish the ties between humans and nature.

The relationship between the intention of evoking an image reflective of a past lifestyle and its land management, hunting, and rusticity and the need to satisfy the contemporary requirements associated with outdoor experiences, such as preservationism, animal rights, photographic safaris, and new communication technologies, produces a tension between the idealization and execution of an enterprise. An example is the way that populations traditionally established their ties with the natural landscape, in opposition to the expectations generated by new projects of commercializing such lifestyles. From this perspective, it is possible to infer that a variety of landscape transfiguration has always been carried out by the villagers, when they brought home elements of the wildlife surrounding them, such as hunting trophies, luck charms, decorations, and healing substances. Nevertheless, the model of rural tourism currently proposed imposes on that population the need to welcome visitors into their everyday lives. Far from going through this conversion without a trace, when villagers are obliged to engage with nature, which has always been "outside," owing to their accommodation of visitors, this does not take place without leaving behind a certain lifestyle change that brings the natural landscape closer to the domestic sphere. Modern life demands that these people rethink, for example, the place of the hunting trophies (or taxidermy displays) that decorate their fireplaces and walls and start identifying the outdoor home of those natural, living trophies. Going from hunting to a photographic safari is, therefore, a significant change of behavior, and one that deeply resonates in the manners of local existence.

References to the past always conjure up idyllic images about what might be done in the future, but they also manifest as barriers to what must be done. For this reason, one of the main (self-)definitions of "renaturalization" initiatives is that they promote an "optimistic" and "positive" view of ecology. The perception of humanity's role as a proactive agent in the process of environmental recovery, and in an environment destroyed by previous human agency, means that such projects represent a privileged locus for anthropological analysis that understands nature as a human coproduction. Perhaps a good example of this can be found in the Faia Brava Reserve. At the time of its acquisition, ruins of old abandoned dovecotes were found near the reserve. These constructions, which are very common in that region of Portugal, had in the past a double function: producing animal excrement to fertilize the poor soil for farming and providing a meat supply, that of pigeons, to the people who lived in the region, especially in times of lack of food. After the reserve was created, the dovecotes were remodeled, and their functionality was partially recovered. It is now providing an increase in the pigeon population in order to feed the eagles, which are in danger of extinction, that live in the rocky cliffs near the Côa River. The incorporation of these dovecotes into the landscape of the reserve also justifies the function of such artifacts in the interaction chains that exist in rewilding projects. Similarly, the vegetable harvest in the reserve provides another food option for rabbits, which are in turn eaten by eagles, foxes, and, occasionally, lynxes.

The "return" of reintroduced or recovered animal species brings a consequent redefinition of animals and also of the environment itself, as seen in the recent return of Iberian wolves to the region. The return of this predator was possible due to several factors: its easy adaptation and transportation, the national protection policy for the Iberian wolf, and the demographic voids that enabled the gradual regeneration of the forests, which create ecological niches that work as a refuge for the wolves' territorial integration. However, the presence of wolves becomes a huge problem for the rural producers when their livestock is eventually attacked and their losses are not compensated for by the state. The presence of wolves, which can be identified by traces left behind even though they are rarely seen, has already been attributed to the rewilding environmentalists. In this way, new myths appear locally, such as the one stating that the wolves are reintroduced into the area during the night via helicopters. Obviously, due to the potential conflict presented by the wolves, the species has never been thought of as a viable candidate for reintroduction. However, in thinking about the rejection of the wolf, the problem brings about a redefinition by the defenders of renaturalization, who observe that the reintroduction of herbivore species would decrease the shepherds' losses by redirecting the predatory attacks on the livestock to the wild fauna.

It is possible to conclude, therefore, that the reconstruction of natural environments happens through the evoking of survival modes and ancestral landscapes, even if it is carried out by human hands. The reintroduction of long-gone animals is, thus, further connected to the recovery of myths, narratives, and images articulated in an anthropic environment. It is possible that the greatest contribution of rewilding thinking is to point to a future perspective of what nature could be. This statement, considered by many to be utopian, gains effectiveness in everyday changes implemented by conservationist initiatives in the field. Although the scale of impact seems microscopic when each specific action is probed, the rewilding initiative demonstrates incrementally that its mission is significant to big questions, like climate change. This is precisely because the objectives it incorporates are concrete and visible to nonexpert eyes. Rewilding Europe showcases a pragmatic approach to dealing with major, global issues, such as global warming and its disastrous consequences for the Earth's populations. Abandoning any interest in reconstructing the past, Rewilding Europe focuses on producing a future for nature that is better than the present scenario. Even if this means triggering images from the past, these only matter as long as they can be useful for planning the forthcoming landscapes.

The reintroduction of the concept of nature as an important category in the social sciences is also occurring at a good moment. Without returning to the ecomaterialist traditions that permeated anthropological theory from the second half of the twentieth century, we are currently observing the reinvention of nature as a concept that can adjust to innovations in the ethnographic field. Taking advantage of the rewilding spirit, we are entering a period in which a culture of creativity allows us to imagine possible futures for anthropology and for the planet.

Guilherme José da Silva e Sá holds a PhD in social anthropology (Graduate Program in Social Anthropology/National Museum of the Federal University of Rio de Janeiro). His main research interests are social anthropology, anthropology of science and technology, anthropology of collectives, human and nonhuman relationships, the nature and culture divide, ethology, intersubjectivity, anthropology of extraordinary experiences, and determinisms. He is an associate professor and researcher at the Department of Anthropology at the University of Brasília (Brazil), where he chairs the research group Laboratory of Anthropology, Science and Techniques (LACT). He was invited researcher at the Laboratoire d'Anthropologie Sociale at the Collège de France, Paris, in 2014. He is currently coordinator of the undergraduate degree in anthropology at the University of Brasília. He is a founding member and vice president of the Brazilian Association of Social Studies of Sciences and Technologies (ESOCITE-Br), a full member of the Brazilian Association of Anthropology, and a member of the Iberoamerican Anthropologists Network (AIBR) and the Société Internationale d'Echnologie et de Folklore (SIEF). His publications include the book *No Mesmo Ramo: Antropologia de Coletivos Humanos e Animais* (In the same branch: Anthropology of human and animal collectives) (Rio de Janeiro, 7Letras, 2013), which won the Marcel Roche Award for the best Latin American scientific work of 2014.

Notes

A preliminary version of this text was originally published in *Vibrant* 14(2) (2017).

1. This is the governmental office responsible for nature and biodiversity conservation administrated by the Portuguese state.
2. A private reserve for nature conservation covering 850 hectares, located between the counties of Pinhel and Figueira de Castelo Rodrigo in Portugal. It was founded in 2003 and is managed by the Transhumance and Nature Association. In 2011 it became part of the Rewilding Europe network, being one of the model rewilding areas in Europe.
3. The term "rewilding" refers to a process of "resavaging" or "renaturalization." I have chosen to preserve the idea of "renaturalization" in this chapter because it highlights the strong character of artificialization that exists within the dynamics of this environmental construction.
4. Guarda is one of the main cities in the region of Beira Alta Interior in Portugal.
5. These are stores where a miscellaneous range of things are sold, from domestic utensils to stationery and clothing.
6. Two notorious Portuguese sheepdog breeds.
7. The Breakthrough Institute is an environmental research center located in Oakland, California. Founded in 2003 by Michael Shellenberger and Ted Nordhaus, Breakthrough Institute has policy programs in energy and climate, economic growth and innovation, and conservation and development.
8. Singularitarianism is a movement defined by the belief that a technological singularity—the creation of superintelligence—will likely happen in the medium future and that deliberate action ought to be taken to ensure that the singularity benefits humans.

References

Bernardina, Sergio Dalla. 2011. *Le Retour du Prédateur: Mises en scène du sauvage dans la société post-rurale*. Rennes: Presses Universitaires des Rennes.
Bonneuil, Christophe, and Jean-Baptiste Fressoz. 2013. *L'Événement Anthropocene*. Paris: Éditions du Seuil.
Danowski, Déborah, and Eduardo Viveiros de Castro. 2014. *Há Mundo por Vir? Ensaio sobre os medos e os fins*. Florianópolis: Cultura e Barbárie.
Descola, Philippe. 2005. *Par-delà Nature et Culture*. Paris: Éditions Gallimard.
———. 2011. *L'Ecologie des Autres: L'anthropologie et la question de la nature*. Paris: Editions Quae.
Donlan, Josh. 2005. "Re-wilding North America." *Nature* 436: 913–14.
Donlan, Josh, J. Berger, C. E. Bock, J. H. Bock, D. A. Burney, J. A. Estes, D. Foreman, P. S. Martin, G. W. Roemer, F. A. Smith, M. E. Soulé, and H. W. Greene. 2006. "Pleistocene Rewilding: An Optimistic Agenda for Twenty-First Century Conservation." *American Naturalist* 168(5) (November): 660–81.
Estes, James A. 2002. "Then and Now." In *Aldo Leopold and the Ecological Conscience*, edited by R. L. Knight and S. Riedl, 60–71. New York: Oxford University Press.
Gonçalves, José Reginaldo Santos. 1996. "A obsessão pela cultura." In *Cultura no Plural*. Rio de Janeiro: CCBB.
Hache, Émilie, org. 2014. *De L'univers Clos au Monde Infini*. France: Éditions Dehors.
Kolbert, Elizabeth. 2015. *A Sexta Extinção: Uma história não natural*. Rio de Janeiro: Intrínseca.
Moore, Jason W. 2017. "The Capitalocene, Part I: On the Nature and Origins of Our Ecological Crisis." *Journal of Peasant Studies* 44(3): 594–630.
Soulé, Michael E., J. A. Estes, J. Berger, and C. M. Del Rio. 2003. "Ecological Effectiveness: Conservation Goals for Interactive Species." *Conservation Biology* 17: 1238–50.
Soulé, Michael E., J. A. Estes, B. Miller, and D. L. Honnold. 2005. "Strongly Interacting Species: Conservation Policy, Management, and Ethics." *BioScience* 55: 168–76.
Soulé, Michael E., and Reed Noss. 1998. "Rewilding and Biodiversity: Complementary Goals for Continental Conservation." *Wild Earth* 8: 19–28.
Stengers, Isabelle. 2015. *No Tempo das Catástrofes*. São Paulo: Cosac Naify.

Chapter 13

Emitting Inequity

The Sociopolitical Life of Anthropogenic Climate Change in Oaxaca, Mexico

Amanda Leppert and Roberto E. Barrios

Introduction

During the last century and a half, human industrialization practices have increased atmospheric concentration of carbon dioxide by 40 per-cent. This CO_2 increase has resulted in a rise in global average surface temperature by 1.4 degrees Fahrenheit, which has led to warming oceans, rising sea levels, and a decline of Arctic sea ice (National Academy of Sciences and the Royal Society 2014). In Mexico, the Secretariat of Environment and Natural Resources has cited the Intergovernmental Panel on Climate Change's (IPCC) fifth report, claiming that anthropogenic climate change has raised the temperature of oceanic waters surrounding the country by 0.79 degrees Fahrenheit in their first seventy-five meters of depth. Additionally, global warming has caused the formation of an area referred to as "the warm pool" close to Mexico's shorelines, where seawater temperatures average 78.8 to 80.6 degrees Fahrenheit (Ponce 2013).

Rising average global temperature, warming ocean waters, and melting ice caps are creating conditions for some parts of the planet that will cause a greater frequency of severe hydrometeorological hazards (e.g., hurricanes, cyclones, tornados, extreme precipitation levels) while triggering drought in other parts; these patterns will threaten human life, infrastructure, and food security. In other instances, rising sea levels, when combined with development-related coastal loss, will force populations to relocate, causing significant societal upheavals. Mexico's federal government has been internationally celebrated for its official recognition of anthropogenic climate change and its move to create a policy framework to mitigate it. In 2012, during the presidency of Felipe Calderón, the Mexican Congress approved the General Law of Climate Change and endorsed the creation of the National Institute of Ecology and Climate Change (INECC).

The General Law and INECC joined existing efforts to promote renewable energy megadevelopment projects that would both help to reduce CO_2 emissions and act as an economic booster in historically marginalized areas. These mitigation and development projects were conceived and implemented through partnerships between federal and state government agencies and multinational energy companies that featured capital investments from Mexican and European players. Furthermore, the projects were conceived within a capitalist framework in which their success was primarily measured through their ability to replicate financial capital, thereby exemplifying green neoliberalism.

In this chapter, we show that, while anthropogenic climate change is a tangible and empirically observable phenomenon (what we call its material life), it is also something that has a sociopolitical life—that is, the varying ways people define climate change as a problem and imagine responses to it. What is more, while anthropogenic climate change's material life has deleterious environmental and societal impacts, so might its sociopolitical life, depending on who is imagining responses to it and how. We also show how ethnographic methods provide a means of documenting the latter and devising policy recommendations that may mitigate these undesirable secondary effects of climate change mitigation. Additionally, we explore how the ethnography of climate change mitigation programs provides data that helps us test a number of assumptions of social theory concerning risk and epistemic and social change.

Background: Climate Change and the Social Theory of Governance

Over the course of several publications, Michel Foucault (1978, 2004) made the case that, beginning in the eighteenth century, Western Europe witnessed a transformation in the ways people imagined the responsibilities of government and sovereign power. Foucault's argument went something like this: unlike the Middle Ages, when sovereigns considered the upkeep, protection, and growth of their estates as their primary responsibility, the late eighteenth century saw the emergence of the care of human populations as biologically living entities as the primary preoccupation of monarchs and governments. This shift in the sovereign's object of concern enabled the rise of a collection of "sciences of man" (e.g., public health, economics, urban planning, anthropology) that focused on the creation of a milieu where human populations could thrive as primarily biological and economic beings (Foucault 2004; Rabinow 2005). Foucault named this emergent modality of power "biopolitics."

The emergence of "man" as an object of concern of the human sciences was roughly contemporaneous with the development of economic liberalism and modern epistemology; the latter being a technique of knowledge making that, according to its practitioners, allowed them to objectively engage the material world by separating objects (things in and of themselves) from subjects (cultural values). Many proponents of modern epistemology claimed it was a mechanism of knowledge production that surpassed "nonmodern" or "primitive" epistemologies whose access to facts and objective reality was inhibited by culturally specific beliefs. An example of ways of knowing and relating to the material world that have been called "nonmodern" is the manner in which many Australian Aboriginal people have historically spoken about human-environment relationships (Povinelli 1995). In the Aboriginal perspective, animals and prominent features of the landscape such as water holes are considered to be sentient and capable of communicating with a divine force called the Dreaming, which precedes the creation of the world and humans. In Aboriginal epistemology, people must relate to the world according to specific ritual and ethical prescriptions, and their failure to do so may be communicated by nonhumans to the Dreaming, which may then enact retribution against people for their transgressions (Povinelli 1995).

Anthropologists have often classified epistemological practices like those of Australian Aborigines as animism, and the anthropological record suggests that similar (but also locality-contingent) ways of speaking about and relating to the material world existed among pre-Columbian populations in the Americas. As far back as 1800 BCE, Mesoamerican (the culture area that manifested across much of today's Mexico and Central America) people upheld the view that the land was part of a supernatural creature that combined reptilian, avian, and feline features, and this mythological "earth monster" also acted as a portal to the spiritual realm of deified ancestors. More recent anthropological case studies of indigenous communities involved in renewable energy projects in Mexico demonstrate that, in certain instances, indigenous people continue to consider deities to be the ultimate owners of the land (Cruz Rueda 2013). Just as in the case of Australian Aborigines, there is a robust ethnographic record that demonstrates Mesoamerican people related to—and, in some instances, continue to relate to—the material landscape not as a thing in itself but as something that was simultaneously material and sacred that required the observation of ritual and ethics when interacting with it.

From the perspective of modern epistemology, the ways Australian Aborigines and pre-Columbian Mesoamericans relate to their environments are expressions of culturally particular beliefs but not matters of fact that transcend their cultural context. A critical reader may suggest that we are conflating cosmology and epistemology, and that the ways indigenous populations and modernist thinkers engage and interpret their surroundings are qualitatively different in a number of key regards. Our argument here, however, is that while practitioners of modern epistemology claim to be able to see the world objectively, they are, in fact, doing things very similar to what Australian Aborigines do when they speak about their human environment relationships; that is, connecting elements of the material world with culturally specific values, meanings, and ethics (see Latour 1993). The power of modern epistemology, then, lies not in being different or superior from "nonmodern" ways of seeing (i.e., cosmologies), speaking about, and relating to the world but in doing the same things while claiming not to (Haraway 1997; Latour 1993). As a number of social scientists and philosophers have observed, cultural values enter the spaces of modern knowledge making (e.g., laboratories) in the form of gender roles, capitalist cost-benefit analyses, copyright laws, ethical guidelines, legal restrictions on experiments, strategic interests of national funding agencies, and local cultures of scientific knowledge making (Franklin 2005; Haraway 1997; Pickering 1995; Latour 1993).

Modernist epistemology is kindred to economic liberalism, which promotes the obliteration of cultural value systems involved in "nonmodern" human-environment relationships in order to make "natural resources" (things in themselves) available for exploitation, extraction, and eventual destruction in the industrial production process. For some anthropologists involved in studies of human ecologies, the emergence of modernist epistemology—which is traced to Robert Boyle's development of the scientific method (Shapin and Shaffer 1985)—is a pivotal moment in the history of human-environment relationships (Descola 2017). Having said this, we also want to clarify that we do not intend to reiterate a narrative that romanticizes "nonmodern" Mesoamericans and Australians as "noble savages" or "ecological Indians." In Mesoamerica, there are examples of historical moments when indigenous populations overtaxed their environments, leading to periods of socioenvironmental crisis. Such was the case of the Central Maya Lowlands in the tenth century CE. Furthermore, Mesoamerican state societies of the second through fifteenth century CE featured economies where material goods were produced and traded at a large scale in market networks that extended from North to South America. At the same time, not all indigenous communities engaged in the same practices of state and economy building, and, in some cases of socioenvironmental collapse (as in the Central Maya Lowlands), the archaeological record shows a transformation of social organization toward village-level life that was less taxing of surrounding environments.

The era of European colonial expansion laid the groundwork for the dramatic global environmental transformations that economic liberalism and modern epistemology made possible. In the region that is today's Mexico, sixteenth-century colonization involved the radical transformation of agricultural production systems and environmental stewardship practices on the part of the indigenous state societies and communities that populated the area (Carmack 2006; García-Acosta 2002, 2018). In the Valley of Mexico and Mesoamerica in general, Iberian colonizers displaced indigenous populations from agriculturally productive lands, interrupted indigenous watershed management systems, introduced cattle ranching, and forced indigenous communities to pay tribute and donate labor to the large estates of newly arrived Iberians (García-Acosta 2002, 2018). The result was a socioenvironmental cataclysm that claimed up to 90 percent of the indigenous population in the Americas during the first half of the sixteenth century. Following the end of the colonial era, Mexico and Central America would witness the introduction of liberal reforms during the late nineteenth century. These reforms were intended to drive economic development by incentivizing the exploitation of indigenous labor and the large-scale cultivation of export crops whose international trade would plug the region into a growing global network of capitalist extraction, production, and circulation (Dore 2006).

Modern epistemology and economic liberalism promoted the dismissal of "nonmodern" human environment relationships that prevented the large-scale extraction of natural resources and the reorganization of communities and populations for the production of export crops and, later on, industrialization. It was this "unleashing" of natural and human resources (i.e., human labor power) from the burden of "traditional" thinking that enabled the environmental destruction (species extinction, deforestation, toxic pollution, rising average global temperatures) that some scholars and environmental activists have termed the Anthropocene (Descola 2017; García-Acosta 2017).

Another scholarly concern related to our analysis is the transition of biopolitical states from what Ulrich Beck called scarcity societies into risk societies during the nineteenth and twentieth centuries (Beck 1992; Collier and Lakoff 2015). As we noted above, the biopolitical state considered the care of the biologically living human population to be its primary preoccupation. Furthermore, economic liberalism came to be seen as one of the key mechanisms through which to provide the resources and commodities necessary to nurture biopolitical societies (Smith 1999). According to

Beck, the key preoccupation of biopolitical states during the nineteenth and early twentieth centuries was the distribution of scarcity. By scarcity distribution, Beck meant the making of decisions dealing with the distribution of limited resources and the means to access these resources (e.g., food, shelter, wages). Through the combination of technological "innovation" (e.g., the green revolution, Fordian mass production) and strategic alliances between "first, second, and third world" nations for the procurement of natural resources—which often involved the conscious dispossession, marginalization, and exploitation of local subaltern populations—Western European nations and the United States managed to create a situation where scarcity became less and less a concern of the majority of their populations. The era of high modernity had arrived.

In the mid-nineteenth century, modern epistemology and economic liberalism seemed to be making good on their promises (at least for the portion of the global population that benefitted from them), but there would soon be signs that something was amiss. Atomic energy, for example, was a technoscientific innovation that the modern biopolitical nation-state made necessary, and which seemed boundless in its application (Masco 2006). The atomic bomb offered to protect the national population from foreign aggression through the threat of mutually assured destruction, while nuclear power production seemed to provide limitless energy to drive industrialization. Nevertheless, physicists involved in open-air atomic testing eventually discovered the effects of radioactive fallout, giving rise to antinuclear proliferation movements (Masco 2006). In the early twentieth century, prior to the development of atomic power, hydrocarbon extraction and consumption also seemed to offer a vast source of energy to drive industrialization and satisfy the needs of scarcity societies, but the global environmental monitoring technologies and consciousness sparked by nuclear proliferation also allowed scientists and the public to begin to recognize the effects of carbon dioxide emissions on the planet's climate (Masco 2009).

In a similar manner, the green revolution promised to eradicate world hunger (another key concern of the scarcity society) through the development of chemical fertilizers, pesticides, genetically modified crops, and industrialized agriculture. However, the public would eventually be forced to wrestle with the dangers some of these technologies posed to human health and environments: toxicity, reduced soil fertility, decreased biodiversity (Fortun 2001). The unexpected consequences of the technologies and practices that economic liberalism deemed necessary and that modern epistemology made possible brought about a transition from the scarcity society to the risk society (Beck 1992). In the risk society, Beck argued, concerns with the distribution of scarcity became overshadowed by concerns with the distribution of risk; that is: who should suffer the effects (toxicity, radiation, and now climate change) of industrialization and modernization? Beck, of course, was writing about the risk society in the mid- to late 1980s, before anthropogenic climate change became a matter of extensive academic and public concern. However, we find it reasonable to list anthropogenic climate change as one of modernity's unexpected emerging risks.

But Beck's preoccupations about risk were not without hopeful hypothesizing for the future. One of the central arguments of *Risk Society* was that, because the risks of modernization (toxicity, radioactivity) were not restricted by national borders, the risk society would give rise to a new social movement—which he referred to as a novel form of cosmopolitanism—that cut across lines of class, race, and national identity as people mobilized to address the socioenvironmental challenges of the time. It is in this optimistic theorizing that we also see another convergence between Beck and Foucault. While Beck anticipated the emergence of a transnational environmental cosmopolitanism, Foucault limited himself to hoping that the epistemic object of "man," which he saw as the culprit of biopolitical violence (i.e., wars in the name of nationalist causes, the sacrifices of people, livelihoods, subjectivities, and nonmodern epistemologies for the sake of modernization and biopolitical well-being) would one day vanish and be replaced with another less virulent organizing *logos* (Foucault 1970).

Mexico and Anthropogenic Climate Change

In what follows, we examine how public-private partnerships between state agencies and renewable energy companies organized as a response to anthropogenic climate change in Northern Latin America (the Isthmus of Tehuantepec in Oaxaca, Mexico, to be exact) provide us with a fruitful context for testing Foucault's and Beck's hopeful hypotheses concerning cosmopolitan environmentalisms and the vanishing and emergence of epistemic objects (*logoi*). We make the case that anthropogenic climate change, in addition to being a scientifically documentable material phenomenon, is something that also has a sociopolitical life; moreover, as part of this latter life, it exists as a phenomenon of discourse and the imagination and, in this latter dimension, takes shape as a situated cultural form subject to interpretation and reconfiguration. Furthermore, the imaginary and discursive manifestations of anthropogenic climate change are not separate from the realm of social theory. Key thinkers such as Beck and Foucault, who were primarily concerned with crisis and modernization (i.e., biopolitics, scarcity and risk societies), imagined specific hopeful futures for humanity whose realization, we insist, should not be treated as a certainty but as tenuous hypotheses that must be tested against the ethnographic record of how people, corporations, and governments are currently responding to global environmental change.

The analysis that follows focuses specifically on a small Zapotec town in the outskirts of the city of Juchitán, Oaxaca, Mexico, where a number of multinational renewable energy companies have worked in collaboration with the Mexican government to develop a number of large-scale wind turbine energy production projects. We have chosen to rename this town with the pseudonym Binnizá, which is the Isthmus Zapotec name for their people, to protect the identities of our interlocutors. In this particular case, we show how, in the context of anthropogenic climate change in Northern Latin America, responses to the risks engendered by modernization are neither leading to the emergence of a transnational environmentalist cosmopolitanism nor featuring a vanishing of biopolitical *logoi,* whether that be "man" or capital. Instead, our ethnographic research indicates that this particular response to anthropogenic climate change is having the effect of increasing socioeconomic inequity among populations directly affected by mitigation programs such as renewable energy megadevelopment projects. Furthermore, our data demonstrates how such megadevelopment projects do not occur within a historical vacuum but manifest in a context of colonial and postcolonial governance that is permeated with overtones of ethnicized and classist discrimination. Finally, in the case of Mexico, large-scale private-public partnerships meant to respond to anthropogenic climate change do not feature a vanishing of "man" as an epistemic object or a dismantling of the biopolitical state. Instead, what we see is a continuation of large-scale energy production whose biproduct is not radioactivity or carbon dioxide but ethnicized inequity. While the scarcity society may have evolved into the risk society, we close by arguing that the risk society is currently morphing into the inequity society at the loci of renewable energy megadevelopment.

Before the Turbines

The Isthmus of Tehuantepec is no stranger to development projects or to external economic forces. Since the colonial era, the region has witnessed multiple attempts on the part of imperial powers and national governments to incorporate it within broader political economic networks. Since the sixteenth century, this part of Oaxaca has been subjected to mining operations focused on mineral extraction, sugar cane plantations, oil drilling operations, hydroelectric construction projects, and now wind turbine installations. Throughout these projects, *Ismeños*—a term used to denote the predominantly indigenous people of the isthmus—have strategically participated in these programs with the intention of maintaining their autonomy and promoting their upward social mobility (Cruz Rueda 2011, 2013; Dunlap 2017; Sellwood and Valdivia 2018).

In the 1990s, the isthmus experienced a series of World Bank, International Monetary Fund (IMF), and US-supported structural adjustment programs and the modification of Article 27 of the Mexican Constitution. The latter altered agrarian land tenure laws to permit the privatization of communal lands known as *ejidos*. The objective of these changes was to make land available for the reproduction of foreign capital (from either other regions of Mexico or international investors) and to transform communal agricultural producers into sources of labor for other industries (Weaver et al. 2012). The structural adjustment economic policies also enforced governmental disinvestment from social programs, deregulation of domestic labor markets, and the privatization of most state-run industries and *ejido* lands under the justification that such measures would increase competitiveness of commodities in the market and minimize government expenditure (Weaver et al. 2012). The beneficial effects of structural adjustment and the neoliberalization process, however, have not manifested uniformly across communities or among all members of communities in the isthmus.

Not only did structural adjustment programs impose ideas about the "best" economic structure or the role of government, they also (re)created inequitable relationships between regions, especially with regards to commodity production and labor. With the signing of the North American Free Trade Agreement (NAFTA), for example, the United States became a destination of agricultural products whose export standards could not be met by many small-scale subsistence farmers, effectively excluding them from emerging opportunities for agricultural development. With diminished subsidies to support subsistence farmers, such as the removal of the state-guaranteed price of corn, many saw themselves forced to abandon agriculture and enter the nonagricultural labor force. Unskilled workers and the subsistence and indigenous communities—the same groups prioritized for green neoliberal development—appear to have been affected the most (Nahmad 2012; Babb 2005).

Nevertheless, the current transnational green neoliberal political economy within which renewable energy megadevelopment projects are implemented differs significantly from these antecedent economic transformations. Although the previous projects were also crafted on the basis of neoliberal ideas about development (i.e., promoting concepts such as individualism, enhancing ability for private investment by allowing the privatization of communal lands with NAFTA, and deregulating protections of the agricultural or mining sectors), many wind turbine projects are sponsored by private international companies that work within the United Nations Framework Convention on Climate Change (UNFCCC) Clean Development Mechanism to incentivize and commodify "green" energy. In this instance, *Ismeños* find themselves in a position where they must negotiate directly with transnational renewable energy companies, something they had not done before during the structural adjustment programs of the 1990s. Official representations of such direct interactions invoke images of equitable partnerships between local communities and energy companies, but our ethnographic experiences indicate that this is not the case and that large power differentials loom over green development encounters. Below, we demonstrate that, unlike the energy produced by wind turbines, the inequities in question do not manifest out of thin air but, rather, are the legacy of colonialism and postcoloniality in the Isthmus of Tehuantepec.

The Struggle for Equity and the Imagined Subjects of Green Development

Chester Karrass earned an MBA from Columbia University and touts himself as the creator of "the most successful negotiation seminar in the United States." On the website he uses to promote his business, he claims forty-five years of experience delivering seminars that arm their participants with an unmatchable "negotiation arsenal" and professes to have trained more than one million professionals, including "salespeople, buyers, corporate leaders, managers, engineers, financial officers, CEOs, and international business people." Karrass would be inconsequential to our discussion of aeolian projects in Mexico were it not for the fact that he wrote a book titled *"In Business as in Life—You Don't Get What You Deserve You Get What You Negotiate."* We begin this section with this reference to Karrass because his seminars and book title propagate a fantasy about business, development, and negotiation in which all people, regardless of gender, class, ethnicity, and race can become equally competent negotiators. The case of Binnizá, on the other hand, shows how colonial and postcolonial legacies of ethnicization and racialization of indigenous communities continue to haunt green megadevelopment projects meant to mitigate climate change.

In 1990, Mexico ratified the International Labor Organization's Indigenous and Tribal Peoples Convention, 1989 (no. 169). The purpose of the convention was to guarantee the right of indigenous communities to assume control of their own institutions, ways of life, and economic development and to maintain and strengthen their identities, languages, and religion (International Labor Organization 1989). The convention recognized that, in many parts of the world, indigenous communities do not enjoy fundamental human rights to the level guaranteed to other sectors of the population of the nation-states they live within and that their own laws, values, and customs have often deteriorated. Article 15 of the convention indicates that rights of indigenous peoples to the natural resources of their lands are to be protected, and that these rights include participation of indigenous communities in the use, administration, and conservation of such resources. Furthermore, in the case where nation-states maintain mineral rights or rights over other resources present on indigenous lands, state governments must establish or maintain procedures focused on the consultation of the affected indigenous communities with the ends of determining whether and how their interests will be negatively affected prior to authorizing any prospecting or exploitation project on their lands. Finally, affected indigenous people must partake of development and extraction projects and receive equitable compensation for any harm suffered as a result of such activities (International Labor Organization 1989).

The establishment of large-scale transnational investment in wind turbine farms in the Isthmus of Tehuantepec dates to the first decade of the twenty-first century, and Binnizá became involved in this megadevelopment program once again in 2014 when it was approached by Eólica del Sur (a multinational renewable energy company supported by Japanese and Mexican investors) for the construction of wind turbines distributed over a space of two thousand hectares. Because Binnizá is recognized as a Zapotec indigenous town and because of Mexico's ratification of Convention 169, negotiations between Eólica del Sur and Binnizáleños had to be conducted via a consultation process. The consultation involved the creation of a council that featured representatives of all three levels of Mexican government (federal, state, and local), energy companies, and local property owners. While Article 15 of Convention 169 is designed to guard over the rights of indigenous communities in the execution of extractive projects, state and energy company officials proceeded in a manner that focused on the rights and compensation of individual landowners, not the community as a whole. The company intended to financially compensate only those people upon whose land wind turbines were installed, effectively excluding both nonlandowning Binnizáleños and landowners whose lands were not directly affected by the project.

Social inequities were certainly present in Binnizá prior to the execution of the wind turbine project, but property owners were apprehensive about the social tensions the inequitably distributed influx of wealth would have upon the town's sociopolitical life. Consequently, landowners used the consultation process to request that state and energy company officials add a community benefits packet to their compensation. Among their requests, they asked that their community health center be updated with the purchasing of diagnostic equipment to help facilitate and expedite town residents' navigation of the national healthcare system, the updating of the town's water drainage system, and the yearly donation of two million pesos to community education, culture, health, and sports programs. State and company authorities agreed to these requests, and the council mandated the creation of a follow-up and monitoring committee to ensure that all parties met their obligations. However, government officials later scrapped the plan to place more diagnostic equipment in the community health center because the community did not have "sufficient need" for such an investment as determined by the low levels of community members on the state-sponsored insurance program Seguro Popular.

The completion of the initial consultation process, however, was not without tension. Binnizáleños involved in the negotiations like Don Eugenio Sanchez recall numerous instances when negotiations between Eólica del Sur and property owners broke down, leading to protests and blockades that impeded access of energy company staff and workers to the project site. The first of these instances was when Binnizá residents discovered that, after agreeing to a specific compensation per wind turbine installed on their lands, Eólica del Sur was paying a much higher sum to property owners in nearby Juchitán. Don Sanchez recollects:

> We accepted the original terms because we wanted to be part of the project. But when we saw what the property owners of Juchitán had asked for, we said, "Well, we were either dumb or we want this to at least have a little business," but in reality, we're not gaining very much. So we decided to protest, and the council representatives would not come. So we complained to the governor, we complained with everyone, and they would not pay any attention to us, but we kept going and going. When they felt pressure because we took action through different means [blocking access to the project site], they finally came. The president of the council came, who was another representative of the governor, the company, the secretariat of SEMARNAT [Secretariat of Environment and Natural Resources], and two or three more came from human rights . . . they came to tell us that, because we had already said that what they had given to Juchitán they had to give us as well, that it didn't matter, it was not a part of the original agreement, but we believed we could come to a new agreement, that was what the council was for, to protect us from the company, to make sure negotiations took place on a neutral ground. "It is wrong," we said, "you are giving forty over there and here you are giving five here, you have to give forty here as well." (Ethnographic interview, 2019)

In the preceding interview excerpt, Don Eugenio recounts how Binnizá's residents felt betrayed by state representatives who staffed the consultation council since Binnizáleños expected them to look after the interests of indigenous community members and to ensure that all negotiating actors engaged one another on a level playing field. Instead, state representatives who were privy to the agreements reached in nearby Juchitán knew very well just how much the energy company was willing to give but left Binnizáleños to their own devices during the negotiation process, which resulted in their settling for much less than they could have. Once Binnizáleños discovered just how much Juchiteco landowners had gained, they attempted a number of strategies to renegotiate their compensation, their idea of fairness being based on the notion that all should receive equal compensation in both Juchitán and Binnizá.

Binnizáleños' initial attempts to resolve the matter amicably were ignored by state authorities, forcing them to take more drastic means like blocking the roads leading to the project site and bringing the installation of wind turbines to a halt. Although Binnizáleños were successful in renegotiating their agreement, they felt the process was not managed in the way stipulated by Convention 169. They felt state representatives were looking out after the interests of the company and not their community. Furthermore, the negotiation process was not one of partnership between the energy company and the community. Instead, the process was organized around the antagonistic principle of "you get what you negotiate," leaving communities to individually battle both the company and state officials for compensation.

The struggle to gain compensation equal to that of Juchitán, however, was not the end of Binnizáleños' tensions with the power company and state government. The installation of wind turbines was to take place on agricultural lands, which required the company to pay a tax to the town's local government for the change of soil use from agricultural to industrial. The municipal president at the time negotiated an amount that landowners considered dismal (3.5 million pesos). The project was deployed over two thousand hectares of land, but the president and company agreed that a tax would only be paid for each eight-square-meter area covered by the bases of the turbines. Once again, the landowners attempted to amicably resolve this matter with state government and the company through talks, but they were once again ignored, pushing them to carry out yet another blockade and halt the construction project. In this particular instance, their efforts proved futile, and the tax amount was not changed.

In the years that have followed the initial consultation, landowners have experienced unexpected impacts of the wind turbine project. The installation of turbines required the building of raised roads that dissected agricultural fields and disrupted water drainage and irrigation patterns. Some fields now flood, while others receive insufficient amounts of water. Landowners also feel they were misled about how much noise the wind turbines would actually produce. Company representatives claimed the wind turbines would make no more noise than a house fan, but landowners disagree. Other complaints on the part of landowners include the incorrect reinstallation of fences that were taken down during wind turbine project and the inadequate replacement of native varieties of trees with cheaper non-native species.

The Temporality of Consultation

From the perspective of Binnizáleños, the process of consultation should be open-ended, allowing them to renegotiate with the state and energy company, especially when unexpected environmental impacts like the disruption of irrigation systems manifest themselves. From the perspective of company and state representatives, the temporality of the consultation is much more limited and ends with the agreement that precedes the initiation of wind turbine construction and installation activities.

In summer 2019, this difference of opinion on the temporal parameters of consultation came to a head during a meeting of the monitoring and evaluation committee that included state and company representatives and landowners. On this occasion, government representatives ran late to the meeting, allowing for landowners to begin an informal conversation with company representatives about their concerns with environmental impacts (disruption of irrigation systems, wind turbine noise, inappropriate replacement of fences and trees) that were not taken into consideration during the initial consultation. Additionally, landowners wanted to request the completion of a baseline soil study to help monitor changes in their lands over the coming decades. At the beginning of the meeting, landowners reintroduced these problems in front of the committee, also demanding an update on a previous agreement allocating state, federal, and company investment for the local outdated drainage system. State representatives responded that these issues were outside those listed in the original consultation, the consultation process had ended with the signed agreement between the landowners and the company, and that it was not to be extended after that, to which the landowners replied that the consultation process had prescribed the creation of a monitoring and evaluation committee that would continue to operate throughout the life of the project, and that these were matters of such a committee's concern.

The landowners requested time be allotted within the meeting's agenda to discuss their concerns, but the representative of the Secretariat of Energy (SENER) objected that such an alteration of the agenda was illegal and corrupt, leading to the property owners walking out of the meeting. With the landowners gone, company and state representatives exchanged stories that depicted Binnizáleños as irrational, corrupt, and inhumane, and complained about locals impeding progress for the state, the nation, and the region. The first story told by a company engineer detailed the case of a landowner from Juchitán who demanded his son be given a job he was not qualified for at the energy company. When the company refused to hire the landowner's son, he blocked the company's access to his land, delaying the installation of some wind turbines. While the company could still conduct its installations on other lands, they opted to hire the landowner's son anyway to diminish tensions with the locals. The point the company engineer wanted to make was that locals were nepotistic, corrupt, and irrational, but it is noteworthy that he used a story about a landowner in Juchitán to express his frustration with landowners in Binnizá, where residents pride themselves in being different from Juchitecos.

A second story told by the engineer did feature events that unfolded in Binnizá. In this second case, the story focused on one of the blockades organized by landowners to exert pressure on the company and state to renegotiate the consultation agreement. The landowners strategically blockaded a substation where energy is fed from the turbines before being exported off-site. There were four company workers at the substation, and one of them was the son of a local landowner. The company knew of a way out of the property that could avoid the blockade, but they did not want the landowner's son to know which route they were taking. Company managers decided to evacuate the other three workers but chose to leave the landowner's son behind for fear he would later reveal their route. The landowner's son was forced to spend four days by himself at the substation without proper sanitation or living conditions. For the company engineer, this case illustrated the inhumanity of Binnizáleños. It is interesting, however, that the engineer chose to pass moral judgement on Binnizáleños for this worker's hardship and not on the company that refused to negotiate with the landowners or ensure the well-being of all its workers equally, regardless of their place of provenance.

After the meeting, Pablo Torres, one of the Binnizáleño landowners, offered his interpretation of what had occurred. He claimed the reason state officials denied their request and accused them of corruption was because the property owners showed up in their traditional sandals and not in suits. Thus, he called attention to the way indigeneity is indexed in this part of Mexico through dress and bodily dispositions and how Binnizáleños' refusal to subject themselves to hegemonic mestizo ideologies of professionalization led the officials to dismiss their requests for further environmental monitoring.

Development and Inequity

In the preceding ethnographic examples, we have shown how the wind turbine project (designed as a means to mitigate climate change) threatened to exacerbate inequities between communities. For example, the attempt to negotiate different compensation packages for Juchitecos and Binnizáleños would have had the effect of enriching some landowners in the region much more than others. Binnizáleños, in contrast, advocated for equal treatment among landowners across communities, insisting that Juchitecos and Binnizáleños should receive equal compensation. But development projects (even ones with the beneficial purpose of reducing carbon emissions), as they have been conducted in the isthmus, promote inequity not only between communities but among community members themselves. Binnizáleños are acutely weary of the creation of inequity, and although they are not opposed to seeking the improvement of their families' standard of living, they also recognize that the gross socioeconomic disparities create sociopolitical tensions. Consequently, Binnizáleños constantly struggle with the tension between personal socioeconomic advancement and the social deterioration that pronounced inequities create.

Binnizáleños often trace the rise of prosperity in their town to the late 1960s and 1970s, but they also recognize how such prosperity increased inequity. During this time, hydroelectric projects were bringing state investment into the region, and a significant number of Binnizáleños began to seek socioeconomic advancement through education. Originally a Zapotec-speaking town, many Binnizáleños experienced discrimination and suppression of their native language in public schools. Nevertheless, many town residents successfully navigated the educational system, became professionals, and brought a positive reputation to the town as a place of learned people. However, one of the deleterious impacts of this rise to intellectual prominence was the abandonment of the Zapotec language, which many Binnizáleños lament today. Over the course of an ethnographic interview, Don Oscar Ramos, the municipal president, explained how relationships among community members changed over the years:

> Look, Binnizá is a very young town. If you compare us to Ixtaltepec, Juchitán, with other towns around Binnizá, it is a distinctly young town. Binnizá had a

> very big development in the last thirty years. A lot of development. The previous generations were dedicated to agricultural labor, to planting, to raising cattle, that's what they were dedicated to. But there was a generation of Binnizáleños who left to study and returned with knowledge, with preparation, and thus began to organize the people. And, for example, the baseball stadium, the sports units, the streets, the drainage system, it was done through *tequios* [indigenous communal labor practices], because there were not many resources. Then, they would arrive prepared and say, "Let's get people together, let's do this," and people would cooperate more. Now when resources [money] are involved . . . because there are resources now, that's the conflict. Many say, "Ah well, if there is money, I have a construction company friend." Right now, we have twenty million pesos that we can't use because [the committees of property owners] are fighting over it. (Recorded interview, July 2019, translated from Spanish by author)

Don Oscar, like several other town residents, stressed the negative impact of economic incentive. Describing how tensions arise among Binnizáleños over how to use the social resources from the turbine project, he juxtaposes how projects were executed through the institution of *tequio* in the past and vocalizes a sense of social disintegration that he fears has occurred over time. Before, he stresses, Binnizá was more "united, without knowledge, but they were united. . . . They went to study and came back with knowledge, so they continued to do *tequios,* but now with the generated knowledge" (recorded interview, July 2019, translated from Spanish by author).

The term *tequio* refers to a form of communal volunteer labor characteristic to the isthmus that was nostalgically invoked by many Binnizáleños when comparing how things were in the past to how things are today in the town's consumer economy. People would congregate upon request for a given period of time to complete a task, whether it be preparing for a party or wedding, building a house, or undertaking larger projects for the community, without economic compensation—although food may be shared as a token of gratitude for their work. Most importantly, many Binnizáleños considered participation in *tequio* as a key ethnic identity marker.

During the late 1960s and 1970s, when the first generation of Binnizáleños left to study, the federal government made obtaining a secondary level of education mandatory. This was not enforced in many parts of the isthmus, as children were needed to help their households with agricultural labor. In Binnizá, however, the requirement was enforced, and parents who did not send their children to school were liable to legal penalization. Unfortunately, once at school, Binnizáleños were exposed to negative conceptualizations of indigeneity. Specifically, speaking Zapotec was banned, and corporal punishment was used to enforce this rule.

Spanish was promoted as the primary language, and Binnizáleños struggled with the ability to be able to fully express themselves in the hegemonic language. Don Ricardo Peralta, a Binnizáleño who lived through this, explained:

> This is what we struggled with in school, we couldn't express ourselves what we knew, those who didn't know Spanish. Little by little, this gave, from the studies, the school, there were these changes. (Recorded interview, July 2019, translated from Spanish by author)

Consequently, the succeeding generation of Binnizáleños did not teach their children Zapotec and instead promoted Spanish within their own households. Those who went on to receive higher education were further pressured to "act like a professional," a phrase that was permeated with ethnocentric assumptions that portrayed mestizo cultural practices as professional attributes. These hierarchies of ethnicity and culture, and the trauma that accompanied them, became the backdrop of power relations against which the negotiation of climate change mitigation programs like the wind turbine project took place.

When individuals left to receive a formalized education, they were seen as bringing "modern knowledge" back to the community, allowing them to progress, "or catch up to" the rest of the nation, in a socioeconomic sense. Indeed, in past development projects, the community remembers not being able to defend themselves as equals among developers to receive the best benefit for their community, because they lacked this "knowledge" of both "how things should be done" and how to do it. The municipal president explained:

> There is a characteristic of Binnizá, in that it has developed a lot. [Binnizá] is very interested that its people prepare themselves. We have many people who are very prepared here in Binnizá, who are in many parts of Mexico with good foreign education as well. People who went abroad to study as well. We have researchers abroad who are originally from Binnizá, so Binnizá is characterized by this . . . being a town of many professionals [as compared to other towns in the Isthmus] . . . but this was created by a generation of Binnizáleños who were attracted to go out and study. (Recorded interview, July 2019, translated from Spanish by author)

During the 1960s and 1970s, Mexican state officials considered this part of the isthmus to be a backward area mired in indigenous "traditionality" and hoped to formally incorporate it into a modern capitalist consumer economy. Therefore, not only did students learn how to benefit from the national market economy as professionals, they also attempted to counteract the ways the region had been historically neglected by the Mexican federal government. Combining their newfound knowledge of how to develop infrastructure with a sense of communal obligation, they worked to make Binnizá the center of "modernity" in the isthmus, which accounts for a willingness of town residents to participate in climate change mitigation programs that have global relevance. However, this process of grassroots modernization also propagated deleterious ideas about indigeneity, influencing many families to suppress Zapotec language as well as several other activities considered indexical of indigeneity (e.g., agricultural livelihoods, foodways). Therefore, subsistence agriculture was traded for cash crops among property owners, and being a *campesino* (person of the fields) came to be seen as a livelihood strategy of the past.

More recently, following indigenous revitalization movements over the last decades, Binnizáleños have worked to redefine professionalization in a way that does not exclude indigenous identity markers. Some town residents focus on the resurgence of the Zapotec language as a key to accomplishing this. As Don Ricardo Mora explained:

> Ah no, it's bad! I want my grandchildren to speak Zapotec. Well, since we lived in a different situation [where speaking Zapotec was denigrated], and for example, if you go to a town close to here, and see the little ones, they are speaking in Zapotec! And to see them, how delightful to hear them! [smiles wide] And [the young Binnizáleños] do not want to.

Don Ricardo is not the only one to feel this way. In fact, some Binnizáleños see Zapotec language suppression as violence against their community. Alberto Rosas, a schoolteacher, explained that, although these social changes began with the agro-industrial transformation of the 1960s and 1970s, they were exacerbated by the current wind turbine projects that enhanced inequity through the exclusion of non-landowners from the lion's share of the compensation money that flooded the region. He describes how those who have disproportionately benefitted from the wind turbine projects manifest a sense of superiority:

> And now the people who work in the wind farms, that generation of the wind turbines, is a generation of the worst of the worst, by the simple fact of working there, they feel great, arrogant. The kids come back to be drug addicts, alcoholics. Vanity, it's a terrible life. Without values, principles, and nothing, they live off wealth, and they come with the illusion they are superior to us, for the simple fact of working [for the energy companies], but there is no identity, they have no identities. (Recorded interview, July 2019, translated from Spanish by author)

When asked why the wind turbine projects have produced this kind of entitlement, Don Alberto explains that it is because the project's benefits are only enjoyed by a small sector of Binnizáleños and that social gaps have grown in the context of green megadevelopment. He insists that the growth of inequity occurs because the turbine projects do not take the social into account and do not focus on the creation of generalized well-being. For example, focusing primarily on the compensation of property owners effectively relegates a large swath of the population to a subordinated socioeconomic status. Don Alberto sees the social transformations beginning in the 1970s as intimately related to what is manifesting currently. For the good of his community, he is adamant that the social must be attended to in green megadevelopment projects and that communal well-being must also be taken into account if these projects are to be beneficial for all.

Although landowners attempted to mitigate the inequity created by the wind turbine project through the creation of a community benefits packet, this packet does not counterbalance the notable increase in revenue they are experiencing in comparison to other community members. Furthermore, the socioeconomic transformation brought about by the turbine projects is bringing new social pressures that are driving further cultural change. The school system, for example, has now required students to learn English as a foreign language that will allow them to be competitive in the eco energy economy. For those looking to become involved in the green energy projects, learning English is desirable because it is the language used for instruction manuals. Scarú Aguayo, a senior in high school, commented that the problem is not that her peers do not want to learn Zapotec but that they have no time to do so in their preparations to be competitive in the new labor market.

Living in the Margins of the Green Economy

Antonio Ruedas is a recent migrant to Binnizá from La Ventosa, another nearby town, who elaborated on the arbitrary representation of indigenous identity, language, and subsistence agriculture as an anachronistic vestige of the past and on the idea that speaking Spanish (and now English) and participating in the national (and now transnational) economy was the future. He learned Spanish approximately two years before our ethnographic study, but because he had previously worked as a mechanic, he was able to work for the energy company motor pool when the turbine projects arrived in La Ventosa. However, he saw his Zapotec monolingualism as a hinderance and did not want his children to learn his heritage language. During the first six years of his children's lives, he did not communicate directly with them. Instead, he relied on their mother to serve as a translator. Even though speaking Zapotec was the norm in La Ventosa at the time, he feared learning the native language would inhibit his children's socioeconomic advancement. With tears in his eyes, he spoke proudly of his success in incorporating himself into the green economy by getting a job as a mechanic for the energy companies, which has allowed him to construct a house. In July 2019, with the completion of the construction phase, his opportunity to work on the wind turbine projects came to an end.

Another Binnizáleño, Don Miguel Aguayo, pointed out that not everyone in Binnizá had the equal capability to fully participate the new green economy. Once a family of subsistence farmers, his parents emigrated from a smaller community in the western part of the isthmus fifty years ago due to increasing hardship. Nevertheless, he considers his family almost *originarios* (original residents) of Binnizá instead of migrants. Don Aguayo notes that his parents thought there was more opportunity in Binnizá, but he observes that, unless you are a landowning "millionaire who had cattle," it was very difficult to advance socioeconomically, and so his father dedicated himself to working as a wage laborer. Describing his father, he says:

> He was always a person who fought a lot; it is on the basis of this effort that they came here looking for a better future, but, well, here, they couldn't "get out" [of poverty]. (Recorded interview, July 2019, translated from Spanish by author)

Don Aguayo describes the difficulty his family had because, although they had emigrated a long time ago, his parents did not have access to formal education and were illiterate. Speaking of himself and his siblings, he explained:

> Us, well thanks to God, we learned something. A little more than them, although we are screwed, we advance a little more. Learn to do other things that they couldn't. (Recorded interview, July 2019, translated from Spanish by author)

Only able to send half of their children to school due to the difficulty to provide for a large family on dismal salaries, his family did not have the same access to resources as other families of professionals. He sees Binnizá as a place for the "rich" and his family having "no future here," as there is a lack of jobs for them to provide for themselves without access to higher education:

> Here in Binnizá there has been . . . there has been a lot of development . . . but there has been a lot . . . a lot of preference for the rich cattle ranchers. As I repeat again . . . here . . . here, the rich monopolize it, the poor are left poor. If

> you imagine that all the help that comes was for the poor, there would be no poverty. We would be more or less the same, but ungratefully there enters the word, selfishness . . . this is greed. The rich stay with all that and then the poor stay, as they prefer to say, without words, they want you quiet. (Recorded interview, July 2019, translated from Spanish by author)

Speaking again on the subject of exclusion, he brings up his belief that his family is systematically excluded by politicians and residents of Binnizá from the economic benefits that green energy development has brought to the town:

> But I'll say it again. Unfortunately, my family didn't receive any of that. I don't know why, I don't know. Maybe because they don't know how to read . . . it gets into politics. And my family, well, [politicians] don't . . . if they support, they don't support, except of course, when the votes come. . . . They're too old [Don Aguayo's parents] to get ahead. For as long as the Lord keeps them alive, well, that's how it is. (Recorded interview, July 2019, translated from Spanish by author)

Conclusion

For a long time, social theorists have seen crisis as a moment that confronts the present order of things and leads to an era where all societal tensions are resolved. The present climate change crisis is no different. Karl Marx, for example, saw the history of social evolution as being one of subsequent upheavals brought about by contradictions within the capitalist mode of production, which he speculated would result in a final revolutionary upheaval (a particular kind of crisis) that led to the communist mode of production, after which the inequities and violences of capitalist society would come to an end. In the case of Mexico, Marxist teleological imaginings of history are exemplified in a mural painted by Diego Rivera on the walls of the country's National Palace in Mexico City (figure 13.1). The mural depicts the history of the nation from the pre-Columbian period to the idyllic future to follow the communist revolution. The painting's dynamic composition zigzags through the three large walls of the palace's main staircase, visually narrating Mexico's history as a series of class crises and struggles, all to culminate at the top of the southern wall, where Karl Marx leads the Mexican people to an idealized future on the upper left corner. This idealized future is one of industrialization, clean skies, and the resolution of the country's identity crisis as *mestizos* reconcile the indigenous and European dimensions of their heritage.

Ulrich Beck, on the other hand, saw the risk society as creating a condition that would move people to form social movements that cut across

Figure 13.1. *The History of Mexico*, mural by Diego Rivera in the Mexican National Palace. © Roberto E. Barrios.

lines of class and national identity. Michel Foucault, for his part, hoped that "man" would one day vanish as an epistemic object, bringing biopolitical societies to an end. More recently, climate change has come to be seen as another crisis that may bring about a resolution to our modernist woes. Gaston Gordillo (2014) has gone so far as to say that climate change may be the final obstacle that brings neoliberalism to its knees. In this latter instance, it is no longer the communist revolution that will bring about the idyllic future imagined by Diego Rivera in the National Palace's walls; instead, it will be the global climate itself. The case of climate change mitigation programs in the Isthmus of Tehuantepec puts these hopeful hypotheses to a test and gives us a very different result. What we see in the case of Binnizá is that, rather than the emergence of a new transnational cosmopolitan environmentalism, climate change is creating a context where neoliberalism is reimagined in the form of green megadevelopment programs. Rather than vanishing, scarcity and biopolitical societies are finding new ways to continue to provide the resources they need—electric power, in this case.

The execution of renewable energy megadevelopment programs is also imbued with positive moral authority. The term "green" or "renewable energy" itself conjures visions of clear skies, environmental harmony, and social progress in the imagination of many consumers. Unfortunately, renewable energy is also subject to fetishization that hides the sociopolitical relations that produce it. In the Isthmus of Tehuantepec, we see an instance where long-standing tensions between national governments and indigenous communities continue to play out over the course of green megadevelopment programs. Furthermore, these programs are structured on the assumption that indigenous communities are to be approached as collections of individual property owners who are left to their own devices to compete for fair compensation. While previous forms of energy production may have created environmentally harmful byproducts (e.g., carbon dioxide), renewable energy programs are now producing exacerbated inequities. Our point in this chapter is not to say that renewable energy programs are undesirable or inherently flawed but that radically different terms for their implementation must be called for. For Binnizáleños, these radically different terms would involve government representatives whose role was not to guard over the financial interests of energy companies during indigenous community consultation processes but to ensure equitable and fair treatment of all communities regardless of their ethnic identities or socioeconomic backgrounds.

Acknowledgments

The authors would like to thank Salomon Nahmad, Jesús Manuel Macías, Ruben Langle, and the Centro de Investigaciones y Estudios Superiores en Anthropología Social Pacífico Sur (CIESAS) for their support and facilitation of this research project.

Amanda Leppert is a graduate student in the Applied Anthropology program at the University of South Florida. She completed her bachelor's degree in anthropology at Southern Illinois University Carbondale under the advisement of Roberto E. Barrios. Her undergraduate research was supported by the SIUC REACH undergraduate grant, which she used to study the social impact of megadevelopment wind projects in the Isthmus of Tehuantepec, Oaxaca, Mexico. In 2020, she was awarded an NSF Graduate Research Fellowship Award.

Roberto E. Barrios is Doris Zemurray Stone Chair of Latin American Studies and professor of anthropology at the University of New Orleans. He has conducted ethnographic research on postdisaster reconstruction during the last twenty-one years, focusing on the ways governmental and nongovernmental recovery policies and practices articulate inherent assumptions about the nature of people, communities, and social well-being, and the ways that disaster-affected populations interpret, navigate, and sometimes contest these assumptions. His ethnographic case studies include Southern Honduras following Hurricane Mitch; New Orleans in the aftermath of Hurricane Katrina; Chiapas, Mexico, after the Grijalva landslides of 2007; Southern Illinois following the Mississippi River floods of 2011; Houston's recovery after Hurricane Harvey; and the U.S. Virgin Islands following Hurricanes Maria and Irma. The results of his work are featured in his book, *Governing Affect: Neoliberalism and Disaster Reconstruction* (University of Nebraska Press, 2017) as well as a variety of peer-reviewed research journals including *Annual Review of Anthropology; Disasters; Identities: Global Studies in Culture and Power; Anthropology News;* and *Human Organization.* He is a founding member and former co-chair of the Society for Applied Anthropology's Risk and Disaster Topical Interest Group.

References

Babb, Sarah. 2005. "The Social Consequences of Structural Adjustment: Recent Evidence and Current Debates." *Annual Review of Sociology* 31: 199–222.

Beck, Ulrich. 1992. *Risk Society: Towards a New Modernity*. Thousand Oaks, CA: Sage.

Carmack, Robert, Janine L. Grasco, and Gary H. Gossein. 2006. *The Legacy of Mesoamerica: History and Culture of a Native American Civilization*. 2nd ed. New York: Routledge.

Collier, Stephen J., and Andrew Lakoff. 2015. "Vital Systems Security: Reflexive Biopolitics and the Government of Emergency." *Theory, Culture & Society* 32(29): 19–51.

Cruz Rueda, Elisa. 2011. "Eólicos e inversión privada: El caso de San Mateo del Mar, en el Istmo de Tehuantepec Oaxaca." *Journal of Latin American & Caribbean Anthropology* 16 (2): 257–77.

Cruz Rueda, Elisa. 2013. "Derecho a La Tierra y El Territorio: Demandas Indígenas, Estado y Capital En El Istmo De Tehuantepec." In *Justicias Indígenas y Estado: Violencias Contemporáneas*, edited by María Teresa Sierra, Rosalva Aída Hernández, and Rachel Sieder, 341–81. Tlalpan, Mexico: CIESAS.

Descola Philippe. 2017. "¿Humano, demasiado humano?" *Desacatos* 54: 16–27.

Dore, Elizabeth. 2006. *Myths of Modernity: Peonage and Patriarchy in Nicaragua*. Durham, NC: Duke University Press.

Dunlap, Alexander. 2017. "Wind Energy: Toward a 'Sustainable Violence' in Oaxaca." *NACLA Report on the Americas* 49(4): 483.

Fortun, Kim. 2001. *Advocacy after Bhopal: Environmentalism, Disaster, New Global Orders*. Chicago: University of Chicago Press.

Foucault, Michel. 1970. *The Order of Things: An Archaeology of the Human Sciences*. New York: Random House.

———. 1978. *History of Sexuality*. Vol. 1. New York: Random House.

———. 2004. *Security, Territory, Population: Lectures at the College de France, 1977–1978*. Edited by Michel Senellart. New York: Picador.

Franklin, Sarah. 2005. "Stem Cells R Us: Emergent Life Forms and the Global Biological." In *Global Assemblages: Technology, Politics, and Ethics as Anthropological Problems*, edited by Aihwa Ong and Stephen Collier, 59–78. Malden: Blackwell Publishing.

García-Acosta, Virginia. 2002. "Historical Disaster Research." In *Catastrophe and Culture: The Anthropology of Disaster*, edited by Susanna M. Hoffman and Anthony Oliver-Smith, 49–66. Santa Fe: School of American Research.

———. 2017. "Presentación: La incursion del Antropoceno en el sur del planeta." *Desacatos* 54: 8–15.

———. 2018. "Los Desastres en Perspectiva Histórica." *Arqueología Mexicana*. 149: 32–35.

Gordillo, Gastón. 2014. *Rubble: The Afterlife of Destruction*. Durham, NC: Duke University Press.

Haraway, Donna J. 1997. *Modest_Witness@Second_Millenium.Female an_Meets_OncoMouse: Feminism and Technoscience*. New York: Routledge.

International Labor Organization. 1989. *C169—Indigenous and Tribal Peoples Convention*. No. 169. Retrieved 12 December 2019 from https://www.ilo.org/dyn/normlex/en/f?p=NORMLEXPUB:12100:0::NO::P12100_ILO_CODE:C169.

Latour, Bruno. 1993. *We Have Never Been Modern*. Cambridge, MA: Harvard University Press.

Masco, Joseph. 2006. *Nuclear Borderlands: The Manhattan Project in Post–Cold War New Mexico*. Princeton, NJ: Princeton University Press.

———. 2009. "Bad Weather: On Planetary Crisis." *Social Studies of Science* 40: 7–40.

Nahmad, Salomon. 2012. "The Impact of World Bank Policies on Indigenous Communities." In *Neoliberalism and Commodity Production in Mexico*, edited by T. Weaver, J. B. Greenberg, W. L. Alexander, and A. Browning-Aiken, 209–24. Boulder: University Press of Colorado.

National Academy of Sciences and the Royal Society. 2014. *Climate Change: Evidence and Causes*. Retrieved 15 March 2020 from https://www.nap.edu/catalog/18730/climate-change-evidence-and-causes.

Pickering, Andrew. 1995. *The Mangle of Practice: Time, Agency, and Science*. Chicago: University of Chicago Press.

Pickering, Andrew, and Kenneth Guzik, eds. 2008. *The Mangle in Practice: Science, Society, and Becoming*. Durham, NC: Duke University Press.

Ponce, Norma. 2013. "Ingrid y Manuel Asociados con el Cambio Climático: Semarnat." Milenio. 30 September. Retrieved 15 August 2014 from http://www.milenio.com/tendencias/Ingrid-Manuel-asociados-cambio-climaticoSemarnat_0_163184144.html.

Povinelli, Elizabeth. 1995. "Do Rocks Listen? The Cultural Politics of Apprehending Aboriginal Australian Labor." *American Anthropologist* 97(3): 505–18.

Rabinow, Paul. 2005. "Midst Anthropology's Problems." In *Global Assemblages: Technology, Politics, and Ethics as Anthropological Problems*, edited by Aihwa Ong and Stephen Collier, 40–54. Malden: Wiley Blackwell.

Sellwood, Scott A., and Gabriela Valdivia. 2018. "Interrupting Green Capital on the Frontiers of Wind Power in Southern Mexico." *Latin American Perspectives* 45(5): 204.

Shapin, Steven, and Simon Shaffer. 1985. *Leviathan and the Air Pump: Hobbes, Boyle, and the Experimental Life*. Princeton, NJ: Princeton University Press.

Smith, Adam. 1999. *The Wealth of Nations*. Books I–III. Penguin.

Weaver, Thomas, James B. Greenberg, William L. Alexander, and A. Browning-Aiken. 2012. *Neoliberalism and Commodity Production in Mexico*. Boulder: University Press of Colorado.

Chapter 14

Disaster and Climate Change

Susanna M. Hoffman

In the last few decades, disasters of both geophysical and technological agency have become alarmingly more frequent and severe across our planet. The effect of this unprecedented development is that ever larger numbers of people are suffering from calamitous events and experiencing escalating conditions of vulnerability. Despite all the modern advances of the current epoch, safety has not increased. It has grown worse.

Since 1998, the world has undergone Hurricane Mitch (1998); Hurricanes Katrina (2005) and Ida (2021); the Haitian Earthquake (2010 and 2021); the Great East Japan Earthquake, tsunami, and meltdown (2011); Hurricane Sandy (2012); Typhoon Haiyan (2013); the Nepalese earthquake (2015); Southeast Asia fires and smog (2015 and 2016); Britain and Ireland floods (2015–16); the El Niño and La Niña of 2016, plus their increased frequency; the Guatemalan, Palomar, and South African droughts (2016); Western United States wildfires (2000–2021); Australian drought, floods, and windstorms (2000–2020); Taiwan's Typhoon Megi and landslide (2016); yet another Mississippi flood (2016); Hurricane Matthew and the North Carolina flood (2016); the worst drought in the Levant and Middle East in a millennia (2016–17); worsening famine in Syria and Sudan (2017); along with countless other smaller-scale and less prominent happenings. All have produced significant losses in life, land, property, habitat, and homeland. While all to a certain extent have also increased the awareness and importance of confronting the risks and episodes of particular regions, the forecast is that similar dire events will accrue and amplify in the future. Storms and other sorts of incidents will grow stronger, last longer, and magnify (EM-DAT 2016).

The disturbing upsurge in disasters, and the vulnerability to them, is due in part to an age-old set of driving factors, albeit today they are often augmented. These include the inherent vicissitudes of embedded environmental features; the actions of humans on land and waterscapes; faulty intentional and unintentional manufacture, construction, arrangement, and assembly; social disparities, poverty, economic and political victimization; and stifling lack of opportunity. The former driving factors are, however, now combined with a number of critical new components, all also well documented by Eriksen in his *Overheating: An Anthropology of Accelerated Change* (2016). They include massive population growth, which has heightened penury, lack of education, marginalization, and depreciated habitat; novel and aberrant demographic processes, including massive worldwide urban migration, burgeoning and overcrowded cities, the drift of populations to coastlines, all of which are inherently hazardous; and the desire for Western lifestyles with concomitant accumulation of goods and demands. The newfangled drivers often interlace with one another. Many of the newly gargantuan cities also sit on coastlines, which are highly prone to earthquakes, tsunamis, cyclones, volcanic eruptions, floods, and mudslides. Enhanced consumerism not only consists of the desire for more goods and services, it includes a concomitant demand for energy and transportation, all of which accelerate depreciation of the environment and advance risk construction. Population movements, whether emanating from poverty, lack of land, or the desire for a better life, have often instead increased hardship, marginalization, and yet further diminished habitat.

A rampant global spread of persons bearing neoliberal agendas with their attendant, often deleterious, actions has also taken place. Heedless of ecozone and culture, aimed only at financial profit, their deeds have often included land grabs, ill-conceived construction, crop conversions, water diversions, and the exploitation of cheap local or replacement labor. Indeed, the spread of neoliberalism and short-sighted development can be viewed as its own sort of firestorm. Perhaps the strangest aspect about the expansion of neoliberalism is that it has not been confined to particular societies, nations, or companies. It so appears to have operated away from and above any of these—I have taken to calling the phenomena "a culture without a people." Its proponents do not share customs, language, homeland, or religion. They simply share an exploiting motivation. Along with their expansion has come the amplification of a dominating intermutual and synergistic economic system that has swept over political factions and subsumed a wide diversity of regions.

Finally, in part due to the above, a globalization of markets has also occurred over the last decades that entails the fabrication, trade, purchase, and sale of products derived from one place and dispatched to a distant other. The items include clothing, cars, machine parts, toys, and, to an ever-expanding extent, food. All have led to unexpected vulnerability to calamity at both ends of the trade routes, and, in fact, all along them. Neither seller, buyer, nor middleman can see, predict, calculate, or comprehend the collapse that takes place when some unanticipated disaster occurs somewhere along the conduit half a world apart. All these new drivers have also been documented by Eriksen (2016).

One contemporary driver, however, is contributing far more than any other to the recent increased frequency and magnitude of disasters: global warming. As Edward O. Wilson states, the rise of annual mean surface temperature caused by pollution on the Earth by 2016 had already reached nearly half of the "2C" threshold above that prior to the birth of the Industrial Revolution. When global atmospheric warming pushes past this marker, Earth's weather will destabilize. Heat records now considered historic will become routine. Severe storms and weather anomalies will become the new normal. The melting of ice shields currently under way will accelerate, bringing to landmasses a new climate and new geography (Wilson 2016: 65–66).

As a consequence, major and minor catastrophes befalling human communities everywhere will simply rise. The process has begun. Warmer seawater is and will generate more ocean-originating cyclones and hurricanes, which upon making landfall will flatten ever larger areas, often populated. Rising sea levels and soaring tides are creating extensive seawater incursions and coastal erosion. Increasing heat inland has caused tornados, once confined to a well-defined "alley," to advance well outside their old corridor to manifest in previously unscathed locales. Rivers are swelling, breaking channels, finding new paths, and inundating previously secure towns as they recently did in North Carolina. Hillsides and mountains will crumble and careen over human concourses as snowpacks, torrential rains, saturation, unstable embankments, and upslope conditions increase. Slow-onset catastrophes like drought and desertification will spread drastically, bringing with them dearth of water, starvation, and abandonment. Heat will kill. Violent snowstorms and cold spells will prove devastating. Some believe destructive earthquakes and volcanic eruptions with their devastating ash and lava deposition might also be connected to climatic changes as Earth's increasing atmospheric temperature disturbs fault lines (McGuire 2016).

Those disasters generally termed "technological" will also increase. Factories and installations that produce or use toxicants and lie on coasts or fault lines will become unstable and unfurl contaminants, as the Fukushima Daiichi nuclear plant did in 2011. Hazardous manufacturing structures located in drought, inundation, or avalanche zones will potentially fissure. Proper maintenance of dangerous facilities will likely decline as regions suffer climate impacts, causing yet more soil and groundwater contamination. Lethal epidemics, like Ebola, swine flu, or COVID-19, which also fall into the domain of disasters, will also conceivably increase. They, too, generally have human action at their genesis, usually some sort of animal-to-human or insect-to-human contact, and further human interaction fans them outward. With global warming, the migration of vectors and carriers will increase (Lafferty 2009). Finally, storms and calamities will bring about far more displacement of human populations seeking new places to settle. The new locales and communities may well be equally or more vulnerable and will create still larger numbers of imperiled people. The process of climate change resettlement has already begun. In future, it will affect legions of persons in myriad locales, causing innumerable economic, political, social, and psychological dilemmas (Oliver-Smith 2009).

Definitions and More

In order to illuminate the unequivocal connection between climate change and disaster, it is necessary to clarify exactly what a disaster is. While most climate change researchers acknowledge that the changing global conditions will cause more catastrophes, few have actually examined what constitutes one. Nor have many considered the pertinence of anthropology in dealing with either, yet anthropology has recently come to the forefront in both the climate change and disaster fields. The emergence has occurred in large part due to the realization that the reason so many climate change adaptation programs and disaster reduction efforts have proven ineffective is that the people's culture, their local knowledge, lifeways, and desires, has not been considered. As a result, anthropologists and the anthropological perspective have become more and more integrated in climate change and risk reduction endeavors today.

In following, then, the first basic tenet in understanding disasters is that there is no such thing as a natural disaster. Paralleling the famous rhetorical saying "If a tree falls in a forest and no one is there, does it make a sound?" in terms of disaster, if there are no people in some way involved, an event would just be an evanescent and unrecognized happening. The statement, however, implies more than just the presence of people. Disasters all involve some form of human interaction at one level or another. The maxim is acknowledged in every field dealing with disaster, from geophysics to engineering to social science, and applies equally to those disasters emanating from climate change as well as other sorts.

There may be natural "triggers" to disasters, but it is what humans did, or chose, or made that results in a catastrophe, even those erroneously called "natural" and clearly technological ones (Squires and Hartman 2006). People have placed themselves in a certain spot. They have altered things or built things in a manner that is perilous. They have acknowledged the embedded characteristics of their environment, the modifications they have made of the environment, and their structures, or they have disregarded them.

Having started with that fundamental principle, let me define what a disaster is: "a process/event combining a destructive agent/force from the physical, modified, or built environment and a population in a socially and economically produced condition of vulnerability that results in the disruption of social needs for physical survival, social order, customary satisfactions, and meaning" (Hoffman and Oliver-Smith 2002: 4). As the definition points out, disasters take place through the conjuncture of two essential factors: a human population and a potentially destructive agent. Neither of these is static. Disasters do not arrive suddenly out of the blue, nor are they mysterious. The factors leading to them evolve over time. Disasters may give the impression that they are demarcated in exact time frames, like the moments of an earthquake or days of a flood, but that is never the full story. Disasters are processual phenomena. They have history and chronology, sometimes quite extended. They have amassed over years, decades, even centuries (Oliver-Smith and Hoffman 1999; Hoffman and Oliver-Smith 2002).

Of course, the intersection of a destructive agent and a human population does not necessarily cause a disaster. Disasters require yet another factor, the third one mentioned in the definition. Disasters take place only in a context of vulnerability. Vulnerability, or the state of being open to injury, is also historically and socioculturally produced. It comes about through the location, infrastructure, sociopolitical organization, and production and distribution systems of a society, and also its ideology. A people's vulnerability is, in fact, such a core ingredient of disaster that it is not merely causal. It conditions the behavior of individuals and organizations throughout the unfolding of the entire disaster scenario, from construction to event, recovery, and possible future mitigation, far more profoundly than the physical force of the destructive agent will. Considering its totalizing influence, and how with global warming it is advancing across more and more populations, vulnerability stands as the key factor in the link between climate changes and disaster (Fiske et al. 2014; Oliver-Smith and Hoffman 1999; Hoffman and Oliver-Smith 2002).

The definition of disaster can also not be entirely separated from the concomitant matter of hazard, another ingredient on the upswing as the climate changes. A hazard can be defined as "the forces, conditions, or technologies that carry a potential for social, infra-structural, or environmental damage. A hazard can be a hurricane, earthquake, or avalanche. It can also be a nuclear facility or a socio-economic practice, such as using pesticides. It can also be an eroding river bank, massive sinkhole, intense heat wave, and devastated growing season. The issue of hazard further incorporates the way a society perceives the danger or dangers, either environmental and/or technological, that they face and allows the danger to enter their calculation of risk" (Hoffman and Oliver-Smith 2002: 4). With advancing shifts, of course, a people may not at all be able to perceive their looming imperilment.

It is important to note that disasters come about in two ways, both of which can be propelled by climate change. Sometimes disasters strike with the sudden impact, as with a violent storm or flash flood. These are called rapid-onset disasters. At other times calamities accumulate over long spans, as with spreading sand or incremental heat accretion. These are called "slow-onset" disasters. Climate change in and by itself is considered a "slow-onset" disaster (Fiske et al. 2014; Fiske and Marino 2020). Nonetheless, it will bring about calamities both unanticipated and abrupt as well as those inching almost imperceptibly forward.

With the disaster definitions in mind, the next matter to elucidate is that humans, in fact, do not dwell in just one environment. They dwell in four, and each must be addressed to understand the interplay of climate and calamity (Hoffman 2017: 194–95). The first environment is the basal terrain in which a people dwell, what is usually referred to by the term "environment," although it might be better termed the "physical plane." The second is a people's "modified" environment. Humans almost never live in a place without altering it. Rather, they sculpt their surroundings. They terrace hillsides, channel streams, lop off mountain tops, and purloin seabeds. In addition, humans erect a third environment, a built one. Upon their physical plane, humans raise houses and temples, pave roads, implant pylons, string bridges, and erect power plants. Their communities spread up and out, all the while superimposing a contrived milieu in which the inhabitants live, eat, sleep, and work. People also reside in a fourth environment, their culture. It is culture that in fact instructs the design and interactions of the other three environments. It proscribes how the physical plane is utilized, the territory modified, what is built upon it, and then how people live (Hoffman 2017: 194–95).

The bottom line is that all four environments are intricately interconnected. A change in one brings about changes in the other. Though environment does not necessarily determine the others, nonetheless, a permutation in the first level, that of the physical plane, generally means an adjustment to the other three: the modifications, built structures, and, in turn, the governing culture. Still, quite frequently, it is the alterations humans have made to their physical plane that exacerbate climate change and bring about full-scale disaster. In southern Louisiana, for example, while it is true that the Mississippi Delta is subsiding into the Gulf of Mexico, it is the shipping channels that the petrochemical industry has gouged out to allow their tankers ready access that have engendered inundation, salinization, and land loss. It is the excessive use of water to irrigate unsuitable crops, such as rice in arid California or pinto beans in sere New Mexico, that has depleted water tables and compounded desertification. The character of the built environment in changing conditions can also create calamity. Deteriorating buildings increase risk of catastrophic failure, and the spread of chemical and other ancillary pollutants back to the physical plane create health and safety risks for occupants (Burton 2012). Simple placement of the built environment intertwines with threatened coasts, such as with the Daiichi Fukushima nuclear facility. A six-story residential building fell in heavy rains in Nairobi, Kenya. The heavy skyscrapers on the southern tip of Manhattan Island in New York already need continuous water pumping due to rising levels of surrounding water. They were already notably affected by wave overflow from Hurricane Sandy, and yet now the area features towering new residential complexes that are occupied not just during office hours but both day and night. Weighty Mexico City is sinking into its foundational swamp.

The overarching cultural environment of a people can also facilitate turning climate change into calamity, particularly, but not limited to, a culture's economic system. A change of physical plane almost inevitably means a loss or readjustment of subsistence, that is, a people's economic base and their potential famine or resettlement. A society's trade systems combined with climate change can also cause calamity. Due to broadening warm zones, inadvertent transportation of various insects has caused the economically destructive diffusion of deleterious structive beetles to North American forests and abandoned communities (Casey and Whittle 2017). The spread of the fetus-threatening Zika virus has alarmed the populations of the Caribbean and South America. Other aspects of a society's traditions can augment the perils of climate change as well. For one, the customs of some cultures are more facile in adapting to changing conditions, while others are intractable and leave their members in more precarious positions as situations mutate. Some cultures engender in their inhabitants a deep sense of place attachment, making it difficult for them to revamp. Perhaps they have revered systems of land tenure and inheritance or sacred territories, or perhaps the land is inhabited by venerated spirits. After the Southeast Asian tsunami, in Ache, Sumatra, people planted stakes atop piles of debris to mark out exactly their once extant gardens. Among other communities, kinship is relatively inconsequential, as is abiding ownership. In some, members adhere to the decisions of leaders, flexible or inflexible, no matter what looms. In still others, individual members are free to do and act as they please, including leave. Some people are schooled to heed warnings from outsiders, others to dismiss. Some cultures display wide consensus among the members. Others cultivate contentiousness. Some see themselves as all similar. Others value individuality. In anthropology, we frequently deal with communities that embody constant participation among members, daily meetings, collective discussions. Among them, agreement and action in order to avert calamity willingly emerges. Within other cultural spheres, especially large-states societies, community is, in reality, a pseudo concept. Perhaps for particular purposes, people strive to embrace a "sense" of community. Community is rather like potential and kinetic energy in physics. It may be triggered by the advent of a dire occurrence, but maybe not, and if activated it is not necessarily universal. Behind such cultural difference lies success or failure when faced with climate change.

Two Examples: One Sudden and Unforeseen; One Slow, Recognized, and Relentless

I offer two examples showing the link between climate change and disaster. One appears as perhaps a rather minor case, but it represents the sort of small, localized, yet highly destructive event that will come to pass worldwide more and more in the future. The other tells of a place where climate change, albeit considerably augmented by human complicity, has already wrought so much damage that the people of the region must leave and resettle elsewhere. Both cases come from the seemingly impervious United States where, despite the warnings of scientists, significant climate change denial continues to prevail.

The Boulder Floods: A Perilous Cocktail of Blissful Denial and Climate Change

Boulder, Colorado, is a town of about two hundred thousand inhabitants sitting at the base of the Rocky Mountains about forty miles from the state capital of Denver. Surrounding the town are a number of smaller communities, most historic in origin, but which in recent years have snowballed with new development, adding greatly to the region's increasingly dense population. Boulder also houses the main campus of the University of

Colorado, which welcomes a yearly student enrollment of around thirty thousand. Due in part to the university, the town is demographically relatively young and quite youthful in its orientation. The inhabitants are by and large environmentally aware and politically progressive. Refillable water bottles bounce on backpacks, an almost universal add-on appendage. Many people ride bicycles rather than drive cars, for which the community features a labyrinth of accommodating bike paths. The city also contains a number of America's most significant climate and atmospheric study centers, including the National Oceanic and Atmospheric Administration (NOAA), the National Center for Atmospheric Research (NCAR), and the National Hazard Center, along with major wind, wave, and other renewable energy research centers. Colorado itself offers a relatively stable environment. The state is devoid of volcanos and earthquakes, and while at times it is quite windy, it rarely has but small tornados, and those are relatively recent phenomena. It is speculated that they are the result of changing continental climate conditions. The state does brave flash floods, sizable blizzards, avalanches of both snow and soil, occasional golf-ball-size hailstorms, significant forest fires, and appreciable aridity and drought.

Nonetheless, although the people of Boulder have largely been quite environmentally mindful, even to the point of acknowledging climate change, until recently they seemed to indulge the common notion that the effects of climate change were geographically distant from them, most likely on the seacoast or in arid deserts to the south, but not close at hand. They did not anticipate what climate change could bring upon them or how disruptive those effects could be.

The first full weekend of September 2013, however, was unusually hot in Boulder. The temperature tied a record of ninety-three degrees Fahrenheit (thirty-four Celsius) for September 8, which fell on a Sunday that year. People were still wearing flip-flops, but then Boulderites do that in the snow. There was talk of a cold front coming in, maybe bringing much-needed rain on Monday, September 9. Climatologists had noticed an unusual amount of moisture in the atmosphere and, indeed, on Monday it began to rain quite hard. Experts were overjoyed. There had been a major drought for months preceding the storm, and most were hoping that the yearly rain level would now rise to the norm. Still, the National Weather Service issued the first flash flood warning on Monday for an area in the mountains behind Boulder left bare from the previous year's forest fires. But at this point no one noticed anything unusual, and no one in town was notified.

The rain continued all Tuesday. By Wednesday, the ground saturation level had been reached due to the amount of rainfall, meaning the foundational soil of the town could absorb no more water. Covertly, upslope conditions were beginning to form, a situation where considerable precipitation occurs despite a lack of moisture in the troposphere, a condition not uncommon in the area in winter but not fall. Upslope conditions are usually marked by cold rain, though residents found the rain oddly warm. Meanwhile, flanking the active Colorado storm was a low-pressure system seated over the neighboring state of Utah to the west. In short, a perfect tempest was building, but the pieces were not linked and the potential not recognized "ahead of time," as the Weather Service later disclaimed.

By late Wednesday, some hiking trails were closed due to mud. The town of Erie, slightly east of Boulder, suddenly had standing water and popping manhole covers. Power lines began to fall in the nearby community of Longmont and along the St. Vrain Creek located to the north of Boulder. Cars were getting stuck. In Boulder itself, streets began to flood. The road up Boulder Canyon into the mountains was almost unpassable. The fact that an event was occurring reached the inkling stage. The University of Colorado advised students to get to higher ground, but police and maintenance providers felt they could manage the situation. Officials began sandbagging overflowing creeks, and everyone thought it would be over by Thursday.

On Thursday, nine inches of rain fell, and three more on Friday. Already by the end of Thursday, buildings were being ripped from foundations. Boulder Creek, which flows through the middle of the town, was, by 1:13 A.M., roaring at a rate of 3,104 cubic feet per second. Homes across the area were taking water in their basements and up to the main floor. The town of Lyons and neighboring Hygiene—yes, I think the name is ironic, too—became inundated islands and remained so for months. Without warning, people were forced to desert their homes. Hundreds had to be airlifted out by helicopter. Most of Boulder's streets had become rivers.

The rain continued three more days, for a total of eight. Altogether, seventeen inches of rain fell, almost the full year's average of twenty. Dozens of roads were washed out, dragging cars along with them. Dry ditches became rivers, most running for many months after. Countless boulders careened down old creek beds. Debris slammed against bridges and washed them away. In the end, water spread over two hundred miles. Boulder County was the worst hit. At least eight deaths were reported, with two persons missing and presumed dead and, at first, hundreds unaccounted for. More than eleven thousand homes were evacuated. The towns of Lyons and Erie were cut off from all forms of ground transportation for a number of months. Several earthen dams along the front range of the mountains burst or were overtopped—mind you, earthen dams were still being used here despite the lesson learned from the 1889 Johnstown flood in the hills of Pennsylvania. Nearly nineteen thousand homes were damaged and over fifteen hundred destroyed. At least thirty state highway bridges were demolished and an additional twenty seriously impaired, with repairs for bridges and roads predicted to cost many millions of dollars. Miles of freight and passenger rail lines were washed out or submerged, including a section servicing Amtrak's iconic California Zephyr (*Boulder Daily Camera* 2013).

But beyond the simple description, there was more to the flood's story. The Rocky Mountains, which form the backdrop to Boulder and the surrounding towns, including the hard-hit Lyons, are not like most mountain ranges. Rather than featuring a series of increasingly rolling and rising foothills building to the highest peaks, the Rockies instead rise abruptly and sharply, much like a massive three-thousand-mile wall. In fact, some of Colorado's highest peaks, called the "Fourteeners," lie in the stretch called "the Front Range," towering directly behind Boulder and within a half-hour's drive of the city. The slopes down from these high peaks, including the very first hills that jut up immediately behind Boulder, bearing the names "Hogback" and "Flatirons," are very steep. Boulder abuts these acute inclines. They are the result of what geologists call the Fountain Formation, a quite young uplift at only about 290,000 years old, comprised of sandstone and gneiss that was formed from the erosion of the mountains laying behind—if that in itself was not a clue. Erosion, dry and water driven, is ever ongoing. As we natives say, for I am a third-generation Coloradan, "The job of mountains is to come down."

I have titled this case study "A Perilous Cocktail of Blissful Denial and Climate Change" because part of the explanation of what became a considerable disaster lies in social causes and the massive denial that led to the event. The area from Denver to Fort Collins, including Boulder, El Dorado, Lyons, Longmont, and Loveland, once an old, sparsely populated cattle ranch and gold-mining region, has experienced extremely fast population growth. There are very few people with any generational history in the area. I am a rarity. My grandfather had shooting matches with Buffalo Bill, but few believe my family has lived in Colorado that long. It is unusual. At best many of the current inhabitants are the firstborn generation of parents who moved to the area. Many are completely new arrivals. Virtual legions of people in the last few years found the district, its geography, its offerings, its climate (not too hot and not too cold), and its lifestyle attractive.

Boulder itself was a sleepy college town when I was a child. Today it has grown from slightly less than twenty thousand in 1950 to over one hundred thousand in 2010. From 1950 through 1970, it doubled every decade. It then grew by a mere ten thousand over each ten-year span, only just recently slowing from that pace. The area is also renowned for attracting what could be termed a "nature-loving," "quasi-hippy" population in search of an alternative or free lifestyle, and believing they are "natural." Most are also definitely "outdoorsy" if not avidly sports minded. Snowboards outnumber persons. People like to ski, climb, raft, hike, and of course bike, even on snow-navigating "fat wheelers" in the winter.

But loving nature, and loving to cavort within it, is far different from understanding it. Few of the many newcomers, despite love of the wild and a desire to exploit its advantages, have endeavored in any depth to understand the environment. Almost none grew up, as I did, with the firm knowledge that mountains, albeit beautiful, are also inherently dangerous. They are powerful, quixotic, and indifferent. They take no heed of the humans traipsing upon them, digging them out for ore or housing developments, or slicing roads through them. Mountains are actually not a toy to be played with, but that has not been the cultural attitude of the new inhabitants. Even in the old days, the first pioneers who only exploited the hills to derive a living, did so with trepidation. Practically no one who lives in Boulder today even knows the history of the name of the town. Well before Colorado became a state in 1876, Boulder was named for the number of boulders that had washed down from the hills and lay strewn about below. That small bit of information might have given yet another hint to what has been happening for eons.

There further exists an extensive history of major floods in Boulder and the surrounding towns, all the way from Denver north to Wyoming. Yet, the current population blithely lacks almost all knowledge of the flood danger and the deadly chronicle of floods, even though numerous written reports exist and old, black-and-white photographs as far back as the old gold-mining days of Boulder and nearby towns buried in flood-deposited mud can be seen hanging on the walls of many of the town's cafés and saloons. In short, even if no one read the stories of past floods, visual clues abounded. Yet, the prevailing attitude in Boulder and in much of Colorado—climate change notwithstanding—continues to elide the history and comprehension of their environment.

As a community, Boulder and the cluster of towns nearby have, indeed, germinated a rather distinct "culture." Not unlike Berkeley, California, it is referred to in jest as "the people's republic of Boulder." More than just having the liberal politics of Berkeley, Boulder has a strong concept of being "at one with nature." I do not mean this pejoratively, as I adhere to much the same feeling, but Boulder is extreme. Gluten dare not enter. Plastic bags can cause an outburst of hysteria, and most would shudder to learn—talk about blinders—that Colorado actually grows sugar, lots of it, and nearby. The geography is treated almost exclusively like a "mother," as I have written, and not a "monster" (Hoffman 2001), though the entire region is virtually crosshatched—veined is the more technical term—with ditches, from tiny, very narrow troughs to wider cracks that we in the west would call "arroyos," to trenches four feet or more wide and deep and lined with rocks. Few among the population, as the habitation has expanded, have apparently questioned how those gullies were formed, as almost all of them appear dry at least most the time, if not always. Nor has the query been broached on how the numerous rocks got in those ditches. Few, furthermore, have ventured to realize that the ridge lying just to the south of Boulder, allowing for a lovely overview of the whole Boulder basin, is actually not hill but a lateral glacial moraine full of dirt and stone detritus, nor do they appreciate that Boulder owns its own glacier, relentlessly melting and moving down the mountain that was for decades the town's main source of water. Glaciers leave behind numerous runoff streams and ditches that carry lots of rubble. Also, the Rockies are arid and unable to hold much moisture. Even the shortest downpours, like the two-hour summer afternoon monsoons that Boulderites thought the September rain was, albeit late, rapidly cause streams to fill and transport considerable debris, such that washouts of roads and streets are common.

There are additional sociocultural factors, but the ones I have already mentioned alone created a serious lack of risk perception leading to the storm. On top of them, the region's growth brought with it a tremendous spread of building probably not fit for the territory, as the fires of the previous year had already demonstrated. Along with far too much urban-forest interface, and, indeed, with some acknowledgment of flood zones, over time developers neglected to take heed of the potential hazard inherent in the ditches and general geological history. Since most of the seemingly inactive trenches had remained dry for years, many were assumed to be of no consequence and were plowed over to make way for highways, parks, apartments, and houses. As a result, a large number of parks, roadways, highway on-ramps, and basements in the town became bathtubs. Hence, expanded denial, increased population, lack of knowledge, naive development, and building for "probability not possibility," as the storm revealed, caused a major surge in vulnerability.

Still, there was another factor involved in the calamity, one quite high, stealthily covert, and very much derived from climate change: global warming. It has long been predicted that, as an aspect of climate change, storms would increase in frequency and power. Already mentioned was that this storm came well after the season for summer monsoons and well before the onset of winter storms. It also involved ingredients common to both yet rarely mixed, upslope conditions, an odd mix of moisture and dryness, and pressure systems, with very hot weather preceding. In addition, and very contributory, before the storm the area experienced a long and serious drought—evidence, long predicted, that the region is and will see increasing aridity.

What these conditions caused is something not often mentioned yet pertinent to ongoing climate change: increased contrast between drought and flood. The whole Denver-Boulder area is becoming hotter and dryer than it was when I was a child, but more importantly, the switch between hot and cold, dry and wet, drought and rain, is no longer a continuum but now a dialectic. Not just true for the state of Colorado but also worldwide, it is one the Earth will have trouble handling, as each extreme entails vastly differing conditions that will implicate more human communities. The Boulder area, as an example, will as a consequence experience more weather systems going upslope, not downslope, and in several variations, not just the mix of winter upslope combined with summer downslope. As well, hot dry air will sit below systems as they move upslope, causing storms to recycle upon themselves much like the single twist of a tornado, heading in an up-down direction instead of a lateral one.

In following, with the advent of such increasing storms, and not yet much discussed in general, is the phenomena called "hovering." Due to climate change and global warming, storms will linger at length and not move readily on, escalating the amount of rain, snow, hail, wind, or whatever destruction they involve and concomitantly impeding the ability of victims to sustain through them. This is partly due to a practice that is not generally done, which is looking high up enough. By and large, climate change studies have focused on changes occurring on the ground or in the near atmosphere, but warming is also affecting something much more elevated some two hundred miles up: the jet stream, and it decidedly added to the Boulder floods.

After crossing the Arctic, the jet stream enters the North American continent around the United States Pacific Northwest, in Washington and Oregon. It then moves east over the Rocky Mountains, makes a sharp turn, and travels down the Rockies to the south, along the eastern edge of the mountains through much of Colorado, up until just about Colorado's border with northern New Mexico. There the jet stream turns east again and moves across the continent to the Atlantic. Due to warming in the Arctic, the jet stream is picking up more moisture than it previously held, and with more moisture in its currents, it is moving slower. Thus, the storms that occur along it do not move as before; rather, they dawdle, or "hover."

Of course, a storm such as the one that brought about the Boulder disaster does not just end when the rain finally ceases. Like a comet, such storms, and the climate change they actualize, have a tail. Due to the damage left behind by Boulder's exceptional storm—the loosened banks, the loss of rock siding, the wider and deeper ditches and creeks—weather experts around Boulder predicted much higher snowmelt and runoff the following winter and, consequently with it, persisting havoc in Boulder and its surroundings. The annual runoff usually begins in May as the air warms up, but the temperature in Boulder already by March was in the seventies Fahrenheit (twenties Celsius). Experts also forecast far more devastating summer "monsoons" the following year. Mountain monsoons are often intense afternoon downpours that last up to several hours in July and August. They are both preceded and followed by the annual destructive fires from summer to autumn, and all of these occurrences were forecast to come not for Boulder not just the year following the flood, but onward.

What is highly disturbing to me is how the Boulder flood became referred to as the "thousand-year flood," or, following protests, the "thousand-year rain." It is declared such in the titles of books about the flood (Prairie Mountain Publishing 2013), in talks and media presentations, and in newspapers, internet sites, and blogs. The population of the region has also adopted the term. Thus named, the people who underwent the disaster have latently assumed in their collective thinking, that a similar storm will not occur in their lifetimes. The very name instills cyclical thinking beneath the America's overt linear view of time. It suggests that a similar tempest will not happen again for another thousand years, as if Boulder operated on a Buddhist time wheel and not a Western progression. But then maybe Boulderites simply operate in this way, along with the younger generation all over the country. A significant number seem to be manifesting a cultural change in America toward viewing themselves as more "natural" and spiritual and the world as more mystical and rhythmic despite their reliance on ubiquitous digital devices. Numerous Tibetan monks reside in Boulder. There exists a Buddhist university in the center of town, and prayer flags flutter on myriad porches. Why not a Buddhist, or Hindu, or Mayan rotating calendar? Of course, climatologists realize that the term means a one-in-a-thousand chance of occurrence again within a certain time span, but the general population does not think that way. They think it indicates a literal thousand years. However, even a one-in-a-thousand chance does not imply that such a storm will not occur again next year, or the year after, or every year, especially with climate change driving the elements.

Isle de Jean Charles: A Story of Advancing Inundation and Dislodgement

The Isle de Jean Charles band of Biloxi-Chitimacha-Choctaw Native Americans was a late-arriving group to the narrow strip of low, fertile land south of the Chickasaw River in the Mississippi Delta of Southern Louisiana when they drifted in around 1832. The place had once before harbored human habitation, probably by the early Hopewell Mound Builders. They had long since disappeared. Those arriving now spoke a Muskogean language and were related to the Iroquois. After first being displaced from their original homeland in Florida and Alabama by French settlers in the 1700s, the people now sought an even more remote sanctuary in the Mississippi Delta in order to escape the Indian Removal Act, also known as "The Trail of Tears." Of the many Native American groups who made up the Muskogean speakers in their original homeland throughout the American southeast, the members of Biloxi-Chitimacha-Choctaw branches were the most renowned agriculturalists. They were also the most democratic, peaceful, well-organized, and least formal in their politic. The new strip of island land this particular group now fled to was considered uninhabitable swampland by white settlers. Yet it was plentiful in vegetation, nurturing estuaries, and wildlife. Once taking roost, all the members held the land communally as they had in their previous home, with individual families responsible for specific fields. Men worked the land, trapped, gathered oysters, and fished, an activity that provided much of the tribe's diet. Women hoed and cooked. Everyone harvested. Afternoons were devoted to collective games and entertainments. Life was bountiful. The group also traded extensively along trade routes extending as far away as Algonquin territory far to the north along the Mississippi River (Underhill 1953). Their life was engrained, their place deep-seated.

Early on, the band had fallen under the geographic and political influence of the French, as had all of Louisiana. One legend states that the island they occupy, which today sits in Louisiana's Terrebonne ("beautiful land") Parish, got its name from a Frenchman named Jean Charles who married one of their native women. The band also adopted the French language, which they continue to speak today. They had started out predominately Choctaw. Over time, the germinal Choctaw incorporated members of the more isolated neighboring Biloxi, the Louisiana Chitimach, and finally people of Acolapissa and Atakapa heritage. Throughout their union forward, the entire blended group lived a highly interactive and kinetic form of community life. While the residents of Boulder, Colorado, share more of a perception of collective culture than reality, the Isle de Jean Charles band, on the other hand, has always vigorously engaged in ongoing interaction and fellowship, partaking in customs, ceremonies, and daily interchange. Included in their traditions has been espousing a single inherited chief who acts as spokesman and arbiter for the whole collective. That custom has continued until today and has allowed them to stay cohesive (www.isledejeancharles.com).

Originally the Isle de Jean Charles land holding consisted of 22,000 acres. It now stands at a mere 320. The land began disappearing due to rising seawater, coastal erosion, and flooding starting in about the 1960s, conveying with it what Nixon has called the "slow violence of climate change" (2011). As is very often the case, however, the devastation of climate change has been greatly augmented by human enterprise. In the Jean Charles case, the initial island has been seriously fouled by three heedless practices: major oil extraction by the petrochemical industry; the gouging of numerous shipping channels to accommodate the oil industry, which has allowed massive water inundation and salinization of the soil; and the construction of dams, dikes, levees, and other flood-control measures by the US Army Corps of Engineers to protect industry and private holdings in ways detrimental to the Isle de Jean Charles people, proving a favorite saying of mine that "one person's protective levee is another person's flood."

In addition, the privatization of land, with its consequent uncontrolled use, allowed for soil-destabilizing logging. Unfortunately, the Isle de Jean Charles territory has always been considered federal property over which the tribe had no control. The Isle de Jean Charles are recognized as a Native American group by the state of Louisiana but not by the United States federal government. The tribe has never had a federal treaty, no deeded reservation, and, thus, no authority over their home; thus logging permits were granted over their territory. Now more than 98 percent of the land has been lost, and, as a result, over 75 percent of the tribe has been dislodged and displaced to nearby towns, cities, and regions farther afield. By 2009, only twenty-five houses remained, down from sixty-three in 2004. The land sank away so rapidly that by 2016 only one substantial garden remained (Maldonado 2014a; Jessee 2016). Beyond advancing climate change flooding, the oil and gas extractions have caused another sort of disaster: severe health issues on an epidemic level (Laska et al. 2005).

There occurred, however, a turning point, which even more vastly accelerated the land loss and disaster. It stemmed from the massive flooding that followed the series of brutal hurricanes that struck Louisiana between 2005 and 2008, all doubtless climate exaggerated. In particular, the island and people suffered acutely from Hurricane Katrina (2005), which savagely glutted the already disappearing dirt. That devastation was then exacerbated by Hurricane Gustav (2008). More recently, in 2016, a multiday torrent arrived from a low-pressure system combined with record amounts of atmospheric water vapor, resulting in two feet of rainfall over the area within three days. In short, the destructive climate change factor for the Isle de Jean Charles occurred from downward atmospheric circumstances as opposed to the upward ones (e.g., the jet stream in the Boulder case), both rather furtive expressions of climate change. The increased atmospheric warming, largely human induced, caused water vapor to increase very near to the land mass, and from it came an unabsorbable, heavy barrage (Mooney 2016). With the salt, the channels, the hurricanes, the climate, and human interference, the Isle de Jean Charles territory, its archaeology, and its history, floundered. Meanwhile, the US government and state of Louisiana, in order to protect the delta and coastal lands from the accumulating erosion ensuing from such events, formed a commission, the Coastal Protection and Restoration Authority, covering much of the endangered territory, but the Isle de Jean Charles lay outside the border of the proposed levee system and was not within its indemnity.

The tribe, with little choice, decided to move. Indeed, they are being called "America's First Climate Refugees," an epithet the tribe resents. As opposed to the naivety of the Boulder, Colorado, citizenry, the Isle de Jean Charles band has long been acutely away of their deteriorating situations and at first adapted indigenous ways to deal with it. Prior to their exclusion in the hurricane protection system, the tribal council had favored accommodating to the changes through restoration of wetlands to stem or reverse erosion and allow them to stay. For example, their original houses were constructed of a mixture of mud and moss with a domed roof covering of palmetto and floors of clay. But recognizing that these dwellings were not flood resistant, the people soon reinvented their built environment to elevated clapboard structures. All along they have shown keen assessment of their physical plane and its character, even as it changed, along with the flexibility to adapt and not just cope. But the exodus of people spurred by the exclusion from the hurricane protection system and the hurricanes themselves proved a pivot point (Maldonado 2014a and 2014b; Maldonado 2019).

Today, and for some years now, as they have noted the seemingly unstoppable loss of their environment, the band has astutely begun to plan their own relocation independently of governing bodies, county, state, and beyond. Their decision to relocate as climate change crept up did not happen instantaneously but developed over time, as houses tilted and then sunk and the fields that sustained them became silted and water-saturated troughs. The number of people gradually dwindled to a small remaining core (Maldonado et al. 2015; Maldonado 2019). Noting the plight, in 2007 the Gulf Coast Comprehensive Restoration Plan offered the tribal members relocation on an individual basis, that is, the plan proposed to buy homes one by one in order that individuals might relocate as they wished. The tribal council countered with the position that, as was their long-standing tradition, they intended to remain together and relocate as a whole. They further resolved to relocate in a fashion that would bring those already dispersed back into the fold. Maldonado and Peterson (2018) call the path taken by tribal leaders and the community members "community-led resettlement." It is a tenet that needs recognition and implementation worldwide in the face of climate change. Community is more than just houses; it is the people, their interface, their kinship, their familiarity, their knowledge.

Heritage communities nationwide and worldwide, like the Isle de Jean Charles band, are searching for support in the forced migration they face due to the disasters climate change is instigating. All are having to so on an ad hoc basis, as there is no body of dedicated laws or programs to which they can turn. The entire-group approach has so far struck governing bodies as far too unwieldy and expensive It often demands unachievable alliances between government and nongovernment agencies and allies. It further bears substantial up-front costs. Land must be obtained, and not merely in terms of singular plots but rather major holdings large enough for many. Homes and facilities need to be constructed. Confronted with these sorts of obstacles, obscuring what is perhaps the truth tantamount to denial, administrating bodies bury their heads in literal rain-drenched or desert-seared sand.

The Isle de Jean Charles band appealed beyond the state to both the US Congress and the United Nations. They have now received a significant grant, a $48 million allocation of US federal tax funds to move the entire community. The grant is a first in enacting climate change relocation. It is being called a "climate resilience" grant. It differs from prior funding in that most of the money the United States has so far provided for climate change issues has been for infrastructure, not human matters (Davenport and Roberson 2016). Still, the "where," "when," and "how" of the Isle de Jean Charles band's move is yet to be determined. The people only know that the next flood will finalize their total exile, so a plan must soon be devised and carried out. The intent is to save as much of their culture as they can, together. Maldonado (2019) calls the objective "cultural triage." As they work toward a proactive community-led resettlement, they hope to provide an exemplary model for other communities who need to choose a new site to sustain their entire communities, to bring back together people who have already been forced apart, and to maintain their family bloodline social connections, lifeways, integrity, and sovereignty (Maldonado and Peterson 2018).

Conclusion

While climate change has great impact for the Earth, its land, seas, and air, my ultimate focus in studying the impact of climate change and its undeniable link to disasters is on the continuity and diversity of human existence and the ability of people to thrive, if not their original their habitats, then in ones in which they can continue holistically as a group with their culture and lifeways.

By and large, climate scientists have focused on the globe's various ecological zones and what is happening to soil, vegetation, water systems, and weather systems. In fact, it is not just one culture that is truly at issue in the matter of climate change and its consequent calamities, it is all of them. To date, while the focus has been largely on the most vulnerable populations, the disenfranchised, poor, and marginal, as the upscale Boulder example shows, the impact of the devastating union of climate and catastrophe is on all people everywhere. The capacity to surmount disasters or adapt to them, to contend with fluctuating or grievous circumstance and prevail, amounts to peril for everyone. Indeed, numerous experts see the matter of whether human lifeways can alter in relation to critical climate phenomena as indicating whether the species retains enough flexibility to adjust and keep thriving or by inattention and not changing, "dig its own grave."

The matter of global warming and its impact on accelerating disasters unquestionably set a critical stage, bringing out existential concerns, action, and arenas of discourse within a society, local and beyond. The hope, as Bergman (2020) points out, is that the intermesh of the two becomes a great motivator of social action, for social action motivates change. People do not sink into inertia when faced with calamitous situations, they react. They throw into stark light not just inequalities but conflicts and struggles over power and over cultural differences within the entire human realm—in a nutshell, the whole spectrum of social matters as well as physical ones. They raise questions of a metaphysical nature as well. All in all, climate-caused disasters present extraordinary examples of the fluid quality of culture, the invention and reinvention of cultural goods, and the areas of harmony, disjuncture, inconsistency, and coherence (Hoffman 2016, 2002). Furthering the matter, we live today in an era of expanding globalization; progressive corporate, state, and international hegemony; the exploitation of resources; and the continuation of global warming, which causes the loss of land and water, resettlement of groups, and demographic and behavioral shifts that force humans to occupy less safe habitats and embrace more perilous technology (Hoffman 2016, 2002)

Stopping and correcting climate change is certainly needed, but along with it, in order to prevent attendant disasters, the other great necessity is to reduce risk everywhere, that is, to end vulnerability. There are numerous ways to expedite human adaptation to the Earth's changing circumstances, to increase capacity, augment flexibility, and heighten safety, including of people's subsistence, social structure, cosmology, cultural sustainability, and potential resettlement as many of the preceding chapters in this volume show. There are ways to advance agency and place-based programs. We cannot change some of the kinds of disasters, but those caused or enhanced by climate change, we can deter. I further posit that the anthropology perspective and approach can help to integrate the human and natural systems. Stopping the global processes that are creating increased vulnerability is like, according to Zygmunt Bauman, trying to prevent weather change itself (2000: 33), but there are ways around the weather. Barnes et al. (2013) emphasize the increasingly critical role of anthropological contributions to discussions on climate change by including not only physical descriptions of the phenomena occurring but also bringing in questions of different cultural groups' receptivity to climate policies, that is, the whole conundrum of what the effects might be on lives and the viability of lifeways. In agreement, Button (2010: 248) contends that we must change the public discourse on what is acceptable to say about disaster, a discourse that so far usually tries to maintain an emphasis on scientific and technical aspects while avoiding other realms, such as values, ethics, policy, and politics of laypeople. That goes for climate change as well. Along this line, Crate (2011) agues for a climate ethnography that is multi-sited, collective, and inclusive of all those concerned in order to trace global processes locally and track how they are being articulated via local knowledge systems. Only such an ethnography will elucidate the convergences and conflicts between the global-to-local conversations and understandings about climate change. Cox and Cox (2016: 324) argue for a correction of the neoliberal and developmental policies, which they call the disciples of creative destruction, that impel both climate change and disaster.

David McDermott Hughes, in his article on climate change in *American Anthropologist* (2013), uses the term "innocence." He means it to refer to ignorance of geography and what climate change is bringing, but he also gives a nod to its meaning in morality. He talks of activists and social justice. Considering the clear and ever-present link of climate change to disasters, present and looming, often appallingly severe, I also see the issue as close to something akin to the criminal charge of "reckless endangerment." It leads not just to damage but also to death. Treating nature as a market or a plaything without understanding the real "nature" of nature; remaining ignorant of geography, weather, and past events; and being guileless about climate change are actions that make for an innocence the world cannot afford.

Susanna M. Hoffman is an internationally recognized expert on disaster. She is the author, coauthor, and editor of twelve books, including *The Angry Earth: Disasters in Anthropology Perspective* (first edition, 1999), *The*

Angry Earth (second edition, 2020), *Catastrophe and Culture: The Anthropology of Disaster* (2002), and *Disaster upon Disaster: Exploring the Gap Between Knowledge, Policy, and Practice* (2020); two ethnographic films, including *Kypseli: Women and Men Apart: A Divided Reality* (1976); and more than forty articles and chapters. She initiated the Risk and Disaster Thematic Interest Group for the Society for Applied Anthropology and is the founder and chair of the Risk and Disaster Commission for the International Union of Anthropology and Ethnographic Sciences. She was the first recipient of the Fulbright Foundation's Aegean Initiative dealing with the Greek and Turkish earthquakes and helped write the UN Statement on Women and Disasters. She is a frequent national and international speaker and also serves on the Task Force on World Food Problems.

References

Bauman, Zygmunt. 2000. *Liquid Modernity*. Cambridge: Polity Press.

Barnes, Jessica, M. Dove, M. Lahsen, A Matheews, P. McElwee, R. McIntosh, F. Moor, J. O'Reilly, B. Orlove, R. Puri, H. Weiss, and K. Yager, 2013. "Contributions of Anthropology to the Study of Climate Change." *Nature Climate Change* 3(6): 541–44.

Bergman, Ann. 2020. "The Time Matter Matters: Disasters as a (Potential) Vehicle for Social Change." In *Disaster upon Disaster: Exploring the Gap between Knowledge, Policy, and Practice*, edited by S. Hoffman and R. Barrios. 313–31. New York: Berghahn Books.

Boulder Daily Camera. 2013. 9, 10, 11, 12, and 21 September.

Burton, Brian. 2012. "Climate Change and the Effects on Commercial Buildings, Building Science Forum©: Potential Impact of Climate Change on Building Envelopes." *Monster Commercial Real Estate*, 1 February.

Button, Gregory. 2010. *Disaster Culture*. Walnut Creek, CA: Left Coast Press.

Casey, Michael, and P. Whittle. 2017. "Spread by Trade and Climate, Bugs Butcher America's Forests." Associated Press, 4 January.

Cox, Stan, and P. Cox, 2016. *How the World Breaks: Life in Catastrophe's Path, from the Caribbean to Siberia*. New York: The New Press.

Crate, Susan, 2011. "Climate and Culture: Anthropology in the Era of Contemporary Climate Change." *Ethnography* 14(1): 145–94.

Davenport, Coral, and C. Robertson. 2016. "Resettling the First American 'Climate Refugees.'" *New York Times*, 2 May.15

EM-DAT (International Disaster Database). 2016. Retrieved 15 February 2021 from https://www.emdat.be/index.php?

Eriksen, Thomas. 2016. *Overheating: An Anthropology of Accelerated Change*. London: Pluto Press.

Fiske, Shirley, S. Crate, C. Crumley, K. Galvin, H. Lazrus, G. Luber, L. Lucero, A. Oliver-Smith, B. Orlove, S. Strauss, R, Wilk. 2014. *Changing the Atmosphere: Anthropology*

and Climate Change; Final Report of the American Anthropology Association Global Climate Change Task Force. Arlington, VA: American Anthropology Association.
Fiske, Shirley, and E. Marino. 2020. "Slow Onset Disasters: Climate Change and the Gaps between Knowledge, Policy, and Practice." In *Disaster upon Disaster: Exploring the Gap between Knowledge, Policy, and Practice*, edited by S. Hoffman and R. Barrios, 139–171. New York: Berghahn Books.
Grigoryan, Armen. 2015. "Technological Hazards: From Risk Reduction to Recovery." *Our Prospective*, United Nations Development Program, 11 and 12 February.
Hoffman, Susanna. 1999. "After Atlas Shrugs: Cultural Change or Persistence after a Disaster." In *The Angry Earth: Disaster in Anthropological Perspective*, edited by Anthony Oliver-Smith and Susanna M. Hoffman, 302–26. New York: Routledge.
———. 2014. "Culture: The Crucial Factor in Hazard, Risk and Disaster Recovery; The Anthropological Perspective." In *Hazards, Risks, and Disaster in Society: A Cross-Disciplinary Overview*, edited by A. Collins et al., 289–304. Amsterdam: Elsevier.
———. 2016. "The Question of Cultural Continuity and Change after Disaster: Further Thoughts." In "Continuity and Change in the Applied Anthropology of Risk and Disaster," edited by A. J. Faas. Special Issue, *Annals of Anthropological Practice* 40(1): 39–51.
———. 2017. "Disasters and Their Impact: A Fundamental Feature of Environment." In *Routledge Handbook of Environmental Anthropology*, edited by Hele Kopnina and Eleanor Shoreman-Quimet, 193–205. New York: Routledge.
Hoffman, Susanna, and A. Oliver-Smith. 2002. *Catastrophe and Culture*. Santa Fe: School of American Research.
Hughes, David McDermott. 2013. "Climate Change and the Victim Slot: From Oil to Innocence." *American Anthropologist* 115(4): 570–81.
Jessee, Nathan. 2016. "Encroaching Gulfs: Memory, Visibility, and Solidarity in Response to Louisiana's Coastal Hazards." Talk delivered at the American Anthropology association Meeting, Minneapolis, MN, 16 November.
Lafferty, Kevin. 2009. "The Ecology of Climate Change and Infectious Diseases." *Ecology* 90(4): 888–900.
Laska, Shirley, G. Wooddell, R. Hagelman, R. Grambling. 2005. "At Risk: The Human Community and Infrastructure Resources of Coastal Louisiana." *Journal of Coastal Restoration* 44: 90–111.
Maldonado, Julie. 2014a. "A Multiple Knowledge Approach for Adaptation to Environmental Change: Lessons Learned from Coastal Louisiana's Tribal Communities." *Journal of Political Ecology* 21: 61–82.
———. 2014b. "Everyday Practices and Symbolic Forms of Resistance: Adapting to Environmental Change in Coastal Louisiana." In *Hazards, Risks, and Disaster in Society: A Cross-Disciplinary Overview*, edited by A. Collins et al., 201–316. Amsterdam: Elsevier.
———. 2019. *Seeing Justice in an Energy Sacrifice Zone: Standing on Vanishing Land in Coastal Louisiana*. New York: Routledge.
Maldonado, Julie, A. Naquin, T. Dardar, S. Parfait-Dardar, and B. Bagwell. 2015. "Above the Rising Tide: Coastal Louisiana's Tribal Communities Apply Local Strategies and Knowledge to Adapt to Rapid Environmental Change." In *Disaster's Impact on Livelihood and Cultural Survival: Losses, Opportunities, and Mitigation*, edited by M. Companion, 239–53. Boca Raton, FL: CRC Press.
Maldonado, Julie, and K. Peterson. 2018. "A Community-Based Model for Resettlement: Lessons from Coastal Louisiana." *The Routledge Handbook of Environmental*

Displacement and Migration, edited by R. McLeman and F. Gemenne, 289–99. New York: Routledge

McGuire, Bill, 2016. "How Climate Change Triggers Earthquakes, Tsunamis, and Volcanos." *The Guardian*, 16 October.

Mooney, Chris 2016. "What We Can Say about the Louisiana Floods and Climate Change." *Washington Post*, 15 August.

Nixon, Rob. 2011. *Slow Violence and the Environmentalism of the Poor*. Cambridge, MA: Harvard University Press.

Oliver-Smith, Anthony. 2009. *Development and Dispossession: The Crisis of Displacement and Resettlement*. Santa Fe: School for American Research.

Oliver-Smith, Anthony, and S. Hoffman. 1999. *The Angry Earth: Disaster in Anthropological Perspective*. New York: Routledge.

Prairie Mountain Publishing. 2013. *A Thousand Year Rain: The Historic 2013 Flood in Boulder and Larimer Counties*. Battle Ground, WA: Pediment Publishing.

Squires, Gregory, and C. Hartman. 2006. *There's No Such Thing as a Natural Disaster*. New York: Routledge.

Underhill, Ruth. 1953. *Red Man's America*. Chicago: University of Chicago Press.

Wallace, Anthony. 1956. *Tornado in Worcester*. Disaster Study #3. Washington, DC: Committee on Disaster Studies, National Academy of Sciences-National Research Council.

Wilson, Edward. 2016. *Half-Earth: Our Planet's Fight for Life*. New York: Norton.

Afterword

Toward Eco-Socialism as a Global and Local Strategy to Cool Down the World-System

Hans A. Baer

For several years, Thomas Hylland Eriksen and his collaborators have discussed how the world system has been overheating due to the drive for profits, economic growth, increased production and consumption, and high dependency on fossil fuels, all of which result in an increase in greenhouse emissions and ultimately contribute to anthropogenic climate change (Eriksen 2016; Stensrud and Eriksen 2019). In this volume, Eriksen with coeditors Susanna M. Hoffman and Paulo Mendes, along with contributors in various countries, make a significant contribution to the relatively young and still burgeoning anthropology of climate change, Their volume delineates strategies to seek to cool down the world system, not so much at the global level but at localized levels. Their book is literally a tour de force focusing on impacts of overheating in various locations, including Northwest Namibia, the Eastern Himalayas of India, New Zealand, French Polynesia, Belém in northeast Brazil, Bangladesh, East Africa, the Austrian Alps, the Elbe River Valley near Dresden, Native American communities of the American Southwest, the remote montane of Portugal, the Isthmus of Tehuantepec in Oaxaca, Mexico, and the state of Georgia in the United States, along with variegated efforts to cool down the ecological and climatic systems of these locales. Included is a chapter depicting disasters consequent to climate change in Colorado and Louisiana. As the various chapters reveal, the efforts to cool down the climatic and ecological systems thus far are meeting with mixed results at best.

Climate Capitalism

There is much debate on how to mitigate climate change. Proposed solutions range from shifting from fossil fuels to alternative energy sources (such as wind, solar, geothermal, biofuel, and even nuclear sources), planting trees, developing more environmentally sustainable technologies, developing and using energy-saving devices, retrofitting buildings with such devices, improving public transport and inducing transitions away from private vehicles, and geoengineering. Many of the specific proposals, albeit not all of them, would be modest steps toward climate change mitigation.

Capitalism has a capacity to turn tragedies of all sorts into profit-making opportunities, thus prompting the development of *disaster capitalism* (Klein 2007; Lowenstein 2015) and, in the case of climate change, *climate capitalism* (Newell and Paterson 2010). Although government officials, politicians, and climate scientists tend to be publicly visible in the climate change mitigation discourse under the UN Framework Convention on Climate Change (UNFCCC), private corporations tend to be less so. Whereas corporations such as the now-defunct Global Climate Coalition were often part and parcel of climate denialism, more and more corporations have come to assert that they are striving to achieve environmental sustainability and reduce greenhouse gas emissions in their business practices. Wright and Nyberg (2015) maintain that corporate environmentalism tends to build on the notion of ecological modernization that stresses the ability to come up with technological innovations that are environmentally friendly.

Ecological modernization, which entails a shift to renewable energy sources, increased energy efficiency, and a numerous array of other techno-fixes, constitutes the overarching agenda of climate capitalism. This is most profoundly seen in the case presented in *Cooling Down* of the wind power development project in the Isthmus of Tehuantepec, which Leppert and Barrio characterize as an illustration of *green neoliberalism* as it involved a public-private partnership between a renewable energy company and the Mexican state. While the project has the potential to reduce greenhouse gases, it actually has exacerbated existing class and ethnic inequities in the region, as is illustrated in a case study of a small Zapotec town, the pseudonymous Bina Za. As Maldonado and Middleton illustrate in their chapter, various Southwest Native American communities are leading the way in their region by turning toward renewable energy sources in a part of the country characterized by plentiful sunshine and powerful winds. In her ethnographic depiction of the highest glacier ski area situated in the idyllic Tirolean Pitztal Valley of Austria, Nöbauer discusses how the ski resort company is turning to an array of environmentally dubious techno-fixes. This includes an All Weather Snowmaker to counter diminishing snowfall, which produces snow even at relatively high temperatures. The irony of high-profile ski areas such as the one at Pitztal is that highly affluent people crisscross the globe to reach them in airplanes that spew greenhouse gas emissions and contribute to the melting of glaciers in numerous places. Fortunately, a large number of environmentalists argue that the downhill ski industry is environmentally unsustainable on a number of fronts, including in its deforestation of mountainsides resulting in destruction of significant carbon sinks and contribution to soil erosion and mudslides.

Existing climate regimes, ranging from the UNFCCC to the European Union's Emissions Trading Scheme (EUETS), to various national emissions trading schemes, have proven ineffective in significantly cutting back on greenhouse gas emissions. In contrast to the Conference of the Parties 15 (COP15) in Copenhagen in 2009, many political pundits, and even some environmentally focused nongovernmental organizations (NGOs), celebrated the Paris Agreement at Conference of the Parties 21 (COP21) in Paris in late 2015. In this effort, the United States and China joined hands with virtually all other nations in agreeing to limit emissions with the parameters of a 2-degree-Celsius, even a 1.5-degree-Celsius, world. However, given the fact that the emission targets that nations have voluntarily pledged would only achieve a 2.7- to 3.5-degree-Celsius world, the Paris Agreement still operates within the parameters of the existing capitalist world system. Thus, large numbers of both social scientists and climate activists are skeptical of the excitement expressed by the UNFCCC delegates and politicians by the outcomes of the Paris assemblage (Lyster 2017). Shortly after his election in late 2016, Donald Trump withdrew the United States from the Paris Agreement.

Parks and Roberts (2010) maintain that the international climate justice movement has an uphill struggle given that many of its actors are opposed to emissions-trading schemes, along with offset schemes such as the United Nations's (UN) Clean Development Mechanism and the Reducing Emissions from Deforestation and Degradation (REDD) program. They favor carbon taxes on the polluters and compensation for the victims of climate change, particularly in the Global South. Climate regimes at international, regional, and national levels have tended to accept the emissions-trading schemes and market mechanisms as axiomatic. For instance, REDD conflicts with the subsistence needs of forest dwellers around the world, such as the Sengwer people in Kenya's western highland studied by Castro in his *Cooling Down* chapter.

Conventional economists contend that market mechanisms, such as emissions-trading schemes and carbon offsetting, will solve the "diabolical problem" of climate change. As a critical anthropologist and historical social scientist, I ask: "How can you expect the system that created the problem to solve the problem?" This question is especially relevant because the problem in this instance—anthropogenic climate change—is not a peripheral feature or unfortunate economic externality of global capitalism. It cannot be easily expunged; rather, it is a significant byproduct of continual expansion of production and the promotion of growing levels of consumption. Nicholas Stern (2015), a mainstream economist who has concerned himself with climate change mitigation for roughly the past fifteen years, argues that various figures have played a significant role in climate action. In his view, they range from Pope Francis to members of royal families to actors, celebrities, sports stars, businesspeople, academics, teachers, and young people. While I would include Greta Thunberg as a student climate activist fitting the bill, it is notably ironic that the carbon and ecological footprints of many of those figures mentioned by Stern contribute to the climate crisis, not to mention the grossly uneven utilization of global natural resources.

Toward an Integrated Critical Understanding of Climate Change

The effort to critically examine and respond to the adverse impacts of climate change on humanity and the ecosystem must be a multidisciplinary effort. It entails collaboration between climate scientists, Earth system scientists, energy analysts, and physical geographers, on the one hand, and social scientists, including anthropologists, archaeologists, sociologists, political scientists, and human geographers, on the other hand (Baer and Singer 2018). Such multidisciplinary endeavors are exemplified in chapters on climate change adaptation in Bangladesh by Siddiqui, Sikder, and Bhuiyan and Aotearoa New Zealand by Schneider and Glavovic.

The reality is that natural scientists and mainstream economists tend to dominate much of the discourse on climate change, as is evidenced by the composition of the Intergovernmental Panel on Climate Change. Rockström (2011: 26–27) advocates moving beyond the "disciplinary status quo" characteristic of the sciences and universities and toward emphasis on "more integrated and problem-solving programmes." Collaboration serves to bring the strengths from various disciplines to what is a monumental but undeniably vital task: understanding and effectively responding to climate change. Climate change research needs to move beyond research centers and universities. It needs to collaborate with communities, particularly those that are being adversely impacted by climate change, as well as NGOs, progressive political parties, women's groups, indigenous communities, and climate action groups that are pushing for effective climate change mitigation strategies informed by a strong sense of social and climate justice.

Several of the chapters in this anthology discuss the significance of traditional ecological knowledge (TEK) in enlightening climate science and other scholars operating in the climate change space. Schnegg found that his ǂnūkhoen informants engage in a pattern of *environmental pluralism* in which they intersperse indigenous, Christian, and scientific discourses to make sense of droughts in their region. The upland villagers with whom Aisher worked the Indian Himalayas are experiencing increasing storms and other extreme weather events that remind people of their need for *conviviality* with powerful supernatural forces. They also are experiencing what some scholars term *solastalgia,* or a sense of loss due to dramatic changes in what once was a more reassuring landscape. As Lauer et al. report in their chapter, the fishers on the French Polynesian island of Moorea have a longer-term interpretation of the periodic crown-of-thorns starfish outbreaks that marine biologists tend to view as adversely impacting coral reefs. This raises the thorny question as to how much credence to give a TEK, including that of Christian fundamentalist perceptions of climate change, in coming to terms in communicating the seriousness of climate change and discussing radical strategies to address it. While getting people to come to terms particularly with the reality of anthropogenic climate change, prompting them to take radical climate action to address it is a much more difficult process. This is because it would require drastic alterations of existing political-economic structures and lifestyles, particularly those of the rich and powerful.

Several of the chapters focus on strategies by which communities and people around the world are seeking to mitigate and even adapt to climate change. Soares discusses the shifting hydric landscape in the city of Belém. The technocratic Una Watershed Project has sought to drain the wetlands, construct soil embankments, and develop a sanitation program. Nevertheless, flooding of the Guarjará Bay constitutes an indicator that nature has a way of reclaiming its own pathways. As Albris found in his ethnographic work in the Elbe River Valley, although peri-urban settlements near Dresden experienced massive flooding in 2002 and 2013, many residents adopted a stance of *defiant acceptance* of the possibility of climate change–related floods. They were not ready to forgo the sense of social cohesion that developed as a result of confronting disaster. As Siddiqui, Sikder, and Bhuiyan discuss in their chapter touching upon migration of Bangladeshi peasants to cities of varying size, it is a process that intricately interweaves the search for job opportunities and urban amenities with escaping climate change–related flooding in low-lying coastal and riverine areas. However, ultimately, while people may be able to adapt to climate change in the short run, any community has a limited amount of resilience. Thus, mitigation or drastically reducing greenhouse gas emissions is the more significant imperative, which raises the question of how much mitigation is possible within the parameters of global capitalism. This is illustrated in Hoffman's chapter on climate change–related disasters as exemplified by the Boulder flood in 2013 and the steady contraction of the Isle de Jean Charles band of Biloxi-Chitimacha-Choctaw Native Americans in the Mississippi Delta. Both of these case studies illustrate how infrastructure projects initially intended to ward off natural forces backfire. Rewilding Europe's plans to introduce large animal species wildlife in the Faia Brava Reserve in Portugal is guided as Sá aptly observes by capitalist principles. It constitutes a collaborative venture involving NGOs, rural landowners, agriculturalists, and banks, and in a loose sense mimics the type of green capitalism espoused by the US-based Breakthrough Institute.

The Need for an Alternative World System

Ongoing global warming and associated climatic and other anthropogenic environmental changes raise the question of how long humanity can thrive into and beyond 2100. While a large section of the international elite has come to recognize the seriousness of climate change, the solutions they propose under the guise of ecological modernization, green capitalism, and existing climate regimes are insufficient to contain catastrophic climate change. As a result, perhaps more than any other environmental crisis, anthropogenic climate change forces us to examine whether global capitalism needs to be transcended and humanity needs to develop a new approach, as some would see it, along eco-socialist lines. Indeed, Dawson (2017: 299) suggests that climate change increases the likelihood of a revolution in the future and argues that "humanity needs a global people's movement to battle climate chaos while generating work for the disenfranchised masses of the world."

Eco-socialism still remains a vision, but one in this age of climate change that merits thoughtful consideration. It entails the following dimensions or desired goals: (1) a global economy oriented to meeting basic social needs, namely adequate food, clothing, shelter, and healthful conditions and resources; (2) a high degree of social equality and social fairness; (3) public or socialized ownership of productive forces at national, provincial, and local levels; (4) representative and participatory democracy; (5) environmental sustainability; and a (6) commitment to a safe climate

(Loewy 2015; Baer 2018a; Saul 2019; Albritton 2019). Ultimately, the shift to eco-socialism in any country would have to be part of a global process that no one can fully envision. Antisystemic movements will have to play an instrumental role in bringing about the political will that will enable the world to shift to eco-socialism. As Magdoff and Williams (2017: 311) observe, any revolutionary upheaval aiming to create a global ecological society "will have to dwarf mobilizations we have seen recently around the world" by linking up many of the struggles seeking to achieve social, economic, and environmental justice.

The transition toward an eco-socialist world system is not guaranteed and will require a tedious, even convoluted path. Nevertheless, while awaiting the "revolution," so to speak, progressive people can work on various system-challenging transitional reforms that open the door to wider socioecological revolution. These include: (1) the creation of new anticapitalist left parties designed to capture the state: (2) the implementation of emissions taxes at sites of production that include efforts to protect low-income people; (3) public ownership of the means of production; (4) increasing social equality, including gender, ethnic, class, and racial equality, within nation-states and between nation-states, and achieving a sustainable global population; (5) the implementation of socialist planning and workers' democracy; (6) meaningful work and shortening of the working week; (7) development of a steady-state economy; (8) the adoption of renewable energy sources, energy efficiency, appropriate technology, and the creation of green jobs; (9) sustainable public transportation and travel; (10) sustainable food production and forestry; (11) resistance to the capitalist culture of consumption; (12) sustainable trade; (13) sustainable settlement patterns and local communities; and (14) demilitarization (Baer 2018a: 201–53). These transitional steps constitute loose guidelines for shifting human societies or countries toward eco-socialism and a safe climate, but it is important to note that these steps would entail global efforts, including the creation of a progressive climate governance regime. Constructing an alternative to global capitalism is the ultimate climate mitigation strategy, even though it will not be achieved anytime soon, if indeed ever. There is the distinct danger that humanity will continue to overheat the planet rather than cool it down, but it is essential that anthropologists play a role in the latter process.

Antisystemic social movements will have to play an instrumental role in bringing about the political actions that will enable the world to shift to eco-socialism. Given the failure to date of established international and national climate regimes to adequately contain the climate crisis, efforts to create a radical climate governance process will have to come from below. Ultimately, the climate justice movement, one that remains quite disparate, will have to form strong alliances with other antisystemic movements, perhaps particularly the anticorporate globalization or social justice and labor movements. Hardt and Negri (2009: 94–95) assert that "only movements from below" possess the "capacity to construct a consciousness of renewal and transformation," one that "emerges from the working classes and multitudes that automatically and creatively propose anti-modern and anticapitalist hopes and dreams." A viable anticapitalist movement will have to address the material impoverishment of much of the world's population, which includes many of the peoples that contributors to this volume have studied. Many parties, ranging from the World Bank to entertainment celebrities, make appeals to "eradicate extreme poverty" or "make poverty history." However, "make wealth, particularly extreme wealth, history," and the eradication of poverty will follow. Personally, I hope that the eco-socialist vision will serve as an integrative focus for antisystemic movements, including the climate justice movement, within nation-states and transnationally, although I recognize how daunting this task will be.

Much of the climate movement is focused on moving beyond fossil fuels, a worthwhile endeavor. However, just as capitalism operated on other forms of energy prior to the Industrial Revolution, capitalism could theoretically operate on renewable sources of energy, a form of *green neoliberalism,* which will require enormous sources to develop and maintain. For example, the Koch brothers have become major investors in windfarm, solar energy, and biofuel projects. Subalterns around the world are increasingly having their land and labor expropriated by mining companies, including ones that are providing resources for renewable energy operations and supposedly green technologies, such as electric cars and autonomous vehicles (Arboleda 2020). Even though some variant of green capitalism might bring down greenhouse gas emissions to some degree, it would not address the social inequities, limited democracy, militarism, threat of nuclear warfare, and global pandemics such as COVID-19 that are byproducts of global capitalism.

Climate justice and social activists face an incredibly daunting task. The next two or three decades, if not the immediate next one, will bring great hardship for much of humanity, exacerbated by the rise of authoritarianism, accompanied by the nexus between corporations and governments constituting surveillance capitalism, including in the US White House and countries in the Global South such as Brazil and the Philippines. As climate change increasingly affects humans and nonhuman beings, the powers that be will be inclined to construct a Fortress World to protect their privileges, borders, and market system. There are no easy fixes to these grim realities, but it is imperative that climate activists become climate justice activists as part of a meta-movement to challenge and transcend global capitalism. I encourage my fellow anthropologists and other social scientists to not only study the global climate movement but become part of it in order to facilitate the cooling down of an overheated planet.

Airplanes, Climate Change, and COVID-19

While a growing number of critical scholars acknowledge that global capitalism constitutes the overarching driver of anthropogenic climate change, one of the smaller elephants in the room has been the aviation industry. The number of airplane flights worldwide has been growing, at least prior to COVID-19, which forced reluctant governments around the world to temporarily restrict the number of flights. Air travel, along with cruise ships, has played a key role in turning a localized epidemic in Wuhan in China into a global pandemic. Prior to this unfortunate event, aircraft flights were contributing 5 to 6 percent of greenhouse gas emissions, not only in the form of carbon dioxide but also nitrous oxide, methane, and ozone (Baer 2020a). Despite repeated claims by airline companies that they were gradually turning to more fuel-efficient and aerodynamic aircraft, these technological innovations were offset by a rise of roughly 5 percent per annum (in keeping with the Jevons paradox, or rebound effect, where the economical use of energy results not in diminished consumption but an overall increase). This rise was even higher for affluent people in China, India, and other developing countries, who started to emulate the habits of their counterparts in developed countries. Airplanes of many sorts (commercial, military, and private) have become sources of tremendous profit and integral components of modernity and the capitalist world system. Furthermore, aviation companies are an excellent example of how corporate profit making is subsidized by public funds.

Airplanes serve to transport both human actors and commodities to keep the world system operating and overheating. However, they do so with dire environmental, climatic, and health consequences (Baer 2020a). The human actors who rely on air travel include businesspeople, politicians, celebrities, the super-rich who own multiple homes in far-flung locations, sports teams, tourists, academics, international university students, other students studying abroad for short-term stints, and even UN climate change conference delegates and observers, environmentalists, and climate activists. The list seems almost endless but, with some exceptions such as low-paid migrant workers, refugees, and rank-and-file military personnel, it consists of relatively affluent people.

Furthermore, air cargo constitutes the underbelly of the airline industry. Its operations often occur at night and at secure inaccessible facilities, bonded warehouses, and multimodal logistics centers, often located some distance from passenger terminals. Corporate globalization has resulted in a growing reliance on air cargo to quickly transport manufacturing components and products. The extractive industry around the world, including in the Atacama Desert of Chile and in Australia, has contributed to air travel by transporting their workers to mining sites (Arboleda 2020: 129–30). However, both extractive industries, such as iron ore and coal, and the petroleum industry are highly dependent upon marine shipping. Paskal (2010: 80) describes sea shipping as the "circulatory system of the global economy" in which about "90 percent of the world trade products are carried at some point."

Last but not least, militarism is highly dependent on aircraft, whether it is in the form of propelling jet fighters and drones or transporting military cargo and personnel around the world to engage in imperialist ventures. Historically, there has been a powerful nexus between the aviation industry—whether aircraft manufacturing or the airlines—and airport construction. This nexus has been strong around the world because of the military significance of aviation, particularly for the United States, but also Britain, Germany, the former Soviet Union and Russia today, and most recently China.

While infectious diseases can be transmitted vis-à-vis ship and train travel, airplane flights have elevated the spread of diseases to a new level. The internal environment of the airplane is an unhealthy one, with little oxygen, germs carried by both crew and passenger, and low-level electromagnetic radiation from flight equipment and x-rays encountered at high altitudes. The outbreak of severe acute respiratory syndrome (SARS) in late November 2002 and lasting to July 2002, which according to the US Centers for Disease Control and Prevention infected more than 8,000 people and killed 774 people as it spread from China to at least twenty other countries, illustrates how air transportation can serve as a rapid transmitter of infectious disease. Tragically, by comparison to SARS, the role of airplanes, as well as cruise ships, in spreading COVID-19 has been exponentially more profound, turning a local epidemic starting out in Wuhan, China, into a global pandemic. Despite the role of airplanes in disseminating SARS and COVID-19, the European Union has permitted airlines to fly with all seats full, thus violating social distancing practices in other walks of life. Developed capitalist societies are the most reliant on air travel, both domestically and internationally. Only time will tell how COVID-19 will adversely affect the health of people in developing or peripheral capitalist countries, such as India, Indonesia, and those in sub-Saharan Africa. Latin America, particularly Brazil, already has emerged as an epicenter of COVID-19. In contrast to the developing capitalist countries, which are quite mixed in terms of the quality of their health infrastructures, the impact upon the former could be devastating in ways that still are presently difficult to ascertain.

Ironically, the coronavirus pandemic has forced governments around the world, in an effort to stem even further spread of COVID-19, to ground the vast majority of international flights as well as many domestic flights. The International Air Transport Association terms the pandemic as "apocalypse now." In Australia, Virgin Airlines is seeking to receive a bailout from the Coalition Party government. Qantas CEO Alan Joyce, reportedly the highest-paid Australian CEO, opposes the bailout for Virgin unless his company obtains one itself on the grounds that Virgin is mostly owned by government-supported foreign airlines. Joyce has been trying to revive Qantas for the moment by pressuring the federal government to allow his airline to fill airplanes for domestic flights, with no empty seats to enable social distancing and making the wearing of masks optional.

A far more preferable strategy would be to allow Virgin to collapse and nationalize Qantas, as it once was, dismiss Joyce, shift to a much-reduced national airline, and create a nationalized solar-powered railway system that greatly improves upon existing state railway systems. Implementing such measures would require tremendous political will, one missing in both the Coalition and Australian Labor Parties, whether in or out of government. While it is far too early to say when humanity will return to some state of normalcy in a post-COVID-19 world, a consolidated airline industry in the wake of airlines that will fall by the wayside, a scenario that has occurred previously in a highly competitive industry, may seek to rise like a phoenix, offering perhaps relatively inexpensive flights, particularly with a low oil price, at least in the short run.

While academics are not generally ranked among the global elites, many in full-time positions, including anthropologists, and particularly those at elite institutions and at higher administrative levels, fall into the ranks of frequent flyers (Baer 2018b). Much of this behavior has been driven by the dictates of the corporate university structure, which seeks to internationalize itself in a competitive bidding war for student numbers, including overseas students, and research funds. This has occurred as governments have reduced funding for particularly public universities. While undoubtedly the vast majority of anthropologists around the world accept climate science, climate change has already adversely affected many of the subjects of their research and will continue to do so as humanity plunges further into the twenty-first century. They often seem to be unaware—or perhaps they compartmentalize their awareness—that their flying may be contributing to a four-degree or more world by the year 2100 if emissions from many sources are not quickly abated in the next few decades. Elsewhere, I have sought to grapple with strategies as to how anthropologists contribute to the cooling down of the planet by reconfiguring the amount of flying that they do in terms of attending conferences, giving guest lectures in distant universities, and conducting fieldwork (Baer 2019, 2020b).

Conclusion

As the chapters in *Cooling Down: Local Responses to Global Climate Change* reveal, anthropogenic climate change has been inducing and will continue to induce severe economic, political, military, sociocultural, and health consequences as the twenty-first century unfolds. Ongoing global warming and associated climate and other anthropogenic environmental changes raise the question of how long humanity can thrive, at least in its present numbers and occupying much of its present places of habitation, into and beyond 2100. The critical anthropology of climate change, the perspective from which I operate, posits that global capitalism has been around for about five hundred years. It has come to manifest so many contradictions, including ecological and climatic crises, that it needs to be transcended to ensure the survival of humanity and animal and plant life on a sustained basis. This points to the need for a critical anthropology of the future that calls for a cooling down of the planet and is informed by an environmental and social need for an alternative world system. Such as system would be committed to social justice, democracy, and environmental sustainability, one that in certain circles is referred to as eco-socialism. However, a robust eco-socialism needs to grapple with and draw upon other anticapitalist discourses, including eco-anarchism and ecological economics (Biehl 2015; Laurent 2020).

In his version of eco-anarchism, Ted Trainer urges the affluent, particularly in the Global North but also the Global South, to adopt a *Simpler Way*, to literally cool down. As anthropologists know from their ethnographic research on indigenous and peasant peoples around the world, without romanticizing them, they have been purveyors of a Simpler Way for eons. We should be addressing to right to not work and ways to overcome the current "work ethic." For the majority of people on the planet, personhood, or the right to personhood, depends enormously on work, on what we do (let alone *la distinction*). Emissions will not change if we keep working/producing as we do and as much as we do. We should also be addressing debt (or the emission of money/debt) in general, and sovereign debt in particular, and therefore the economic growth imperative.

Hans A. Baer is principal honorary research fellow in the School of Social Political Sciences at the University of Melbourne. He has published twenty-three books and some two hundred book chapters and academic articles on a diversity of research topics, including Mormonism; African American religion; sociopolitical life in East Germany; critical health anthropology; medical pluralism in the United States, United Kingdom, and Australia; the critical anthropology of climate change; and eco-socialism. Baer has published five climate change–related books: *Global Warming and the Political Ecology of Health* (with Merrill Singer, Left Coast Press, 2009), *Global Capitalism and Climate Change* (AltaMira, 2012), *Climate Politics and the Climate Movement in Australia* (with Verity Burgmann, Melbourne University Press, 2012); *The Anthropology of Climate Change* (with Merrill Singer, Routledge, 2nd edition, 2018), *Democratic Eco-Socialism as a Real Utopia: Transitioning to an Alternative World System* (Berghahn Books, 2018), and *Climate Change and Capitalism in Australia: An Eco-Socialist Vision for the Future* (Routledge, 2022).

References

Albritton, Robert. 2019. *Eco-Socialism for Now and the Future: Practical Utopia and Rational Action*. Cham, Switzerland: Palgrave Macmillan.

Arboleda, Martin. 2020. *Planetary Mine: Territories of Extraction under Late Capitalism*. London: Verso.

Baer, Hans A. 2018a. *Democratic Eco-socialism as a Real Utopia: Transitioning to an Alternative World System*. New York: Berghahn Books.

———. 2018b. "Grappling with Flying as a Driver to Climate Change: Strategies for Critical Scholars Seeking to Contribute to a Socio-ecological Revolution." *Australian Journal of Anthropology* 29: 298–315.

———. 2020a. *Airplanes, the Environment, and the Human Condition*. London: Routledge.

———. 2020b. "The Elephant in the Sky: On How to Grapple with Our Academic Flying the Age of Climate Change." *Anthropology Today* 35(4): 21.

Baer, Hans A., and Merrill Singer. 2018. *The Anthropology of Climate Change: An Integrated Critical Perspective*. 2nd ed. London: Earthscan at Routledge.

Biehl, Janet. 2015. *Ecology or Catastrophe: The Life of Murray Bookchin*. Oxford: Oxford University Press.

Dawson, Ashley. 2017. *Extreme Cities: The Peril and Promise of Urban Life in the Age of Climate Change*. London: Verso.

Eriksen, Thomas Hylland. 2016. *Overheating: An Anthropology of Accelerated Change*. London: Pluto.

Hardt, Michael, and Antonio Negri. 2009. *Commonwealth*. Cambridge, MA: Harvard University Press.

Klein, Naomi. 2007. *The Shock Doctrine: The Rise of Disaster Capitalism*. New York: Metropolitan Books.
Laurent, Eloi. 2020. *The New Environmental Economics*. Cambridge, UK: Polity.
Loewy, Michael. 2015. *Ecosocialism: A Radical Alternative to Capitalist Catastrophe*. Chicago: Haymarket Books.
Lowenstein, Antony. 2015. *Disaster Capitalism: Making a Killing Out of Catastrophe*. London: Verso.
Lyster, R. 2017. "Climate Justice, Adaptation and the Paris Agreement: A Recipe for Disasters?" *Environmental Politics* 26: 438–58.
Magdoff, Fred, and Chris Williams. 2017. *Creating an Ecological Society: Toward a Revolutionary Transformation*. New York: Monthly Review Press.
Newell, Peter, and Matthew Paterson. 2010. *Climate Capitalism: Global Warming and the Transformation of the Global Economy*. Cambridge: Cambridge University Press.
Parks, Bradley C., and J. Timmons Roberts. 2010. "Climate Change, Social Theory and Justice." *Theory, Culture & Society* 27(2–3): 134–66.
Paskal, Cleo. 2010. *Global Warring: How Environmental, Economic, and Political Crises will Redraw the World Map*. New York: Palgrave Macmillan.
Rockström, Johan. 2011. "Science's Role and Responsibility." In *Bankrupting Nature: Denying Our Planetary Boundaries*, edited by Andres Wijkman and Johan Rockström, 19–35. London: Earthscan.
Saul, Quincey, ed. 2019. *The Emergence of Ecosocialism: Collected Essays by Joel Kovel*. New York: 2Leaf Press.
Stensrud, Astrid B., and Thomas Hylland Eriksen. 2019. *Climate, Capitalism and Communities: An Anthropology of Environmental Overheating*. London: Pluto Press.
Stern, Nicholas. 2015. *Why Are We Waiting? The Logic, Urgency, and Promise of Tackling Climate Change*. Cambridge, MA: MIT Press.
Thornett, Alan. 2019. *Facing the Apocalypse: Arguments for Ecosocialism*. London: Resistance Books.
Trainer, Ted. 2010. *The Transition to a Sustainable and Just World*. Sydney: Envirobook.
Wright, Christopher, and Daniel Nyberg. 2015. *Climate Change, Capitalism, and Corporations: Processes of Creative Self-Destruction*. Cambridge, UK: Cambridge University Press.

Index

Note: Page references noted with an *f* are figures.

Aboriginal epistemology, 315
acceptance, defiant. *See* defiant acceptance
acorns, loss of, 274
Acosta, Virginia, 98
actions in response to change, 150–53, 151*f*
activism, 1, 370
adaptation, 8, 167, 179; animal presence, 295 (*see also* animal presence); Aotearoa New Zealand (ANZ), 169; attitudes to responses, 148–50, 150*f*; climate, 249–51 (*see also* Elbe River Valley); climate change, 219; coastal hazard risk, 167, 168; Coromandel Peninsula, Aotearoa New Zealand, 180–92, 189*f*; defiant acceptance and, 263; institutional setting for, 173–77; narratives, 168 (*see also* Aotearoa New Zealand [ANZ]); responses, 172; strategies, 201; tourism, 230
affluence, 14
AFOLU sector (Kenya), 205
Africa, 201; global warming in, 201; Intergovernmental Panel on Climate Change (IPCC), 201, 202; vulnerability in, 201. *See also* East Africa
agriculture, rewilding strategies, 302
Agua Caliente Band of Cahuilla Indians v. Coachella Valley Water District (2015/2017), 277, 278
Aguayo, Don Miguel, 333
airline travel, COVID-19 effect on, 371–73
Airs, Waters, Places (Hippocrates), 3
Aisher, Alexander, 26
Alaska, 34
Aldrich, Daniel, 263
All Weather Snowmaker, 231, 238, 365
alpine cryosphere environments, 230; challenges in, 232–38; high alpine environments, 226–29; human-environmental dynamics, 224, 225; rock falls, 234
alternative world systems, 368–571
Amah Mutsun Tribe, 277. *See also* Tribal nations
Amazon: climate change in the urban, 92–93; Delta, 27
American Anthropological Association, 3
American Anthropologist, 359
American Community Survey, 142*f*
The Angry Earth: Disasters in Anthropological Perspective (Oliver-Smith/Hoffman), 8
animal populations, 272
animal presence: anthropology and rewilding, 304–6; ethnographies, 293–396; Faia Brava Reserve (Portugal), 292; reconstruction of a natural heritage, 306–10; reintroduction of predators, 299, 304, 307, 308, 309; rewilding concept, 297–99; Rewilding Europe network, 300–302, 303; rewilding strategies, 302–3
animism, 315
the Anthropocene, 2, 3, 9, 78, 223, 238, 298, 305, 306
anthropogenic climate change, 319–20
anthropology: climate, 6–8; climate change and, 342; contribution of, 4–6; of energy, 6; global diversity, 8–11; and rewilding, 304–6
Anthropology and Climate Change (Crate/Nutall), 7
anticipation, 8

antisystemic movements, 369
Aotearoa New Zealand (ANZ), 167–69; adaptation, 169, 180–92, 189*f*; beachfront property, 184*f*; climate change denial, 182–85; climate change leadership, 181–82; erosion, 185*f*; ethnographies, 169–71; geography of Coromandel Peninsula, 171–73; institutional setting for adaptation, 173–77; local and regional council collaboration, 191; management of, 178; natural hazards, 177–80; political economy in, 181
apartheid, 39
Arctic sea ice decline, 313
Arizona (United States), droughts in, 273
artificialization, 299, 306
Arunachal Pradesh (India), 51, 54–56
assemblage, 9
Asset Management Plans, 176
atmospheric pollution, 7
atomic energy, 318
attachment to place, 137–38, 144–48, 345
attitudes to adaptation responses, 148–50, 150*f*
Australia: coral reefs, 71; crown-of-thorns starfish (COTS) outbreak, 71, 72; disasters, 339; Great Barrier Reef, 76
Australian Aborigines, 315, 316
Austria: alpine cryosphere environments, 232–38; glaciers in, 223–25; high alpine environments, 226–29; human-environment interactions, 229–32; modernity and, 228; performing identity through snow, 228–29; Pitztal glacier ski resort, 226–28; retreating glaciers, 225; snowmaking in, 236, 237
Austrian Alps, 223, 232. *See also* glaciers
autonomy, 174

Bachelard, Gaston, 50, 60
Baer, Hans, 7, 8
baixadas (land occupation), 90
Baldy, Cutcha Risling, 275
Bangladesh, 366; access to livelihoods, 118–19; Chattogram, 115; climate change in, 113–14; Dhaka, 115; drinking water access, 122–23; electricity access, 122–23; Great Britain and, 124; growth hubs of, 126; housing, 119–22; migration and, 116–18; Natore, 116, 119, 123; Rajshahi, 116, 117, 119, 121, 123; sanitation access, 122–23; Sirajganj, 116, 120, 122
Barnes, Jessica, 5, 8
Barre regime (Somalia), 209, 210, 211
Barriball, Philippa, 181
barriers, symbolic, 9
BASIS Collaborative Research Support Program, 202, 215, 217
Bates, Henry Walter, 90, 103
Bateson, Gregory, 6, 12
Bauman, Zygmunt, 3, 359
Beaton, Sheena, 188
Beatty, Andrew, 60
Beck, Ulrich, 3, 317, 318, 319
behaviors: control, 156*f*, 160; outcomes, 157*f*; Perceived Behavioral Control, 160; Theory of Planned Behavior, 154, 160
Belém, Brazil, 27; climate change in the urban Amazon, 92–93; disasters, nature, and culture, 97–98; environmental memory and disasters, 95–97; Una Watershed Project, 92, 93–95, 98, 100, 103; urban floods in, 92, 93, 98–103; urban transformations in, 90–92
belief systems, 81. *See also* knowledge
Biloxi-Chitimacha-Choctaw Native Americans, 353–57, 368
Binnizáleños, 325, 326, 327, 328, 329, 330
biomass energy, 278
biophysical processes, 298
birds of prey, 308
Boulder, Colorado, flooding, 346–53
Boyle, Robert, 316
Brazil, 25, 27; Amazon Delta, 27; Belém, 90–92 (*see also* Belém, Brazil); Galo Channel, 100; Guajará Bay, 94; São Joaquim Channel, 100; Visconde de Inhaúma Channel, 101; water supplies, 91
Breakthrough Institute, 305, 306
Brondizio, Eduardo, 92
Brown, Lester, 204
Brundtland, Gro Harlem, 11
building materials, 121

calamities, vulnerabilities of, 341
Calderón, Felipe, 314
California tribes, 274, 277, 278, 279, 280. *See also* Tribal nations
Camus, Albert, 51, 52
Canada: Inuit in, 137; Labrador, 137

canals, 100
Canterbury earthquake sequence (2010–11), 175, 179
capitalism, 5, 305, 306; climate, 364–66; disasters, 364; glaciers and, 230
capitalist nature, 231
Capitalocene, 305
carbon, 8, 280; emissions, 11; footprints, 12; offsetting, 365
carbon dioxide. *See* CO_2
CARE-Somalia, 208, 211
Carson, Rachel, 15
Cartesian dualism, 6
catastrophes, 342. *See also* disasters
Central European floods, 251
Centre de Recherches Insulaires et Observatoire de l'Environnement (CRIOBE), 70, 72, 75, 81
channels, 102*f*
Chattogram, Bangladesh, 115
China, 7, 9, 10; airline travel, COVID-19 effect on, 371–73
Choctaw, 353–57
Christianity, 38–39
Civil Defence Emergency Management (CDEM) Act 2002, 175
civilizational collapse, causes of, 3, 4
clan warfare, 58
climate: adaptation, 249–51 (*see also* Elbe River Valley); disruption, 272; relationships and, 3; stressors, 270
climate anthropology, 6–8
climate capitalism, 364–66. *See also* capitalism
climate change, 1; actions in response to change, 150–53, 151*f*; adaptation, 134–35, 179, 219 (*see also* climate change); anthropogenic, 319–20; anthropology and, 342; attachment to place, 137–38, 144–48; attitudes to adaptation responses, 148–50, 150*f*; atypical dangers in, 233–34; Bangladesh, 113–14; categories of, 2, 3; Coromandel Peninsula, Aotearoa New Zealand, 182*f*; COVID-19, 371–74 (*see also* COVID-19); definition of, 17; denial, 182–85; disasters, 339–42 (*see also* disasters); discussions, 158–60; displacement, 135–37; East Africa and, 201 (*see also* East Africa); Eastern Himalayas, 49–54 (*see also* Eastern Himalayas); ecological perturbations (*see* ecological perturbations); effects of location, 138–39; effects on tribes, 271–73; Elbe River Valley (*see also* Elbe River Valley); Ethiopia, 213–19; expectations about future change, 147–48, 158; future effects of on Tribal nations, 273–75; glaciers and, 223–25 (*see also* glaciers); global, 52; global topic of, 2; importance of stories, 59–60; inaction, 177; indigenuity responses, 275–80; intentions to move, 153, 154–58, 157*f*; Kenya, 203–7; lack of climate action, 13–16; leadership (New Zealand), 181–82; localizing, 131–39; migration, 135–37; mitigation, 92, 366; narratives of, 25–28; New Zealand, 167–69 (*see also* New Zealand); in Oaxaca, Mexico, 313–14 (*see also* Oaxaca, Mexico); political solutions to, 1; and progress, 2; research, 366; rewilding and, 292 (*see also* Faia Brava Reserve [Portugal]); rise in sea levels, 133–34; scientific knowledge of, 29; Shishmaref, Alaska, 253; social theory of governance and, 314–19; study designs, 139–41; study methods, 143–44; systemic changes to, 5; understanding of, 366–68; in the urban Amazon, 92–93
Climate Change Attitude Survey (2015), 142
Climate Change Response (Zero Carbon) Amendment Bill (2019), 188
climate refugees, 207–13, 356. *See also* Tribal nations
climate refugia, 50
climatology, 26
CO_2 (carbon dioxide): emissions, 11, 13, 314; increase of, 303
Côa River Valley (Portugal), 294, 303
coastal hazard risk, 167, 168, 169, 175, 182*f*, 186; management, 175; natural hazards, 177–80
Coastal Hazards Policy, 186
Coastal Management Strategy/Coastal Hazards Policy, 173, 186
Coastal Protection and Restoration Authority, 356
coastal tribes, 273. *See also* Tribal nations
collaboration, 191, 366
Collaborative Research for Moorea's Fishery, 70

colonialism, 38, 179, 202, 204, 271, 274, 281, 317, 320. *See also* Indigenous communities
Colorado disasters, 346–53. *See also* Boulder, Colorado
commodities, 11
communal lands *(ejidos)*, 321
communism, 334
community, sense of, 346
community-based natural resource management (CBNRM), 36
Concow Maidu people, 270. *See also* Tribal nations
conservation, 80
consultation, Mexico and, 326–28
consumption, 8, 13
control, behaviors, 156*f*, 160
conviviality, 59, 367
coral bleaching, 77, 78
coral reefs, 26; Australia, 71; crown-of-thorns starfish (COTS) outbreak, 70–73 (*see also* ecological perturbations); diversity in, 74; fish populations on, 74–76
Coromandel Peninsula, Aotearoa New Zealand, 167–69; adaptation, 173, 180–92, 189*f* (*see also* adaptation); beachfront property, 172*f*, 184*f*; climate change, 182*f*; climate change denial, 182–85; climate change leadership, 181–82; decision-making processes, 183; erosion, 185*f*; ethnographies, 169–71; geography of, 171–73; institutional setting for adaptation, 173–77; local and regional council collaboration, 191; management of, 178. *See also* Aotearoa New Zealand (ANZ)
corporate power, 1
COVID-19, 342, 370, 371–74
Crate, Susan, 7, 8
credit market, carbon, 280
crown-of-thorns starfish (COTS) outbreak, 65–68; culling, 72; effect on fish abundance, 74–76; eradication efforts, 73; glut, bloom, 76–79; knowledge spaces, 82–83; local ecological knowledge (LEK), 73–74; management, 80, 81; perceptions of, 76–79; start of, 70–73; tensions between fishers and scientists, 80
Crutzen, Paul, 3, 305
cryospheres, 224, 225. *See also* glaciers
Cuba, 208
Cultural Fire Management Council (CFMC), 276
culture, 17, 97
Cultures of Energy (Strauss/Rupp/Love), 7
Cyclone Cook (2017), 184
Cyclone Oli, 76

Daiichi Fukushima nuclear facility (Japan), 345
damages: economy of repairs, 258–61; flooding, 251, 256 (*see also* flooding; insurance); reconstruction, 259
Damara people (ǂnūkhoen), 34, 40. *See also* Namibia
Darfur, Sudan, 202
decision-making processes, 93, 183
defiant acceptance, 249–51, 367
deforestation in Kenya, 203
democracy, 178
demographics, 142*f*, 145*f*. *See also* populations
denial, climate change, 182–85. *See also* climate change
Denver, Colorado, 352. *See also* Boulder, Colorado
depreciated habitats, 340
Derg, the, 213, 214, 215
Descola, Philippe, 97
Deutsche Gesellschaftfür Internationale Zusammenarbeit (GIZ) GmbH, 115
development, 90; green development and Mexico, 322–26; megadevelopment projects, 320, 322; Mexico and, 314; Oaxaca, Mexico, 328–32. *See also* urban transformations
Development Bank of Saxony (Sächsische Aufbaubank, SAB), 256
Dhaka, Bangladesh, 115
Diamond, Jared, 4
dikes, protests against, 254
Diné, 272. *See also* Tribal nations
disasters, 50, 339–42; Biloxi-Chitimacha-Choctaw Native Americans, 353–57, 368; Boulder, Colorado, 346–53; capitalism, 364; definitions, 342–46; environmental memory and, 95–97; examples of, 346–57; human dwellings and, 344; nature, and culture, 97–98; slow-onset, 344; socionatural, 91; triggers for, 343; vulnerabilities and, 340, 343

displacement, 1, 8, 135–37
disruption, climate, 272
ditches, 102
diversity, 8–11, 18
dogs (in Portugal), 295
Douglas, Mary, 102
Dove, Michael, 8
drainage, 93, 95, 98, 102*f*. *See also* flooding
Dresden, Germany, 251. *See also* Elbe River Valley
drinking water, access to, 122–23
DRM (Direction des Ressources Marines), 72, 73
droughts, 339; in Arizona (United States), 273; in Ethiopia, 213–19. *See also* disasters
Dulu-Kungu Dojung, 56, 58
Dust Bowl (United States), 206

earthquakes, 27, 175, 179. *See also* disasters
East Africa, 201; Ethiopia, 204, 205, 213–19; Kenya, 203–7, 219; Somalia, 207–13, 219; Sudan, 204, 205
Eastern Himalayas: climate change in, 49–54, 51–54; genealogy of a storm, 56–59; significance of climate change in, 52; storms in, 54–56; weather in, 53
Ebola, 342
ecological modernization, 364
ecological perturbations, 65–68; COTS effect on fish abundance, 74–76; crown-of-thorns starfish (COTS) outbreak, 70–73 (*see also* crown-of-thorns starfish (COTS) outbreak); differences in local ecological knowledge (LEK), 79–81; glut, bloom (COTS), 76–79; knowledge spaces, 82–83; local ecological knowledge (LEK), 73–74; Moorea (French Polynesia), 68–70; tensions between fishers and scientists, 80
ecology, 6
economic costs, flooding, 251
economic development, 296
economic liberalism, 316
economic systems, 2
eco-socialism, 363; climate capitalism, 364–66; COVID-19 and, 371–74; understanding of climate change, 366–68; world systems, 368–571
ecosystems, health of, 299. *See also* animal presence
education, 340
effects of location, 138–39
ejidos (communal lands), 321
Elbe River Valley, 249–51, 367; defiant acceptance, 261–64; economy of repairs, 258–61; floods and climate change in, 251–54; protests against dikes and walls in, 254; uncertainty and flooding, 254–58
electricity access, 122–23
Eliot, Charles, 203
El Niño, 7, 339
emissions, 18; airplanes, 371–73; carbon, 11; CO_2, 11, 13, 314
employment in Pitztal glacier ski resort, 226–28
energy, 278; anthropology of, 6; atomic, 318; exploitation of, 4; renewable, 322; sources, 364; wind, 320–22; wind power, 364
energy sacrifice zones, 270, 271
English language, 332
environmental degradation, 1
environmental memory and disasters, 95–97
environmental pluralism, 19, 26, 29–31, 367; definitions of, 33–34; linking ways of knowing, 32–33; northwestern Namibia, 34–36; patterns of, 41–42
environments, connections to, 344
epidemics, 342
Eriksen, Thomas, 340, 341, 363
erosion, 121, 185*f*
estuaries, 172
Ethiopia, 202, 204, 205, 213–19. *See also* Africa
ethnographies, 15, 293–396
Europe: Austria, 223 (*see also* Austria); flooding in, 252; rural exodus in, 304
European Commission (EC), 12
European Environment Agency, 252
European Union (EU), 207, 230
European Union's Emissions Trading Scheme (EUETS), 365
evacuations, 140. *See also* storms
evolution, 14
excessive success, 14
exploitative treatment, 67
exports, oil, 12
exposure, 167
extreme weather, 1

Fagan, Brian, 3

Faia Brava Reserve (Portugal), 292; anthropology and rewilding, 304–6; ethnographies, 293–396; navigation of, 296–97; reconstruction of a natural heritage, 306–10; rewilding concept, 297–99; Rewilding Europe network, 300–302, 303; rewilding strategies, 302–3
famine, 213–19, 339. *See also* disasters; Ethiopia
Figueira de Castelo Rodrigo (Portugal), 293
fire, 99; outbreaks, 293; Serra da Estrela Natural Park (Portugal), 294; use of to resist climate change, 276
Firestorm Inc., 276
Fischer, Ernst, 253
fishing, 67; COTS effect on fish abundance, 74–76; food fisheries, 78; Moorea (French Polynesia), 69, 70; strings of fish, 75*f*; tensions between fishers and scientists, 80
flooding, 67, 121; Belém, Brazil, 90–92, 98–103 (*see also* Belém, Brazil); Boulder, Colorado, 347–53; Central European floods, 251; control systems, 94; defiant acceptance, 261–64; economy of repairs, 258–61; Elbe River Valley, 249–51 (*see also* Elbe River Valley); Europe, 252; flood-proofing, 151; hurricanes, 138; insurance, 256, 258–61; inundation zones, 254, 257, 264; living close to rivers, 254; management, 258, 261; management of, 27; New Zealand, 175; preparation for, 252, 253; reconstruction, 259; Una Watershed Project, 98 (*see also* Una Watershed Project); uncertainty and, 254–58; US Army Corps of Engineers, 355; vulnerable populations and, 95, 96. *See also* disasters
flood-prone areas, living in, 254–58. *See also* flooding
Florida (United States), 135, 137
food: caravans, 203; fishing (*see* fishing); in Namibia, 41; security, 273, 281; sovereignty risks, 273; traditional foods of Tribal nations, 271, 274, 281
food-for-work scheme, 209
footprints, carbon, 12
forests: conservation, 207; management, 292, 293
Fortress World, 370
fossil fuel, revolution of, 4
Foucault, Michel, 51, 314, 319, 335
Four Corners region (southwest US), 271
France, 70, 71. *See also* French Polynesia
Fransfontein, Namibia, 26, 35, 36, 40, 41
French Polynesia, 26. *See also* Moorea
future change, expectations about, 147–48, 158

Galaal, Muusa H. I., 212
Galo Channel (Brazil), 100
Gedge, Ernest, 203
gender, 8
genealogies, 50, 51, 56–59
General Law of Climate Change (Mexico), 314
Georgia (United States), 132; coastal map of, 133*f*; coastal residents, 132; Hurricane Matthew (2016), 132, 134, 139*f*, 143, 145; intentions to move, 153, 154–58, 157*f*; migration, 140; populations, 141*f*; reasons for migration, 146–47, 147*f*; reasons people live on the coast, 144, 144*f*
geotextiles, 234
geothermal, 278
German Democratic Republic (GDR), 259
Germany: colonial rule in Namibia, 34–35; Elbe River Valley, 249–51 (*see also* Elbe River Valley)
giant spider conch, 74
giant triton, 74
Gila River Indian Community Water Rights Settlement Act (2004), 273
Gill, Peter, 216
Gizaw, Shumete, 216
glaciers, 29, 54; alpine cryosphere environments, 232–38; Austria, 223–25 (*see also* Austria); capitalism and, 230; high alpine environments, 226–29; human-environment interactions, 229–32; modernity and, 228; performing identity through snow, 228–29; Pitztal glacier ski resort, 224*f*, 225, 226–28; reliability of, 234–38; removing textiles, 236*f*; retreating, 225; rock falls, 234; ski resorts, 223–25; snowmaking, 236, 237; vanishing, 232
global capitalism, 5
global climate change, 52, 223
Global Climate Coalition, 364

global diversity, 8–11
globalization, 69, 181; of markets, 340; modernity and, 228
global population, 15
Global South, 11, 12, 93, 365
global warming, 79, 137; in Africa, 201; increase of carbon dioxide, 313; in India, 51
Global Warming and the Political Ecology of Health (Baer/Singer), 7
God, role of, 41, 42. *See also* religion
Goudie, Sandra, 184, 187, 188
governance, social theory of, 314–19
Great Barrier Reef (Australia), 76
Great Barrier Reef Marine Park Authority, 71
Great Britain: Bangladesh and, 124; floods (2015–16), 339; Kenya and, 203, 204, 206; New Zealand and, 173, 174
Great East Japan Earthquake (2011), 339
Green Capitalism, 305
green development and Mexico, 322–26
green economy, Oaxaca, Mexico and, 332–34
green growth, 15
green neoliberalism, 364
Green Party, 175
green revolution, 318
green technology, 15
Guajará Bay (Brazil), 90, 94. *See also* Belém, Brazil
Gump Research Station, 70

habitats, depreciated, 340
Haitian Earthquake (2010), 339
Haraway, Donna, 50, 81
Hardin, Garrett, 205, 211
Harper, Mary, 209
Harris, Marvin, 16
Hastrup, Kirsten, 8
Hau'ofa, Epeli, 66
Hauraki Coromandel Climate Action (HCCA), 188
Hazard Mitigation Plans, 176
hazards, definitions of, 343, 344. *See also* disasters
health of ecosystems, 299. *See also* animal presence
hegemonies, 330
heritage communities, 357. *See also* Tribal nations
high alpine environments, 226–29
Himalayas, 25, 26; climate change in, 49–54 (*see also* Eastern Himalayas)
Hippocrates, 3
Hiran Refugee-Reforestation Project, 208
The History of Mexico (Diego Rivera), 334, 335*f*
Hoffman, Susanna, 99, 363
holistic leadership by Tribal nations, 269–70. *see also* Tribal nations
holistic research, 16
holistic worldviews, 31
Hopi, 272. *See also* Tribal nations
Hornborg, Alf, 8
housing, 119–22
Hughes, David McDermott, 359
Human Choice and Climate Change: An International Assessment (Malone/Rayner), 7
human dwellings and disasters, 344
human-environmental dynamics, 224
human-environment interactions, 34, 229–32
human industrialization, 313
humanitarianism, 212
human rights abuses, 215
humidity, 52
hunting, 309
Hurricane Gustav (2008), 355
Hurricane Katrina (2005), 135, 136, 137, 339, 355
Hurricane Matthew (2016), 132, 134, 139*f*, 143, 145, 339
Hurricane Mitch (1998), 339
hurricanes, 78, 131. *See also* disasters
Hurricane Sandy (2012), 135, 138, 339, 345
hybridization, 98
hydroelectric construction projects, 321
hydroelectric storage infrastructure, 270
hydrographic basins, 100

IAUES World Congress, 99
Iberian colonizers, 317
Iberian lynxes, 308
Iberian Peninsula, 292
Ice Age, 7
igarapé (natural entity), 102
Imperial British East Africa Company (IBEAC), 203
imports, 11
In Business as in Life—You Don't Get What You Deserve You Get What You Negotiate (Karrass), 322

Independent Commission on International Humanitarian Issues (1985: 82), 209
India: Arunachal Pradesh, 51, 54–56; climate change in, 51–54; global warning in, 51; Kurung Kumey, 51
Indian Ocean, 52
Indigenous and Traditional Knowledges (IK/TKs), 269, 275
Indigenous communities, 269; indigenuity responses, 275–80; population of, 270, 271; racialization of, 322. *See also* Tribal nations
indigenous issues, 1; New Zealand, 173, 174 (*see also* Māori); Tribal nations (*see* Tribal nations)
indigenous knowledge, 31, 32, 65
individualism, 6
Industrial Revolution, 15, 341, 370
inequality, 1, 2
inequity: mitigation, 332; Oaxaca, Mexico, 328–32
Institute for Nature and Biodiversity Conservation/Institute for Nature Conservation and Forests (ICNB/ ICNF), 292
insurance, flooding, 256
integrated approach, 32
Intentional Labor Organization, 323
Interamerican Development Bank (IDB), 94, 95
Intergovernmental Panel on Climate Change (IPCC), 11, 59, 113, 201, 202, 313, 366
International Displacement Monitoring Centre (IDMC), 113
International Labor Organization's Indigenous and Tribal Peoples Convention (Mexico [1989]), 322
International Monetary Fund (IMF), 321
Inuit in Canada, 137
Iñupiat, 34
IPCC AR5 Working Group II assessment (2013–14), 183
IRD (Institut de Recherche pour le Développement), 71
Ireland floods (2015–16), 339
Isle de Jean Charles band of Biloxi-Chitimacha-Choctaw Native Americans, 353–57, 368
Isthmus of Tehuantepec (Mexico), 320, 323
Iwi Management Plans, 176, 178

Jansson, Kurt, 214
Japan: Daiichi Fukushima nuclear facility, 345; postdisaster communities in, 263
Jemze Pueblo (New Mexico, US), 279
jobs, 118. *See also* livelihoods
Johannes, R. E., 65

Kaika, Maria, 101
Karrass, Chester, 322
Karuk, 276. *See also* Tribal nations
kaumātua (elderly Māori of standing), 170
Kenya, 202, 203–7, 219. *See also* Africa
Kenya Land Commission, 206
key informant interviews (KIIs), 116
Khoekhoegowab language, 34, 35
Kiage, Lawrence, 205
Kikuyu highlands (Kenya), 203, 204, 205
Klein, Naomi, 1
knowledge: coexistence of, 36; indigenous, 31, 32, 65; Indigenous and Traditional Knowledges, 269; lack of information, 32; linking ways of knowing, 32–33; local ecological knowledge (LEK), 65, 66, 67, 68; Māori, 193; *Mātauranga Māori* (ancestral Māori knowledge), 169; models, 66; modern, 330; scientific, 30, 34; traditional, 275; traditional ecological knowledge (TEK) (*see* traditional ecological knowledge [TEK]); ways of knowing, 30
Kunene region (Namibia), 34. *See also* Namibia
Kurung Kumey (India), 51
Kyoto Protocol (1997), 10, 11

labor force, reasons for migration, 118
labor movements, 370
Labrador, Canada, 137
land degradation, 214
Land Information Memoranda, 175
land occupation *(baixadas)*, 90
landowners (Mexico), 325, 326, 327, 332
languages: Damara people (ǂnūkhoen), 40; Khoekhoegowab, 34, 35; Ovaherero people (Namibia), 40
La Niña, 339
Latif, Abdul, 117
Latour, Bruno, 9, 97
Lauer, Matthew, 26
laws: flooding insurance, 261; General Law of Climate Change (Mexico), 314

Leach, Glenn, 182
Lévi-Strauss, Claude, 5
Likert scale, 143
lime juice fight, 71. *See also* crown-of-thorns starfish (COTS) outbreak
livelihoods, 124; access to, 118–19; loss of due to flooding, 251; in Pitztal ski resort, 226–28
Living in Denial (Norgaard), 7
local, primacy of, 16–17
local ecological knowledge (LEK), 65, 66, 67, 68; claims of, 82; crown-of-thorns starfish (COTS) outbreak, 73–74; differences in, 79–81; discrepancies between fishers and scientists, 79; Polynesian fishers, 77 (*see also* crown-of-thorns starfish (COTS) outbreak)
Local Government Act (LGA), 175
Local Government Funding and Financing Productivity Commission draft report (2019), 188
localizing climate change, 131–39; actions in response to change, 150–53, 151*f*; attachment to place, 144–48; attitudes to adaptation responses, 148–50, 150*f*; discussions, 158–60; expectations about future change, 147–48, 158; intentions to move, 153, 154–58, 157*f*; study designs, 139–41; study methods, 143–44
local weather patterns, 30
location, effects of, 138–39
logging, 355
long-term cycles, 25
Long Term Plan (LTP [2018–28]), 186
Lutheran Protestants, 38. *See also* religion

MacNeice, Louis, 59
maintenance, ski resorts, 233
maladaptation risks, 202
Maldives, 136
Malone, Elizabeth, 7
management: coastal hazard risk, 175; crown-of-thorns starfish (COTS) outbreak, 80, 81; flooding, 27, 258, 261; forests, 292, 293; resource, 10, 204; watersheds, 317
manufacturing, 341, 342
Māori, 169, 170; *kaumātua* (elderly Māori of standing), 170; knowledge, 193; local action from, 180; *Mātauranga Māori* (ancestral Māori knowledge), 169; relationships with, 178, 190; rights, 193; settlement in New Zealand, 173, 174; Waitangi Tribunal (1975), 174. *See also* Aotearoa New Zealand (ANZ); New Zealand
Ma'ow, Hussein, 210, 212
Maren, Michael, 209
marginalization, 340
Marino, Elizabeth, 253
markets: carbon, 8; globalization of, 340; systems, 370; treating nature as, 359
Marx, Karl, 334
Marxism, 204, 334
Mātauranga Māori (ancestral Māori knowledge), 169
Mayan empire, 4
Mayer, Helena, 187, 187*f*
McNeill, John, 3
Mead, Margaret, 6
Mean Higher High Water (MHHW), 143, 159
megacities, 115. *See also* Bangladesh
megadevelopment projects, 320, 322
Mendes, Paulo, 363
Mengistu Haile Mariam, 213
Mesoamerican people, 315, 316, 317
Mexican Constitution, 321
Mexico: anthropogenic climate change and, 319–20; climate change and, 313, 314; consultation and, 326–28; General Law of Climate Change, 314; green development and, 322–26; Isthmus of Tehuantepec, 320, 323; landowners, 325, 326, 327; National Palace, 334, 335; North American Free Trade Agreement (NAFTA), 321, 322; Oaxaca (*see* Oaxaca, Mexico)
Micronesia, 66
Middle Ages, 314
middle-class, 2
migration, 1, 8; access to livelihoods, 118–19; climate change, 135–37; cycles of, 116; drinking water access, 122–23; electricity access, 122–23; Georgia (United States), 140; housing, 119–22; Hurricane Katrina (2005), 137; intentions to move, 153, 154–58, 157*f*; poverty as reason for, 125; reasons for, 146–47, 147*f*; responses, 138; rural-to-urban, 116; sanitation access, 122–23; social motivations for, 117; surveys, 142; urban, 340

Millennium Development Goal (MDG): Bangladesh Progress Report 2008, 116
mitigation, 8; in Bangladesh, 113–14; climate change, 92, 366; inequity, 332; Mexico and, 314; rural flow, 116
Moapa River Indian Reservation (Nevada, US), 279
models, knowledge, 66. *See also* knowledge
modernist epistemology, 316, 317, 318
modernity, 90, 94, 101, 318; tourism and, 228
modernization, 27, 90; ecological, 364
modern knowledge, 330
Monga, 115
monoculture planting, 272
monsoons, 53
Montesquieu, 3
Moore, Jason, 305
Moorea (French Polynesia), 67, 68; crown-of-thorns starfish (COTS) outbreak, 73–74 (*see also* crown-of-thorns starfish (COTS) outbreak); ecological perturbations, 68–70; knowledge spaces, 82–83; map of, 69*f*; tensions between fishers and scientists, 80
Moorea Coral Reef LTER, 70
Mora, Don Ricardo, 331
mother-places *(aaney-nyoku)*, 57
mountain landscapes, 229. *See also* glaciers
Mount Kenya (Kenya), 203, 205, 206
movements: antisystemic, 369; labor, 370
multispecies assemblages, 60

Namibia, 9, 19, 25; environmental pluralism, 29–31 (*see also* environmental pluralism); food in, 41; Fransfontein, 26, 35, 36, 40, 41; northwestern, 34–36; patterns of pluralism in, 41–42; rain in, 41; weather in, 29; ǁgamo!nâb, 36
narratives: adaptation, 168 (*see also* Aotearoa New Zealand [ANZ]); of climate change, 25–28; oral, 59
Naso, 74, 75, 78
National Center for Atmospheric Research (NCAR), 347
National Climatic Data Center, 273
National Geographic, 208
National Hazard Center, 347
National Institute of Ecology and Climate Change (INECC), 314
nationalism, 2
National Oceanic and Atmospheric Administration (NOAA), 143, 347
National Palace (Mexico), 334, 335
The National Republican Guard (GNR), 294
National Science Foundation, 73
national voter identity card (ID), 120, 123
Native American communities, 364. *See also* Tribal nations
Natore, Bangladesh, 116, 119, 123
natural disasters, 343. *See also* disasters
natural entity *(igarapé)*, 102
natural hazards, 181; adaptation to, 249–51; Aotearoa New Zealand (ANZ), 177–80. *See also* disasters
natural heritage, reconstruction of, 306–10
naturalization, 299. *See also* animal presence; rewilding
natural resources, 16, 366
nature, 97
Nature Conservancy Fire Learning Network, 276
nature reserves, 295. *See also* animal presence; Faia Brava Reserve (Portugal)
Navajo Nation, 271, 272, 277. *See also* Tribal nations
neoliberalism, 180, 181, 340, 364
neo-Malthusianism, 205
Nepal, 9, 54. *See also* Eastern Himalayas
Nepalese earthquake (2015), 339
networks, social, 32
New Ireland, 74
New Mexico tribes, 275. *See also* Tribal nations
New Orleans, Louisiana, 135, 136. *See also* Hurricane Katrina (2005)
New York Times, 76
New Zealand: adaptation, 180–92, 189*f*; beachfront property, 184*f*; climate change denial, 182–85; climate change leadership, 181–82; Coromandel Peninsula, Aotearoa, 167–69; erosion, 185*f*; ethnographies, 169–71; geography of Coromandel Peninsula, 171–73; institutional setting for adaptation, 173–77; local and regional council collaboration, 191; natural hazards, 177–80; storms, 183
New Zealand Coastal Policy Statement (NZCPS [2010]), 175, 176*f*, 181

New Zealand First, 175
New Zealand National Party, 174
New Zealand Transport Agency (NZTA), 186
Njega of Ndia, Chief, 206
nongovernmental organizations (NGOs), 72, 73, 74, 77, 92, 300, 365, 366
Norgaard, Kari, 7
North American Free Trade Agreement (NAFTA), 321, 322
Northern California Indian Development Council, 276
Norway, 11–13
Nuttall, Mark, 7, 8
Nyishi, 51, 57

Oakland, California, 99
Oaxaca, Mexico, 313–14; anthropogenic climate change and, 319–20; consultation and, 326–28; development and inequity, 328–32; green development and, 322–26; green economy and, 332–34; social theory of governance and, 314–19; before turbines, 320–22
occupationscapes, 224, 225, 229. *See also* ski resorts
Oceania, 66
ocean temperatures, 77, 79
Ogaden War (Somalia), 208, 211
oil: exports, 12; platforms (Norway), 13. *See also* fossil fuel, revolution of
One-Way Analysis of Variance, 143
oral narratives, 59
Orlove, Ben, 7, 8
Oromiya Zone (Ethiopia), 216, 217
outbreaks, fire, 293
outcomes of behaviors, 157*f*
Ovaherero people (Namibia), 40
overheating, impact of, 363
Overheating: An Anthropology of Accelerated Change (Eriksen), 340

Pacific islands, 66
Pacific Ocean populations, 173
Paiute, 273, 274, 279. *See also* Tribal nations
Palauan fishers, 65
Panama Beach, Florida, 135, 137
pandemics, 17
Pankhurst, Atula, 213
Paris Agreement at Conference of the Parties 21 (COP21), 365
parrotfish, 74
pastoralists, 39, 207–13
path dependence, 13
patronage networks, 210
Pauli, Julia, 35
Pentecostal churches, 38. *See also* religion
Peralta, Don Ricardo, 330
Perceived Behavioral Control, 160
permafrost, 232–38. *See also* glaciers
permanent residency (Bangladesh), 119–22
pesticide, 73
petroleum industries, 271
Philippines, 65
photographic safaris, 309
physical planes, altercations to, 344, 345
pine nut populations, 272
Pitztal glacier ski resort, 224*f*, 225, 226, 238; employment in, 226–28; snowmaking, 237
place, attachment to, 144–48, 345
Pleistocene rewilding, 297, 298. *See also* rewilding
The Poetics of Space (Bachelard), 50
policy of villagization (Ethiopia), 215
political movements, 1
political solutions to climate change, 1
pollution, 7, 345
Polynesia, 25, 26; ecological perturbations in, 65–68; Moorea (French Polynesia), 67 (*see also* Moorea [French Polynesia])
populations, 90; animal, 272; animal presence, 293 (*see also* animal presence); Bangladesh, 113 (*see also* Bangladesh); Colorado disasters, 349, 350; emergence of human, 314, 315; Georgia (United States), 141*f*; global, 15; growth, 4; inequity among, 320; movements, 340 (*see also* migration); Pacific Ocean, 173; pine nut, 272; postlogging effects, 272; pre-Columbian, 315, 316
Portugal, 11–13, 368; anthropology and rewilding, 304–6; dogs in, 295; Faia Brava Reserve, 292 (*see also* Faia Brava Reserve [Portugal]); Figueira de Castelo Rodrigo, 293; renaturalization, 302, 306; Rewilding Europe network in, 303; Serra da Estrela Natural Park, 294
postdisaster solidarity, 263
postlogging effects, 272
poverty, as reason for migration, 117, 125

pre-Columbian populations, 315, 316
predators, reintroduction of, 299, 304, 307, 308, 309. *See also* wolves
Prescribed Fire Training Exchange (TREX), 277
Principal Components Analysis (PCA), 143, 154
Private Protected Area (Portugal), 292
products of knowledge practices, 82
progress, climate change and, 2
protection, flood, 261–64. *See also* flooding; insurance
protests, 187*f*
public understanding of science (PUS), 32
purification, 97, 98
Pyramid Lake Paiute Tribe, 274

Qualtrics Panels, 140
Quiroste Valley Cultural Preserve, 277

racialization of Indigenous communities, 322
racism, 67
racist exclusion, 1
rahui organizations, 80, 81
rain, 37, 41, 98. *See also* flooding; weather
Rajshahi, Bangladesh, 116, 117, 119, 121, 123
Ramos, Don Oscar, 328, 329
rapid disturbances, 78
Rayner, Steve, 7
REDD (Reducing Emissions from Deforestation and Forest Degradation), 11, 12, 280, 365
Reef Check Polynésie, 72
Refugee and Migratory Movements Research Unit (RMMRU), 115
refugees, 50; climate, 208 (*see also* Somalia); United Nations High Commission for Refugees, 209
Le regard éloigné (Lévi-Strauss), 5
regions, 18
reintroduction of predators, 299, 304, 307, 308, 309
reliability, snow, 231, 234–38
religion: Christianity, 38–39; role of, 41, 42
renaturalization, 298, 302; in Portugal, 306; support of, 307. *See also* animal presence; rewilding
renewable energy, 322
repairs, economy of, 258–61. *See also* flooding
repopulation programs, 293
Republic of Somaliland, 212
research, climate change, 366
resettlement, 213–19, 342
residency (Bangladesh), 119–22
resilience, 8, 18
resource management, 10, 204
Resource Management Act 1991 (RMA), 175, 177
Resource Management (Energy and Climate Change) Amendment Act (2004), 181
responses, migration, 138
restoration, rewilding, 292. *See also* Faia Brava Reserve (Portugal)
retrofitting buildings, 364
returns on investments in energy (EROI), 4
revitalization movements, 331
rewilding, 230, 296; anthropology and, 304–6; concept of, 297–99; strategies, 302–3
Rewilding Europe network, 292, 296–97, 300–302; in Portugal, 303; rewilding strategies, 302–3
rights, Māori, 193
risks, 8; adaptation to, 249–51 (*see also* adaptation); flood inundation zones, 254; food sovereignty, 273; inundation zones, 264; society, 319, 320
Rivera, Diego, 334, 335*f*
river erosion, 121
rivers, living close to, 254. *See also* flooding
RMA Regional Policy Statements (RPS), 176
Road to Survival (Vogt), 15
rock falls, 234
Rocky Mountains (United States), 351. *See also* Boulder, Colorado
Roman empire, 4
Rosa, Hartmut, 3
Routledge, W. Scoresby, 203, 204
Royal HaskoningDHV, 186
Ruedas, Antonio, 332
rural exodus in Europe, 304
rural-to-urban migration, 116

SADC (Southern African Development Community), 40
safe havens, 57
safety of ski resorts, 233
Sahelian nomads, 9

Samoa, 74
Sanchez, Don Eugenio, 324, 325
Sanchez, Kathy, 275
sand dune mobility, 272
sanitation, 27, 93, 95, 122–23. *See also* flooding
Santa Ynez Band of Chumash Indians, 279. *See also* Tribal nations
São Joaquim Channel (Brazil), 100
Savannah, Georgia, 134, 140. *See also* Georgia (United States)
Saxony, Germany, 249–51. *See also* Elbe River Valley
Schnegg, Michael, 26
Schneider, Dieter, 258
science, claims of, 82
scientific discourses, 39–41
scientific knowledge, 34
sea ice decline, 313
sea levels, 29, 67, 92, 103; adaptation to changes in, 134–35; in Georgia (United States), 132; intentions to move, 153, 154–58, 157*f*; in New Zealand, 172; rise of, 147, 148*f*, 213, 313
seawater, warming, 341. *See also* global warming
Secretariat of Energy (SENER [Mexico]), 327
security, food, 273, 281
sedentarization, 95
SEMARNAT (Secretariat of Environment and Natural Resources), 324
septic tanks, 94
sequestration, carbon, 8, 280
Serra da Estrela Natural Park (Portugal), 294
settlements (Bangladesh), 119–22, 125
settler colonization, 270
severe acute respiratory syndrome (SARS), 372
sewage, 27
Sewage Company of Pará, 94
sexism, 67
shanties, 120, 125. *See also* housing
shelter, types of, 121. *See also* housing
Shishmaref, Alaska, 253
Shoreline Management Plans (SMPs), 186
Siberia, 79
Silent Spring (Carson), 15
Singer, Merrill, 7, 8
Sirajganj, Bangladesh, 116, 120, 122
Sisk, Caleen, 278
ski resorts, 365; alpine cryosphere environments, 232–38; glaciers, 223–25 (*see also* glaciers); human-environment interactions, 229–32; maintenance, 233; modernity and, 228; performing identity through snow, 228–29; Pitztal glacier ski resort, 224*f*, 226–28; removing textiles, 236*f*; retreating glaciers, 225; snowmaking, 236, 237
slavery, 40, 132
slow-onset disasters, 344
smog, 7
smoke, 7
snow, 234; All Weather Snowmaker, 231, 238, 365; alpine cryosphere environments, 232–38; performing identity and, 228–29; reliability, 231, 234–38; snowmaking, 236, 237; tourism, 229 (*see also* ski resorts); as white gold, 228
snowmaking, 236, 237
social institutions, relationship to climate and, 3
social life, relationship to climate and, 3
social networks, 32
social norms, 155*f*, 160
social theory, perspectives from, 3–4
social theory of governance and climate change, 314–19
social welfare, 12
Society for Applied Anthropology (SfAA), 8
sociocultural approach, 32
socionatural disasters, 91
soil erosion, 54
solar wind, 278
solidarity, postdisaster, 263
Solomon Islands, 74
Solukhumbu District (Nepal), 54
Somalia, 202, 219; climate refugees, 207–13; food-for-work scheme, 209. *See also* Africa
Southeast Asia fires and smog (2015 and 2016), 339
southwestern United States: climate change effects on tribes, 271–73; Four Corners region of, 271; future effects of climate change on, 273–75; population of Indigenous communities, 270, 271; Tribal nations, 269–70 (*see also* Tribal nations). *See also* United States
South Wollo Zone (Ethiopia), 217

Soviet Union, 208
Spanish language, 330
spatiality, 14, 138
spearfishers, 78
Srinivas, M. N., 25
starfish. *See* crown-of-thorns starfish (COTS) outbreak
State Water Resources Control Board (CA-SWRCB), 278
Steffen, Will, 3
Steingraber, Sandra, 213
Stern, Nicholas, 366
Steward, Julian, 6
Stoermer, Eugene, 3
Stoltenberg, Jens, 12
stories, importance of, 59–60
storms, 339; anxiety and, 50; in the Eastern Himalayas, 49–54, 54–56 (*see also* Eastern Himalayas); genealogies of, 56–59; hurricanes, 131, 138; New Zealand, 183. *See also* disasters
strategies: adaptation, 201; eco-socialism, 363 (*see* eco-socialism); rewilding, 302–3
study designs, 139–41, 140*f*
study methods, 143–44
Sudan, 204, 205, 339
surveys, migration, 142
sustainability, 10, 11
Sustainable Development Goals (SDGs), 126
swine flu, 342
Switzerland, 238
Syria, 339

Tahiti, 69, 75
Tainter, Joseph, 3
Talum village, 56, 57, 60
TAMA (There Are Many Alternatives), 5
taramea. *See* crown-of-thorns starfish (COTS) outbreak
technological disasters, 343. *See also* disasters
technonature, 231
Tegg, Dennis, 191
temperament, relationship to climate and, 3
temperatures: Boulder, Colorado, 346–53; effects on southwestern US, 271, 272; global warming and, 137; impact of overheating, 363; increase in carbon dioxide, 313; rising ocean, 77; shifting weather patterns (Ethiopia), 216; ǁgamo!nâb Namibia, 36, 37
temporality, 14
Tesuque Pueblo (New Mexico, US), 276
Tewa Women United, 275
textiles, removing, 236*f*
Thames-Coromandel District Council (TCDC), 168, 176, 177, 183, 187, 187*f*, 188, 192, 193
themes, 18
Theory of Planned Behavior, 154, 160
This Changes Everything (Klein), 1
thousand-year floods, 353. *See also* flooding
Thunberg, Greta, 14, 366
tilapia, 101
Time magazine, 76
timescales, weather conditions and, 31
TINA (There Is No Alternative), 5
Tohono O'odham Nation, 277. *See also* Tribal nations
tourism, 223; adaptation, 230; alpine cryosphere environments, 232–38; definitions of, 230; employment in Pitztal glacier ski resort, 226–28; modernity and, 228; rewilding strategies and, 302; snow, 229 (*see also* ski resorts). *See also* glaciers
traditional ecological knowledge (TEK), 26, 27, 275, 276, 367. *See also* knowledge
traditional foods of Tribal nations, 281
traditions, 81
Transhumance and Nature Association (ATN), 292
trenches, 102
Tribal nations, 269–70; Choctaw, 353–57; climate change effects on tribes, 271–73; future effects of climate change on, 273–75; indigenuity responses, 275–80; Isle de Jean Charles band of Biloxi-Chitimacha-Choctaw Native Americans, 353–57; population of Indigenous peoples, 270, 271; Pyramid Lake Paiute Tribe, 274; traditional foods of, 271, 274, 281
Trump, Donald, 365
tsunamis, 345. *See also* disasters
Tuvalu, 67
2018 Fourth United States National Climate Assessment (NCA), 269
Tybee Island, Georgia, 132. *See also* Georgia (United States)

Typhoon Haiyan (2013), 339
Typhoon Megi (2016), 339

Una Basin, 27, 90. *See also* Belém, Brazil
Una River, 91. *See also* Belém, Brazil
Una Watershed Project, 92, 93–95, 98, 100, 103, 367
uncertainty, 8
uncertainty and flooding, 254–58
UN Food and Agriculture Organization (FAO), 214
UN Framework Convention on Climate Change (UNFCCC), 364, 365
United Nations (UN), 11; Clean Development Mechanism, 365; Declaration on the Rights of Indigenous Peoples, 278; Framework Convention on Climate Change (UNFCCC), 322; High Commission for Refugees, 209; Reducing Emissions from Deforestation and Degradation (REDD), 365
United States, 6; budget for climate change, 357; Colorado (*see* Boulder, Colorado); Dust Bowl, 206; Florida, 135, 137 (*see also* Florida [United States]); Georgia, 132 (*see also* Georgia [United States]); North American Free Trade Agreement (NAFTA), 321, 322; slavery, 132; Tribal nations, 269–70, 355 (*see also* Tribal nations); wildfires, 339
United States Agency for International Development (USAID), 208, 209
upward social mobility, 321
urban migration, 340
urban planning, 103
urban transformations (Belém, Brazil), 90–92
urban vulnerability: access to livelihoods, 118–19; in Bangladesh, 113–14; drinking water access, 122–23; electricity access, 122–23; housing, 119–22; migration and, 116–18; overview of, 114–15; sanitation access, 122–23
US Army Corps of Engineers, flooding, 355. *See also* flooding
US Centers for Disease Control and Prevention
US Forest Service, 276
US Geological Survey (USGS), 143

varimax rotation, 143
Vetlesen, Arne Johan, 16
Vietnam, 34
Viliui Sakha, 79
villagization, policy of (Ethiopia), 215
Visconde de Inhaúma Channel (Brazil), 101
Vogt, William, 15
von der Leyen, Ursula, 12
voter identity card (ID), 120, 123
vulnerabilities, 8, 50, 167, 179, 193; access to livelihoods, 118–19; in Africa, 201; of calamities, 341; disasters and, 340, 343; drinking water access, 122–23; electricity access, 122–23; forms of, 114–15; housing, 119–22; migration and, 116–18; sanitation access, 122–23; urban, 113–14 (*see also* Bangladesh)

Waikato Regional Council (WRC), 176, 183
Waitangi Tribunal (1975), 174
walls, protests against, 254
water: access to drinking, 122–23; retention basins, 254; storage infrastructure, 270; supplies (Brazil), 91. *See also* flooding
watershed management, 317
The Water Tower of Asia, 51–54
ways of knowing, 29–31; and environmental pluralism, 33–34 (*see also* environmental pluralism); linking, 32–33; weather changes, 36–41. *See also* knowledge
weather: climate change (*see* climate change); conditions and timescales, 31; destabilization of, 341; in the Eastern Himalayas, 53; El Niño, 7; extreme, 1; forecasting, 34; local weather patterns, 30; in Namibia, 29; shifting weather patterns, 216; uncontrollability of, 31; ways of knowing, 29–31; ways of knowing weather changes, 36–41
well-being, 13
wells, water, 122
wetlands, draining, 94
White, Leslie, 6, 16
white gold, snow as, 228
white sand, 173
Wildavsky, Aaron, 102
wildfires, 339
Wilhite, Hal, 8

Wilson, Edward O., 341
wind power, 364
wind turbine installations, 320–22, 327, 331
Winters v. US (1908), 277, 278
Wolf, Edward, 204
wolves, 294, 308, 309. *See also* animal presence
World Bank, 207, 321
world systems, 368–571
World War II, 6

Yazzie, Melanie, 275
Yurok, 276, 277. *See also* Tribal nations
Yurok Forestry/Wildland Fire, 276

Zapotec language, 328, 330, 331, 332, 333
Zwaan, Eric, 187*f*
Zwaan, Nancy, 187*f*